KU-236-350

Electrical
Networks

Electrical Networks

A. Henderson
Delft University of Technology

Edward Arnold
A division of Hodder & Stoughton
LONDON   MELBOURNE   AUCKLAND

© 1990 VSSD

First published in Great Britain 1990

British Library Cataloguing in Publication Data
A CIP Catalogue record for this book is available from the
British Library.

ISBN 0 7131 3637 5

All rights reserved. No part of this publication may be reproduced or
transmitted in any form or by any means, electronically or
mechanically, including photocopying, recording or any information
storage or retrieval system, without either prior permission in writing
from the publisher or a licence permitting restricted copying. In the
United Kingdom such licences are issued by the Copyright Licensing
Agency: 33–34 Alfred Place, London WC1E 7DP.

Typeset in Times and Univers by VSSD, Delft, The Netherlands.
Printed and bound in Great Britain for Edward Arnold, the
educational, academic and medical publishing division of Hodder and
Stoughton Limited, Mill Road, Dunton Green, Sevenoaks,
Kent TN13 2YA by J. W. Arrowsmith Ltd, Bristol

Preface

The book begins with simple basic concepts and the principal circuit theorems, which form a good link to the knowledge of the starting student; initially the mathematics is not difficult either. Very soon the controlled sources (including opamps) are introduced.

Subsequently the whole theory is extended to alternating currents including complex voltages, currents and impedances, while the mathematics becomes increasingly complicated. The subsequent chapters discuss transformers, three-phase systems, Fourier analysis, the complex frequency, poles and zeros, two ports including filters and networks with switches (transient response). Then an extensive chapter on computer aided design follows; it turns out that the problems dealt with before can be solved by computer.

Each chapter finishes with a large number of problems with increasing difficulty. The book concludes with the answers to the problems.

In the field of network theory one cannot avoid a great influence of mathematics, but in many places I explain the physical background of mathematical results.

In order to avoid calculations that are too complex I have in most cases used simple values for the network elements.

I wish to express my appreciation to Professor K.M. Adams and Professor P. Dewilde for the many informative conversations I had with them and to my colleague W. Buijze for his critical remarks and because he was so kind as to offer me a number of instructive problems.

Finally I wish to thank J. Schievink (VSSD) for his splendid and friendly cooperation and his suggestions.

Delft, December 1989   A. Henderson

Contents

Symbols

		unit
Admittance matrix	$\mathcal{Y}$	S
Admittance	Y	S
Ampere turns	(AT)	A
Amplitude and modulus of the boltage	|V|	V
Amplitude and modulus of the current	|I|	A
Angular frequency	ω	rad/s
Apparent power	|S|	VA
Argument of the impedance Z	arg Z	rad
Average, real power	P	W
Bandwidth	B	Hz or rad/s
Capacity	C	F
Cascade matrix	$\mathcal{K}$	
Coefficient of coupling	k	
Complex frequency	λ	s^{-1}
Complex power	S	VA
Conductance	G	S
Conductancematrix	$\mathcal{G}$	S
Conjugate complex of Z	Z*	Ω
Constant charge	Q	C
Constant, or complex current	I	A
Constant, or complex voltage	V	V
Coupled flux	Φ	Vs
Dampingexponent	σ	s^{-1}
Determinant of the matrix	det $\mathcal{Z}$ or |$\mathcal{Z}$|	
Detuning	d	
Differentiating operator	p	s^{-1}
Effective value of the current I	I_{eff}	A
Electric field	E	V/m
Energy	W	J
Force	F	N
Frequency	f	Hz
Hybrid matrix	$\mathcal{H}$	
Impedance matrix	$\mathcal{Z}$	Ω
Impedance	Z	Ω
Inductor	L	H
Integrating operator	p^{-1}	s
Leak	σ	

Magnetic field	H	A/m
Magnetic fluxdensity	B	Vs/m^2
Magnetic resistance	R_m	A/Vs
Momental charge	q	C
Momental current	i	A
Momental power	p	W
Momental voltage	v	V
Mutual induction	M	H
Period	T	s
Permeability of the vacuum	μ_0	Vs/Am
Phase	φ or p	rad
Quality of an inductor	Q	
Reactance	X	Ω
Reactive power	Q	VA$_r$
Relative permeability	μ_r	
Resistance matrix	$\mathcal{R}$	Ω
Resistance	R	Ω
Reverse cascade matrix	$\mathcal{J}$	
Reverse hybrid matrix	$\mathcal{G}$	
Specfic resistance	ρ	Ω/m/mm^2
Specific conductance	γ	S/m/mm^2
Susceptance	B	S
Time constant	τ	s
Voltage or current ratio	H	
Voltage ratio in decibel	G	dB
Voltage	$V_{ab} = V_a - V_b$	V

Magnetic coupled inductors

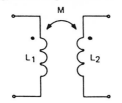

Transformer

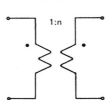

Gyrator

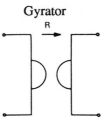

Switch: make contact

Switch: break contact

Transistor

12   Symbols

Operational amplifier

Voltage source

Current source

Controlled voltage source

Controlled current source

Resistor or conductor

Inductor

Capacitor

Impedance or admittance

Branch current

Mesh current

Non-lineair resistor

Diode

1

d.c. currents and d.c. voltages

1.1 Current, potential, voltage and resistance

The unit of electric *current* is the *ampere*. If a *charge* of 1 *coulomb* (1 C) is passed through the cross-section of a conductor in 1 *second* (1 s), the current will be 1 *ampère* (1 A). The current is indicated by I or i, so if Q is charge and T is time we have

$$I = \frac{Q}{T}. \tag{1.1}$$

We will use capital letters if the quantity is constant and lower case letters if the quantity varies with time. In the latter case for the current we get:

$$i = \frac{dq}{dt}. \tag{1.2}$$

The *potential* V_A (measured in *volt*, V) at a point A is equal in magnitude to the *energy* (measured in *joule*, J) required to take a charge of 1 C from a point where the potential is regarded as zero (V = 0) to point A.

The *voltage* V_{AB} between two points A and B is the energy that has to be supplied to take a charge of 1 C from point B to point A. So we have:

$$V_A = V_B + V_{AB},$$

thus

$$V_{AB} = V_A - V_B. \tag{1.3}$$

It follows that

$$V_{BA} = -V_{AB}. \tag{1.4}$$

We assign to each wire in a network a direction and we call this the *positive current direction*. We use this as a reference. One can freely choose this positive current direction. Once chosen, the sign of the current is known.

In Figure 1.1 a wire is drawn with the positive current direction I_1. It is possible that there is no current in the wire, but we still maintain the arrow with the indication I_1, I_1 being zero. If there is a current in the wire of 7 A from left to right we write $I_1 = 7$ A; if there is a current in the wire of 5 A from right to left we write $I_1 = -5$ A.

Figure 1.1

In Figure 1.2 a situation is represented in which the current direction changes. Between 0 and t_1 seconds the current i_3 is positive, between t_1 and t_2 the current i_3 is negative.

For the polarity of a voltage we have a similar situation. The arrow of the the positive current direction has the same meaning as the signs + and − in the *positive voltage polarity* (see Figure 1.3.a).

If $V_2 > 0$ the left hand terminal has a higher potential than the right hand terminal. If $V_2 < 0$ the potential of the left hand terminal is smaller than that of the right hand terminal. As has been already explained we can also denote a voltage between two terminals by two indices. This is beneficial in the case of big networks. The indices give the incident nodes. The first index is the positive terminal of the reference voltage (see Figure 1.3.b). Thus we can omit the notation + and − and also the notation V_{AB} without difficulties.

Finally, if there is no confusion we can omit the double arrow (see Figure 1.3.c).

The relation between voltage and current in a conductor is given by *Ohm's law*:

$$V = RI, \tag{1.5}$$

in which R is supposed to be linear. R is called *resistance*. In Figure 1.4 it is shown how the positive voltage is related to the positive current.

We see that the positive current flows from plus to minus through the resistor. In other words we can say that *the current and the voltage are interrelated (belong to each other).*

The resistance of a wire with a constant cross-section is

$$R = \frac{\rho l}{A} = \frac{l}{\gamma A} . \tag{1.6}$$

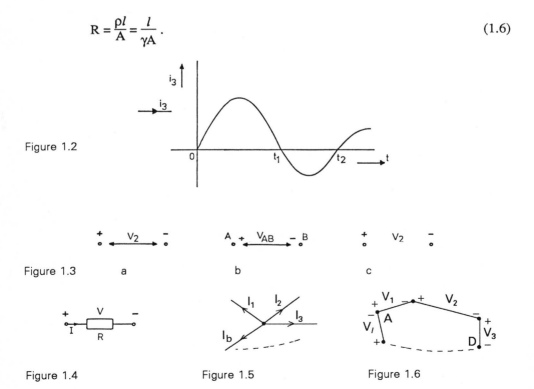

Figure 1.2

Figure 1.3   a   b   c

Figure 1.4   Figure 1.5   Figure 1.6

In this equation l is the length (m), A is the cross-section in square metres (m^2), ρ is the specific resistance (Ωm) and γ the specific conductance ((Ωm)$^{-1}$). The values of ρ and γ depend on the material used. ρ and γ also depend on the temperature. This leads to non-linearity which we will discuss at the end of Chapter 10.

The *conductance* G is the inverse of the resistance:

$$G = \frac{1}{R} . \tag{1.7}$$

So Ohm's law can also be written as

$$I = GV. \tag{1.8}$$

We will assume that the connecting wires have no resistance, i.e. R = 0.

If R = 0 we speak of a *short circuit*; if G = 0 we speak of *open terminals*.

Kirchhoff's current law is:

$$\sum^{b} I_n = 0 \qquad n = 1,2,...,b. \tag{1.9}$$

This equation refers to a node, where b wires (*branches*) coincide (see Figure 1.5). This formula also means that the sum of the currents flowing into the node equals the sum flowing out.

Positive currents are those which leave the node. *Kirchhoff's voltage law* is:

$$\sum_{m=1}^{l} V_m = 0 \qquad m = 1,2,...,l. \tag{1.10}$$

This equation refers to a *loop* in which there are l voltages (see Figure 1.6).

Following the loop the drop in voltage is chosen as positive. The use of the node indications with figures or letters, together with the corresponding double voltage index results in a clear formula:

$$V_{AB} + V_{BC} + V_{CD} + V_{DA} = 0.$$

So:

$$V_{AD} = V_{AB} + V_{BC} + V_{CD}. \tag{1.11}$$

Note the position of the indices in the left- and right-hand parts. Equation 1.11 is called the *index rule*. Not all nodes necessarily have a branch between them. The voltage law may also contain voltages of *fictitious* branches.

1.2 The voltage source and the current source

The voltage V_1 between the terminals of an *accumulator* with a varying *load* (see Figure 1.7) and the load current I_b have the idealised plot of Figure 1.8.

The analytic expression for $V_1 = f(I_b)$ is:

$$V_1 = V - R_i I_b. \tag{1.12}$$

R_i is the *internal resistance*.

With the aid of Kirchhoff's voltage law we can now draw the circuit of Figure 1.9.
I_S is the *short circuit current*. The element V is the (ideal) *voltage source*.
If we divide both terms of equation 1.12 by R_i we get

$$\frac{V_1}{R_i} = \frac{V}{R_i} - I_b \qquad (1.13)$$

and together with Kirchhoff's current law this gives the network of Figure 1.10.

We have set $\frac{V}{R_i} = I$ and this is called the (ideal) *current source*.

As a collective name we will use the expression *source intensity* or *source strength* for a voltage source voltage and a current source current.
If we connect a resistor R to a voltage source V the current through that resistor is $I = \frac{V}{R}$ with $R \neq 0$. This leads to the important rule:

A voltage source must not be short-circuited.

If we connect a conductor G to a current source I the voltage across that conductor is $V = \frac{I}{G}$ with $G \neq 0$. This means:

The terminals of a current source must not be be opened.

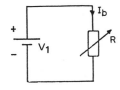

Figure 1.7

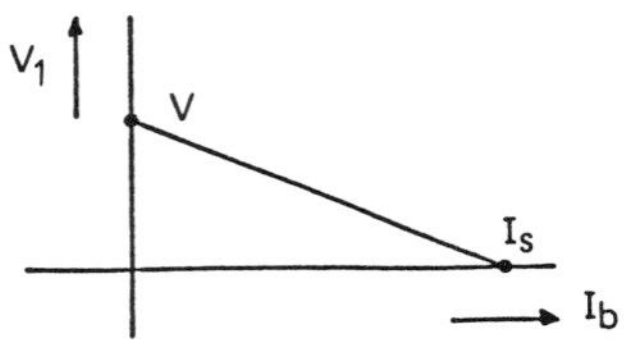

Figure 1.8

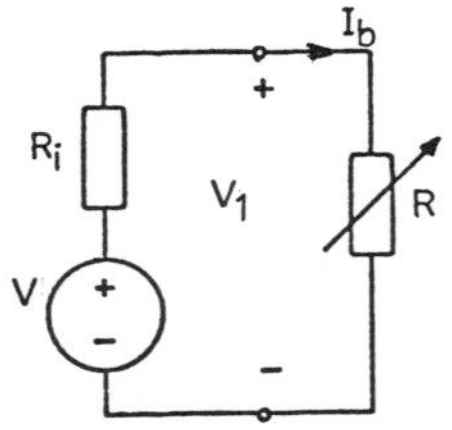

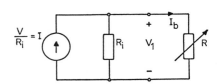

Figure 1.9

Figure 1.10

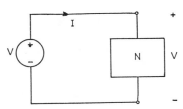

Figure 1.11

A voltage source with intensity zero is a *short circuit* and a current source with intensity zero is *open terminals*.

Two voltage sources in parallel are forbidden if the sum of the source strenghts is not zero (for in this case Kirchhoff's voltage law is not valid). Two voltage sources in parallel are not forbidden if the sum of the source strengths is zero. In that case, however, the source currents cannot be calculated.

Two current sources in series are forbidden if the sum of the source strengths is not zero (for in this case Kirchhoff's current law for the node between the two sources) is not valid. Two current sources in series are not forbidden if the sum of the source strengths is zero. In that case the source voltages cannot be calculated.

In network theory it often happens that there is analogy between two formulas, between two elements or between two circuits. For instance, one Kirchhoff's law turns into the other if one substitutes voltage for current and vice versa. We therefore say that the current law is *the dual* of the voltage law and vice versa.

The dual character is also found in

- voltage – current

- open nodes – short circuit

- resistance – conductance

In the following chapters we shall often meet this phenomenon of *duality*.

1.3 Energy and power

The voltage V_{AB} between two points A and B is defined as the work needed to move a unit charge (1 C concentrated in a point) from point B to point A.

If the charge is Δq the work is therefore

$$\Delta W = (V_A - V_B)\Delta q = V_{AB}\Delta q, \tag{1.14}$$

in which V_A and V_B are the potentials of the points A and B. If V_{AB} is constant (d.c.) and if the work is done in a time Δt, the average power is

$$P = \frac{\Delta W}{\Delta t} = V \frac{\Delta q}{\Delta t},$$

in which $V = V_{AB}$.

For $\Delta t \to 0$ we obtain

$$P = VI. \tag{1.15}$$

So the power, in the case of d.c., is the product of voltage and current.

Energy is expressed in joule (J), power in watt (W).

Power can be consumed or supplied. If a current I flows through a network N with two terminals (also called a *one-port*) and if the polarity of the voltage V is such that I flows from + to – the power consumed is positive (Figure 1.11).

The voltage source transports (positive) charge from minus to plus and so delivers electrical energy to N (this energy is supplied by the chemical or mechanical system outside the network).

The network N consumes this energy. We have pointed out before that the voltage and the current of a two-terminal network *belong to each other* if the current flows from plus to minus through the network. So we derive the following rule:

The energy $P = VI$ consumed is positive if current and voltage belong to each other.

We note that the current I does not necessarily have to be supplied by a voltage source; it can also be a current source or in general any other network.

1.4 Connection of resistors

One can easily show, using both Kirchhoff's laws and Ohm's law, that for the series connection (Figure 1.12) the total resistance is the sum of the separate resistances.

$$R = \sum_{k=1}^{n} R_k. \tag{1.16}$$

The dual situation is the parallel connection (see Figure 1.13). We find

$$G = \sum_{k=1}^{n} G_k. \tag{1.17}$$

1.5 Voltage and current division

A formula which is often used is the formula of voltage *division*. Consider the network of Figure 1.14. We calculate V_2, and find $I = \dfrac{V}{R_1 + R_2}$ and $V_2 = R_2 I$.

So

$$V_2 = \frac{R_2}{R_1 + R_2} \cdot V. \tag{1.18}$$

This is called *voltage division*.

The dual of voltage division is the *current division* (Figure 1.15). We find

$$V = \frac{I}{G_1 + G_2} \text{ and } I_2 = G_2 V.$$

So

$$I_2 = \frac{G_2}{G_1 + G_2} \cdot I. \tag{1.19}$$

Sometimes this formula is written with resistances. With $G_1 = \dfrac{1}{R_1}$ and $G_2 = \dfrac{1}{R_2}$ we obtain

$$I_2 = \frac{R_1}{R_1 + R_2} \cdot I. \tag{1.20}$$

Note the index in the numerator.

1.6 The solution of larger networks

In general the solution of a network problem involves the computing of all voltages and currents in the network. As an example we shall solve the bridge circuit and attach a current to each branch (Figure 1.16).

Kirchhoff's current law for the nodes A, B, C and D gives:

$$I_1 = I_2 + I_5 \tag{a}$$
$$I_2 = I_3 + I_4 \tag{b}$$
$$I_1 = I_3 + I_6 \tag{c}$$
$$I_6 = I_4 + I_5 \tag{d}$$

Not all of these equations are independent. It follows from formulas (a) and (b) and from (c) and (d) that:

$$I_1 = I_3 + I_4 + I_5.$$

Kirchhoff's voltage law for the loops 1, 2 and 3 gives:

$$V_{AD} + V_{DC} + V_{CA} = 0 \tag{e}$$
$$V_{AB} + V_{BD} + V_{DA} = 0 \tag{f}$$
$$V_{BC} + V_{CD} + V_{DB} = 0 \tag{g}$$

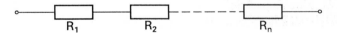

Figure 1.12

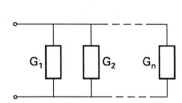

Figure 1.13

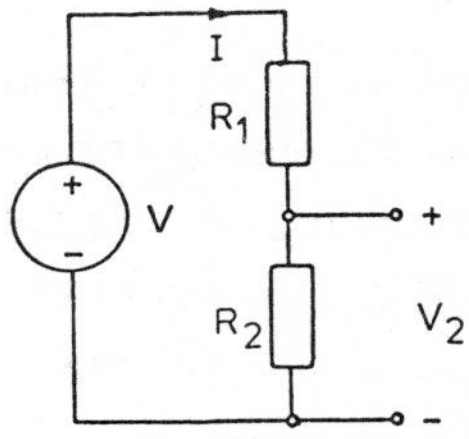

Figure 1.14

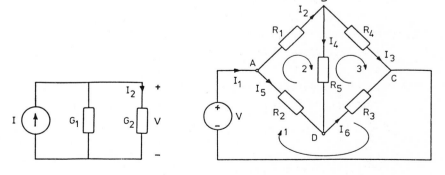

Figure 1.15

Figure 1.16

These three loops, each not enclosing smaller loops, are called *meshes*.
Kirchhoff's voltage law applied to the loop ABCD results in:

$$V_{AB} + V_{BC} + V_{CD} + V_{DA} = 0. \tag{h}$$

Formula (h) is dependent, because adding (f) and (g) yields (h).
Finally we apply Ohm's law:

$$V_{AB} = R_1 I_2, \tag{i}$$
$$V_{AD} = R_2 I_5, \tag{j}$$
$$V_{BD} = R_5 I_4, \tag{k}$$
$$V_{BC} = R_4 I_3, \tag{l}$$
$$V_{DC} = R_3 I_6. \tag{m}$$

The number of equations we obtain with this so-called *branch method* is not convenient and is also unnecessarily large. We shall now discuss two more systematic methods.

1.7 The mesh method

We again examine the bridge circuit and attach a so-called *mesh current* to each mesh. This is a fictitious current which *circulates* in a mesh (see Figure 1.17).
Consequently a *branch current* is the difference between two mesh currents if that branch is the separation of two meshes:

$$I_4 = I_2 - I_3. \tag{1.21}$$

In an outer mesh the mesh current is equal to the branch current, e.g. I_1 in the source and I_2 in R_1. Now apply the voltage law to the three meshes. We write the intensity of the source in the left-hand part of the equation and the voltages across the resistors in the right-hand part:

$$V = R_2(I_1 - I_2) + R_3(I_1 - I_3)$$

or, in another sequence

$$V = (R_2 + R_3)I_1 - R_2 I_2 - R_3 I_3. \tag{α}$$

The first term in the right-hand part is the voltage created by the mesh current in the sum of the resistances in the mesh, the other terms are negative and are created by the 'counteracting' currents in the adjacent meshes.
Further note that the source voltage is positive, if, following the mesh direction, the plus terminal is left. In a similar way for the other meshes we find:

$$0 = -R_2 I_1 + (R_1 + R_5 + R_2)I_2 - R_5 R_3, \tag{β}$$

$$0 = -R_3 I_1 - R_5 I_2 + (R_4 + R_3 + R_5)I_3. \tag{γ}$$

With the three equations (α), (β) and (γ) we can solve the three mesh currents. All branch currents (and thus all branch voltages) are then known. In matrix notation the three mesh equations become:

$$\begin{bmatrix} V \\ 0 \\ 0 \end{bmatrix} = \begin{bmatrix} R_2+R_3 & -R_2 & -R_3 \\ -R_2 & R_1+R_5+R_2 & -R_5 \\ -R_3 & -R_5 & R_4+R_3+R_5 \end{bmatrix} \begin{bmatrix} I_1 \\ I_2 \\ I_3 \end{bmatrix}.$$

With

$$\mathcal{V} = \begin{bmatrix} V \\ 0 \\ 0 \end{bmatrix} \quad \mathcal{R} = \begin{bmatrix} R_2+R_3 & -R_2 & -R_3 \\ -R_2 & R_1+R_5+R_2 & -R_5 \\ -R_3 & -R_5 & R_4+R_3+R_5 \end{bmatrix}$$

and $I = \begin{bmatrix} I_1 \\ I_2 \\ I_3 \end{bmatrix}$,

we get Ohm's law in *matrix* form

$$\mathcal{V} = \mathcal{R}I. \tag{1.22}$$

Note that this formula can only be written in this sequence. On the principal diagonal of $\mathcal{R}$ there are only positive terms. The other terms are not positive (if we choose all mesh currents turning clockwise or anti-clockwise).

The symmetry is remarkable: If we place a two-sided mirror on the principal diagonal both subject and image of the non-positive terms are equal!

For a second example we turn to a network with a voltage source and a current source (Figure 1.18). Choose the voltage V across the current source with arbitrary polarity and choose three mesh currents I_1, I_2 and I_3.
We find

$$-V \quad = \tfrac{5}{2}I_1 - \tfrac{1}{2}I_2, \tag{a}$$

$$V - 9 = -\tfrac{1}{2}I_1 + \tfrac{3}{2}I_2, \tag{b}$$

$$9 \quad = I_3. \tag{c}$$

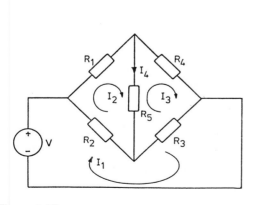

Figure 1.17

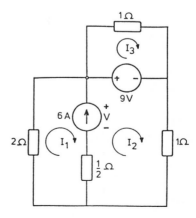

Figure 1.18

As an additional equation we have

$$6 = I_2 - I_1. \tag{d}$$

The terms of (a), (b) and (c) are voltages, those of (d) are currents.
Solution gives

$$I_1 = -5 \text{ A} \qquad I_2 = 1 \text{ A} \qquad I_3 = 9 \text{ A} \qquad V = 13 \text{ V}.$$

Note that the current law is apparently not used in the mesh method. This is due to the fact that each mesh current automatically satisfies the current law: Each mesh current enters a node and leaves the same node.

In the above we have chosen all mesh currents clockwise. This is not necessary. One mesh current can be chosen clockwise and the other anti-clockwise.

The mesh method is limited to *planar* networks, these are networks which can be drawn on a flat surface without crossing branches (Figure 1.19).

In Figure 1.19 the so-called *graphs* of some networks are drawn; the sources and resistors have been replaced by lines connected to nodes.

The lines are called *branches*. The graph is drawn in such a way that a discontinuity in the line is a node. In this way one is often forced to draw curvilinear branches. All three graphs of Figure 1.19 are planar. The graph of Figure 1.19(b) contains two crossing branches but it is the same graph as shown in Figure 1.19(c). It is also usual in graph-theory to give each branch a direction; we then speak of *directed graphs*.

A method that can also be used to solve non-planar networks is the so-called *node method*.

1.8 The node method

Consider the network of Figure 1.20.
One node is at the potential zero (*earth*); for the other nodes we write the current law:

$$\text{Node 1:} \qquad 5 = 1 \cdot V_{10} + 2 \cdot V_{12}.$$

But we have

$$V_{12} = V_{10} - V_{20} \text{ (index rule).}$$

Hence

$$5 = 3V_{10} - 2V_{20}. \tag{α}$$

We can reason out this equation as follows:
The left-hand part contains the source current, going to the node. The first term in the right-hand part is the sum of the conductances adjacent to the considered node, multiplied by the voltage of the node, measured with respect to earth. The other terms are not positive and are created by the 'counteracting' voltage of the adjacent nodes. (Note the dual reasoning with respect to the mesh method).
For the other nodes we find:

Node 2: $0 = -2V_{10} + 9V_{20} - 4V_{30},$ (β)

Node 3: $10 = \quad 0 \quad - 4V_{20} - 6V_{30}.$ (γ)

In matrix form:

$$\begin{bmatrix} 5 \\ 0 \\ 10 \end{bmatrix} = \begin{bmatrix} 3 & -2 & 0 \\ -2 & 9 & -4 \\ 0 & -4 & 6 \end{bmatrix} \begin{bmatrix} V_{10} \\ V_{20} \\ V_{30} \end{bmatrix}.$$

In general

$$I = \mathcal{G}\mathcal{V}. \tag{1.23}$$

Note that the voltage law is used in the index rule. If the node voltages have been solved, all branch voltages are known and thus all branch currents. Now consider the network again with a voltage source and a current source (Figure 1.18) and solve it with the node method (Figure 1.21)

We choose the current I in the voltage source.

Node 1: $6 - I = \frac{3}{2} V_{10} - V_{20},$ (1)

Node 2: $I = -V_{10} + 2V_{20},$ (2)

Node 3: $-6 = \qquad 2V_{30},$ (3)

Additional equation: $V_{10} - V_{20} = 9.$ (4)

Solution gives:

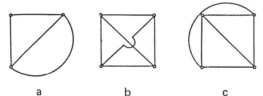

Figure 1.19   a   b   c

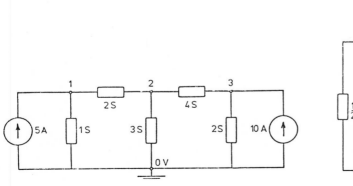

Figure 1.20

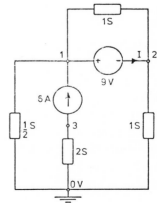

Figure 1.21

$$V_{10} = 10 \text{ V}, \quad V_{20} = 1 \text{ V}, \quad V_{30} = -3 \text{ V} \quad \text{and} \quad I = -8 \text{ A}.$$

Instead of the voltages V_{10}, V_{20} etc one can write V_1, V_2, etc while it is assumed that the voltages are always measured with respect to the zero potential.

1.9 The current law for a cut-set

Consider Figure 1.22.

In this graph of a network the branch currents are given. For nodes 1, 2 and 3 it holds that:

$$-I_1 - I_2 + I_4 = 0,$$

$$I_1 - I_3 + I_6 = 0,$$

$$I_2 + I_3 + I_5 = 0.$$

Adding results in $I_4 + I_5 + I_6 = 0$ and these terms happen to be the currents crossing the intersecting line A-B. The set of branches, cut by such an intersecting line is called a *cut-set*.

If we remove all branches from a cut-set the graph falls apart into two separate parts; if we replace one arbitrary branch from the cut-set, we get a connected graph again.

So Kirchhoff's current law is also valid for a cut-set. The current law for a node is to be regarded as a particular case, namely for a *node-cut-set*, for example the intersecting line C-D. (Note that one node is also part of a graph).

1.10 Superposition

Consider the network of Figure 1.23.

We calculate the current I. For example, we can do this with the mesh method :

$$V_1 = 12I_1 + 6I_2,$$

$$V_2 = 6I_1 + 9I_2,$$

$$I = I_2 + I_2.$$

It follows that $I = \frac{1}{24} V_1 + \frac{1}{12} V_2$. We find a function of two variables

$$I = f(V_1, V_2). \tag{1.24}$$

This function is *linear*, because it does not contain constant values, second and higher order powers, roots, etc. The network can be drawn as follows (Figure 1.24).

V_1 and V_2 are input quantities. They are called *excitations*. I is the output quantity, often called *response*. The system is determined by the network.

One can deduce (see Appendix I) that the principle of superposition is valid for such a linear network:

A voltage or a current in a network can be found by making all but one source strengths zero, then by finding the relevant voltage or current and further executing this for all sources and finally adding all results.

In practice the superposition rule is seldom used in networks with more than two sources.

Example (see Figure 1.25).
Find the current I with the superposition rule.

Solution
In the first situation we take the current source intensity zero, i.e. open nodes. In the second situation the voltage source intensity is set at zero, which means a short circuit (Figure 1.26).
We find $I' = 2$ A, $I'' = 1$ A. So $I = I' + I'' = 3$ A.

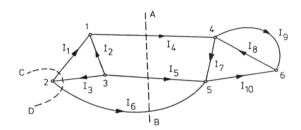

Figure 1.22

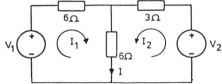

Figure 1.23

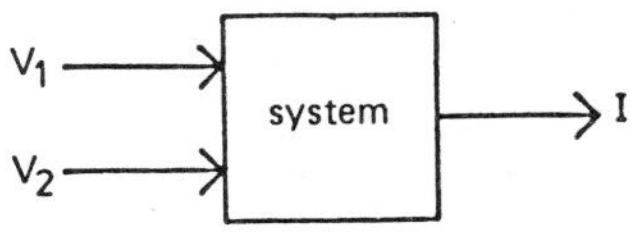

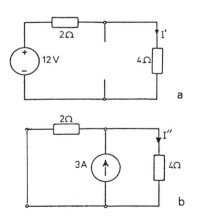

Figure 1.24

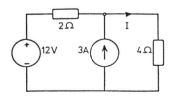

Figure 1.25

Figure 1.26

1.11 Tellegen's theorem

In Figure 1.27 the graph of a network is drawn, and in it the branch currents and node potentials are indicated.

Now calculate the product of the voltage and the current of each branch and add the results. For each branch we let the current 'belong to the voltage', i.e. the current in each branch flows from plus to minus.

$V_{12}I_1 + V_{12}I_2 + V_{13}I_3 + V_{32}I_4$

$$= (V_1 - V_2)I_1 + (V_1 - V_2)I_2 + (V_1 - V_3)I_3 + (V_3 - V_2)I_4$$

$$= V_1(I_1 + I_2 + I_3) + V_2(-I_1 - I_2 - I_4) + V_3(I_4 - I_3).$$

This expression is zero because according to the current law we have:

Node 1: $I_1 + I_2 + I_3 = 0,$

Node 2: $-I_1 - I_2 - I_4 = 0,$

Node 3: $I_4 - I_3 = 0.$

So for this network:

$$\sum_{n=1}^{4} V_n I_n = 0.$$

In this formula n is the branch number, V the branch voltage and I the branch current.

One can prove (see Appendix II) that this equation in general holds for a network with b branches:

$$\sum_{n=1}^{b} V_n I_n = 0. \tag{1.25}$$

This is *Tellegen's theorem*.

The theorem also holds for non-linear networks. *It even holds for two networks with the same graph, in which the branch voltages in one network are taken and the currents in the other.*

Example (see Figure 1.28).

Both networks have the same graph. Figure 1.29 shows the graphs of both networks, in which each branch has been given a number.

Calculating gives the following table:

Branch	Network a		Network b		V_aI_a	V_aI_b
	Voltage	Current	Voltage	Current		
1	12	−5	22	−10	−60	−120
2	5	5	10	10	25	50
3	7	7	12	4	49	28
4	7	−2	12	6	−14	42
					0	0

It is very important to use the rule that for each branch the voltage and the current belong to each other. In the sixth column the product of the branch voltages and the branch currents are both taken in network a; in the last column the branch voltages in network a and the branch currents in network b have been taken. In both cases summation results in zero.

If the branch voltages and the branch currents concern the same network, Tellegen's theorem is the same as the law of power balance. The unity then is 1 W (only in the case of d.c.!).

If the branch voltages and the branch currents do not concern the same network or if they concern two different situations in one network we cannot speak of power.

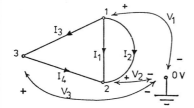

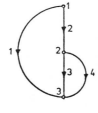

Figure 1.27

Figure 1.29

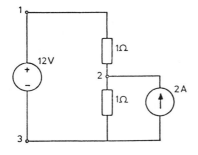

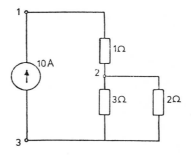

Figure 1.28   a

b

1.12 Two-ports

In Figure 1.30 the block scheme of a *two-port* is drawn.

Within the rectangle there is a network consisting of resistors and possibly of sources, interconnected by wires (without resistance). Both terminals on the left are often called *input gates*, *input terminals* or in short *input*. In the right-hand part is the *output*.

There often are only resistors within the rectangle in a certain configuration. The two-port is then *passive*. Often a source is connected to the input, whereas the output is connected to a resistor.

The two-port theory is an extensive subject in network theory. Here we shall merely treat some basic properties. In Chapter 8 we shall discuss the two-ports further. Two terminals of a network together form a *port* if the current entering one terminal equals the current leaving the other terminal. This is called the *port condition*.

In some situations the port condition is not met. In Figure 1.31 a situation is drawn in which it is not at all clear whether the port condition has been met. In general $I_a \neq I_b$ and $I_c \neq I_d$. We further note that we often choose the directions of the currents as drawn in Figure 1.30. Sometimes, however, we deviate from this rule.

As an example of a two-port we shall consider Figure 1.32.

Relations between the port quantities (V_1, V_2, I_1, I_2) are derived with the aid of the mesh method.

$$V_1 = 3I_1 + 2I_2 - I_3,$$

$$V_2 = 2I_1 + 8I_2 + 6I_3,$$

$$0 = -I_1 + 6I_2 + 8I_3.$$

Elimination of I_3 leads to

$$V_1 = \frac{23}{8} I_1 + \frac{11}{4} I_2,$$

$$V_2 = \frac{11}{4} I_1 + \frac{7}{2} I_2.$$

In matrix notation

$$\begin{bmatrix} V_1 \\ V_2 \end{bmatrix} = \begin{bmatrix} \frac{23}{8} & \frac{11}{4} \\ \frac{11}{4} & \frac{7}{2} \end{bmatrix} \begin{bmatrix} I_1 \\ I_2 \end{bmatrix}$$

or $\mathcal{V} = \mathcal{R}I$ with $\mathcal{R}$ as resistance matrix:

$$\mathcal{R} = \begin{bmatrix} \frac{23}{8} & \frac{11}{4} \\ \frac{11}{4} & \frac{7}{2} \end{bmatrix} = \begin{bmatrix} R_{11} & R_{12} \\ R_{21} & R_{22} \end{bmatrix}.$$

We find

$$R_{21} = R_{12}. \tag{1.26}$$

This symmetry in the resistance matrix of a passive two-port is due to the validity of the so-called *reciprocity theorem*. See Appendix III.

We now replace the voltage source V_2 in Figure 1.29 by a short circuit (see Figure 1.33). We choose V for V_1.

To calculate I_2 in the short circuit we may use the mesh method, and after some calculating find $I_2 = 1.1$ V. We now change the short circuit and voltage source and thus get Figure 1.34.

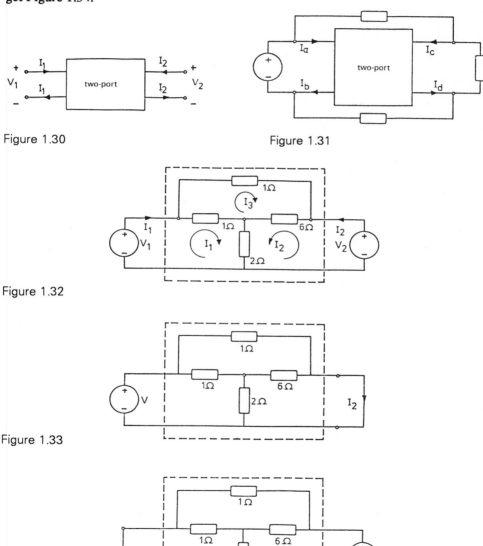

Figure 1.30

Figure 1.31

Figure 1.32

Figure 1.33

Figure 1.34

Once more we calculate the current in the short circuit and find that $I_1 = 1.1$ V. We have found a remarkable result: If we interchange an ampere meter by a voltage source the deflection of the meter remains the same (the ampere meter has to be ideal, i.e. with resistance zero).

The dual rule says: If we interchange a voltage meter (its conductance must be zero) by a current source the deflection of the meter remains the same.

These results are also due to the reciprocity theorem.

1.13   Thévenin's and Norton's theorems

Consider the network of Figure 1.35. The network within the rectangle contains voltage sources, current sources and resistors.

We want to find the current I in the resistor R. According to Thévenin's theorem one can replace the network inside the rectangle by a voltage source V_T in series with a resistor R_T. This is called *Thévenin's equivalence*, see Figure 1.36.

The voltage source intensity V_T is found by determining the open voltage $\overset{o}{V}_{ab}$ (that means take away R) between the terminals a and b. We have

$$V_T = \overset{o}{V}_{ab}. \tag{1.27}$$

The resistance R_T is found by determining the resistance at the terminals a and b after having removed the resistor R and after having set all internal source intensities at zero:

$$R_T = R_{ab}. \tag{1.28}$$

The dual of Thévenin's theorem is *Norton's theorem*. The network inside the rectangle of Figure 1.35 is then replaced by the *Norton equivalence* (see Figure 1.37).

I_N is found by determining the *short circuit current* I_s between the terminals a and b in Figure 1.35:

$$I_N = I_s. \tag{1.29}$$

The conductance G_N equals the conductance one measures at the terminals a and b after having taken away R and having set the internal source intensities at zero. So we find:

$$G_N = \frac{1}{R_T} \tag{1.30}$$

and

$$I_N = \frac{V_T}{R_T}, \tag{1.31}$$

so

$$R_T = \frac{\overset{o}{V}_{ab}}{I_s}. \tag{1.32}$$

This introduces a second method to determine R_T.

Finding the current I in the network of Figure 1.35 is no longer difficult after determination of one of the equivalences. With the Thévenin equivalence we find

$$I = \frac{V_T}{R_T + R} \qquad (1.33)$$

and with the Norton equivalence

$$I = I_N \frac{R}{R_T + R} . \qquad (1.34)$$

In Appendix IV both theorems are proved.

Example (Figure 1.38).
We wish to find I. Take away $R = 18\ \Omega$ and determine the open voltage $\overset{o}{V}_{ab}$ (Figure 1.39). We find

$$I_1 = \frac{1}{3}\ A.$$

So

$$\overset{o}{V}_{ab} = 12 - 6 \cdot \frac{1}{3} = 10\ V = V_T.$$

It further follows from Figure 1.39 when we set the source intensities at zero (here short circuits) that

$$R_{ab} = \frac{3 \cdot 6}{3 + 6} = 2\ \Omega = R_T.$$

With these results the Thévenin equivalence has been found, see Figure 1.40.

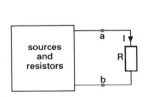

Figure 1.35

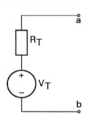

Figure 1.36

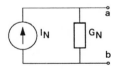

Figure 1.37

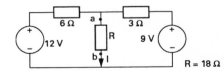

Figure 1.38

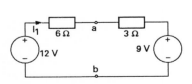

Figure 1.39

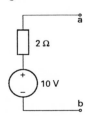

Figure 1.40

If we finally bring back the resistor R = 18 Ω it follows that I = $\frac{10}{20}$ = $\frac{1}{2}$A.

The Norton equivalence is also known, because from the above:

$$I_s = \frac{V_T}{R_T} = \frac{10}{2} = 5 \text{ A} = I_N \quad \text{and} \quad G_N = \frac{1}{R_T} = \frac{1}{2} \text{ S} \quad \text{(see Figure 1.41).}$$

We finally note that R_{ab} often does not consist of resistors in series or in parallel for more intricate networks. In order to find R_{ab} in those cases, one has to connect a voltage source or a current source to the terminals a and b and subsequently calculate the source current or the source voltage with one of the methods known. In that case the resistance R_{ab} is the quotient of source voltage and source current.

1.14 Maximum power transfer

Consider the network of Figure 1.42. The part to the left of terminals a and b is the *non-ideal voltage source* or the Thévenin equivalence of a network. The part to the right of nodes a and b is the *load*.

Given V_T and R_T we calculate the load R so that in that load R will be dissipated a power that is as large as possible. We find

$$I = \frac{V_T}{R_T + R}, \quad \text{so} \quad P = I^2 R = \frac{V_T^2 R}{(R_T + R)^2},$$

$$\frac{dP}{dR} = V_T^2 \frac{(R_T + R)^2 - R \cdot 2(R_T + R)}{(R_T + R)^4} = V_T^2 \frac{R_T^2 - R^2}{(R_T + R)^4}$$

$$= V_T^2 \frac{R_T - R}{(R_T + R)^3}.$$

This expression is zero if $R_T = R$. So we have

$$R = R_T. \tag{1.35}$$

In order to learn if the extremum found in this way is a maximum we have to calculate the second derivative.

$$\frac{d^2P}{dR^2} = V_T^2 \cdot \frac{(R_T + R)^4 \cdot -2R - (R_T^2 - R^2) \cdot 4(R_T + R)^3}{(R_T + R)^8},$$

$$\frac{d^2P}{dR^2}\bigg|_{R=R_T} = V_T^2 \cdot \frac{16 R_T^4 \cdot -2R_T}{(2R_T)^8} < 0,$$

so the extremum is indeed a maximum. We find the plot in Figure 1.43.
For R = 0, P = I^2R = 0 and for R $\to \infty$

$$P = \frac{V_{ab}^2}{R} = \frac{V_T^2}{R} = 0.$$

Maximum power transfer is very important if the available power of the source is very small, while one wants to transfer a part as large as possible to the load. Just think of telephony, in which the microphone is to be regarded as a voltage source with a resistor in series while the telephone at the end of the transmission line is the load.

Note that 50 % will be dissipated in R_T in the case of maximum power transfer. In high power technology, however, one strives for as small a loss as possible, so R_T is made minimal.

1.15 Star-delta transformation

Three resistors in *delta connection* are given (Figure 1.44).
A, B and C are *resistances*. This circuit is identical to three resistors in star according to Figure 1.45.
It turns out that

$$D = \frac{AB}{A + B + C} \qquad (1.36)$$

and there are analogous equations for E and F. See Appendix V.
If, on the other hand, the star circuit of Figure 1.46 is given, which one wants to alter into the delta connection of Figure 1.47, one can use the same equations, provided one takes *conductances* instead of resistances.

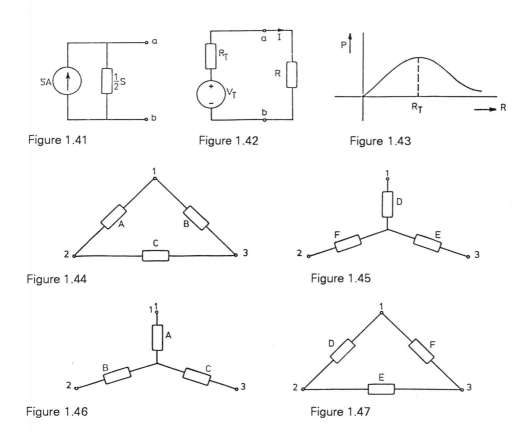

Figure 1.41   Figure 1.42   Figure 1.43

Figure 1.44   Figure 1.45

Figure 1.46   Figure 1.47

Example (Figure 1.48).

We want for the equivalent star circuit. We find thatthe resistances are $\frac{2\cdot5}{10}\,\Omega$, $\frac{5\cdot3}{10}\,\Omega$ and $\frac{2\cdot3}{10}\,\Omega$ (Figure 1.49).

Starting with Figure 1.49 we now ask for the equivalent delta circuit, and so we take conductances instead of resistances (Figure 1.50).

The conductances of the delta circuit become

$$\frac{\frac{5}{3}\cdot1}{1+\frac{2}{3}+\frac{5}{3}}=\frac{1}{2}\,\text{S}, \quad \frac{\frac{5}{3}\cdot\frac{2}{3}}{1+\frac{2}{3}+\frac{5}{3}}=\frac{1}{3}\,\text{S}, \quad \frac{1\cdot\frac{2}{3}}{1+\frac{2}{3}+\frac{5}{3}}=\frac{1}{5}\,\text{S}$$

with which the circuit of Figure 1.51 is found, which is the same as the circuit of Figure 1.48.

1.16 Controlled sources

The sources we have met so far had an independent strength, i.e. the strength is always the same no matter which currents or voltages are present in the network. In electronics, however, one has developed devices which made it necessary to introduce so-called *controlled sources* in network theory. Such a source has a strength that depends on a current or a voltage elsewhere in the network. The *transistor* is an example of this (Figure 1.52). This is a two-port with b and e (base and emitter) as input terminals and c and e (c = collector) as output terminals.

Note

An electronic circuit only works if connected to it is a so-called *bias voltage* equipment. Often these bias voltages and currents are not further indicated, neither have they been drawn in Figure 1.52. The input current I_b and the output current I_c should be considered as small increases of the bias currents. Both port voltages V_{ce} and V_{be} are also positive variations of the bias voltages. Below we shall consider the four port quantities as (small) d.c. currents and d.c. voltages.

A simple substitute is shown in Figure 1.53.

Practicable values are $\alpha = 100$, $R_b = 100\,\Omega$, $R_e = 25\,\Omega$. Here we have a *current-controlled current source*, i.e. a source of which the intensity is dependent. The symbol is drawn in Figure 1.53. The source current is zero when the input current I_b is zero and is, for instance, 100 mA when $I_b = 1$ mA (with $\alpha = 100$).

Such a source together with the control unit is called a *transactor*.

1.17 Transactors

a. *The current-current transactor* (iit) (Figure 1.54).

Here we have a current-controlled current source. The input is a short circuit, so

$$V_1 = 0.$$

Therefore it is not allowed to connect a voltage source to the input. Further, the output current is α times the input current, so

$$I_2 = \alpha I_1.$$

If I_1 and α are not zero, a load must have been connected to the output. In matrix notation the above equations become

$$\begin{bmatrix} V_1 \\ I_1 \end{bmatrix} = \begin{bmatrix} 0 & 0 \\ 0 & -\dfrac{1}{\alpha} \end{bmatrix} \begin{bmatrix} V_2 \\ -I_2 \end{bmatrix}.$$

The matrix introduced is known as the *chain matrix* of the two-port:

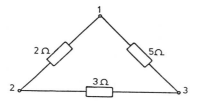

Figure 1.48

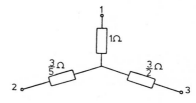

Figure 1.49

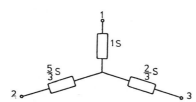

Figure 1.50

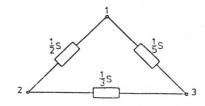

Figure 1.51

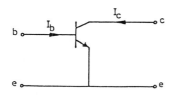

Figure 1.52

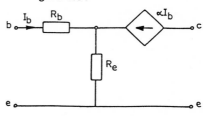

Figure 1.53

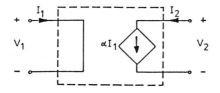

Figure 1.54

$$\mathcal{K} = \begin{bmatrix} 0 & 0 \\ 0 & -\dfrac{1}{\alpha} \end{bmatrix}. \tag{1.37}$$

The current I_1 and the voltage V_2 are determined by the network in which the transactor is connected. From the above it follows that α has no dimension. The meaning of the negative sign will be explained in Chapter 8.

b. *The voltage-voltage transactor* (vvt) (Figure 1.55).
This is a *voltage-controlled voltage source*. We have

$$I_1 = 0,$$

$$V_2 = \mu\, V_1,$$

$$\mathcal{K} = \begin{bmatrix} \dfrac{1}{\mu} & 0 \\ 0 & 0 \end{bmatrix}. \tag{1.38}$$

μ has no dimension.

c. *The voltage-current transactor* (vit) (Figure 1.56).
This is a *voltage-controlled current source*.

$$I_1 = 0,$$

$$I_2 = GV_1.$$

The conductance matrix is

$$\mathcal{G} = \begin{bmatrix} 0 & 0 \\ G & 0 \end{bmatrix}. \tag{1.39}$$

G has the dimension of conductance.

d. *The current-voltage transactor* or *current-controlled voltage source* (ivt) (Figure 1.57).

$$V_1 = 0,$$

$$V_2 = RI_1.$$

The resistance matrix is

$$\mathcal{R} = \begin{bmatrix} 0 & 0 \\ R & 0 \end{bmatrix}. \tag{1.40}$$

R has the dimension of resistance.

The transactors are active devices: The power delivered to the input is zero (the voltage or the current is zero), while a non-zero power may be given by the output. Further, they are linear and non-reciprocal. From (1.40) we clearly see that the current-voltage transactor is non-reciprocal: $R_{21} \neq R_{12}$. The four transactors are not all elementary (Figure 1.58).
We have $V_2 = R_A I_1$, $I_4 = G_B V_3$ and $V_3 = V_2$.
Thus

$$I_4 = G_BR_AI_1.$$

So

$$I_4 = \alpha I_1$$

with

$$\alpha = G_BR_A.$$

Cascading a current-voltage transactor and a voltage-current transactor is a current-current transactor.

1.18 Special networks

Consider the circuit of Figure 1.59.

This circuit contains a current-current transactor. The current through the resistor is $(1 - \alpha)I_1$, so the voltage becomes

$$V = (1 - \alpha)RI_1.$$

The substitute or input resistance of this one-port becomes

$$R_i = (1 - \alpha)R.$$

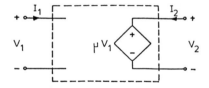

Figure 1.55

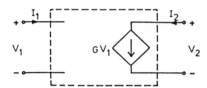

Figure 1.56

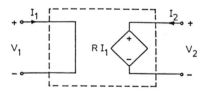

Figure 1.57

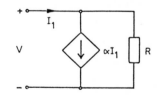

Figure 1.59

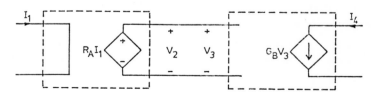

Figure 1.58

For $\alpha = 1$, $R_i = 0$.

For $\alpha < 1$, $R_i > 0$.

For $\alpha > 1$, $R_i < 0$.

So with a transactor and a positive resistance we can make a *negative resistance*.

Next consider the circuit of Figure 1.60.

Here we have made a two-port by connecting two two-ports in parallel. We have

$$I_1 = GV_2,$$

$$I_2 = -GV_1.$$

With $R = \frac{1}{G}$ it follows that

$$\left. \begin{matrix} V_1 = -RI_2 \\ \\ V_2 = RI_1 \end{matrix} \right\} \tag{1.41}$$

with the resistance matrix

$$\mathcal{R} = \begin{bmatrix} 0 & -R \\ R & 0 \end{bmatrix}.$$

Such a two-port is called a *gyrator*. Its symbol is shown in Figure 1.61.

Although the gyrator consists of active components it is *passive* itself.

R is called the *gyration resistance*.

We shall not treat the properties of the gyrator further here, but mention only one aspect.

If we connect a resistor R_o to the output we have $V_2 = -R_oI_2$. Together with the gyrator formulas (1.41) we find

$$V_1 = R \cdot \frac{V_2}{R_o} = \frac{R^2}{R_o} I_1,$$

with which the input resistance becomes

$$R_i = \frac{R^2}{R_o}.$$

With $R = 1\ \Omega$ we find

$$R_i = \frac{1}{R_o}.$$

A resistor of $100\ \Omega$ on the input of a gyrator with a gyration resistance of $1\ \Omega$ means an input resistance of $0.01\ \Omega$.

So we might call the gyrator a resistance inverter, here a positive one. In *a.c.* theory the concept *impedance* is introduced (see Chapter 2). Thus we are led to the expression *positive impedance inverter*. We also know the *negative impedance inverter* (NII) (see Figure 1.62).

We find:

$$V_1 = R_1I_2 \quad \text{and} \quad V_2 = R_2I_1. \tag{1.42}$$

If we connect a load R_o to the output of the NII then $V_2 = -R_oI_2$, so

$$V_1 = R_1I_2 = \frac{-R_1V_2}{R_o} = \frac{-R_1R_2I_1}{R_o} \,,$$

this means that the input resistance is

$$R_i = -\frac{R_1R_2}{R_o} \,.$$

For $R_1 = R_2 = 1 \; \Omega$ we get

$$R_i = -\frac{1}{R_o} \,. \tag{1.43}$$

The input resistance is the negative inverse of the load.

The *negative impedance convertor* (NIC) is drawn in Figure 1.63.
We find:

$$V_1 = \mu V_2 \quad \text{and} \quad I_2 = \alpha I_1. \tag{1.44}$$

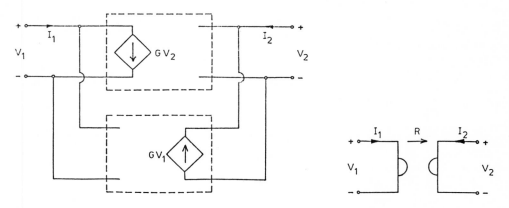

Figure 1.60

Figure 1.61

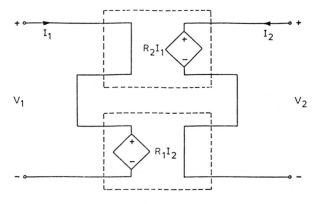

Figure 1.62

Connecting R_o to the output results in $V_2 = -R_oI_2$, so

$$V_1 = \mu V_2 = -\mu R_oI_2 = -\mu\alpha R_oI_1.$$

Hence the input resistance is $R_i = \dfrac{V_1}{I_1} = -\mu\alpha R_o$.

For $\mu = \alpha = 1$ we get

$$R_i = -R_o. \tag{1.45}$$

The input resistance is the negative of the load.

As an example of a network with a transactor look at the network of Figure 1.64.

Question: find the current I.

With the mesh method we find

$$1 = 2I_1 - I,$$

$$0 = -I_1 + 8I + 5I_1.$$

It follows that $I = -0.2$ A.

1.19 Thévenin's theorem in networks with transactors

We shall try to find the current I in the network of Figure 1.64 with Thévenin's theorem.
So we remove the resistor of 6 Ω (Figure 1.65).
We find

$$1 = 2I_1 \implies I_1 = \tfrac{1}{2}\,A,$$

$$\overset{\circ}{V}_{ab} = V_T = I_1 - 5I_1 = -2\ V.$$

We now short-circuit terminals a and b and calculate the short circuit current (Figure 1.66).

$$1 = 2I_2 - I_s,$$

$$0 = -I_2 + 2I_s + 5I_2.$$

It follows that $I_s = -\tfrac{1}{2}\,A.$

(We have changed I_1 into I_2, because in general this current will take another value while short-circuiting a and b.) We find

$$R_T = \frac{-2}{-\tfrac{1}{2}} = 4\ \Omega,$$

with which the Thévenin equivalence has been found (Figure 1.67).

Should we calculate R_{ab}, the intensities of the independent source of 1 V and the dependent current source of $5I_1$ being set at zero, it follows that $R_T = \tfrac{3}{2}\ \Omega$, which is an incorrect result.

The reason is that in the equation $R_T = R_{ab}$ (1.28) the intensities of the controlled sources must not be made zero (see Appendix IV). In order to find the correct value of R_T in the second method we consider the network of Figure 1.68 in which

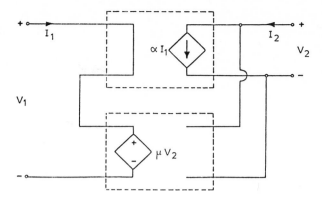

Figure 1.63

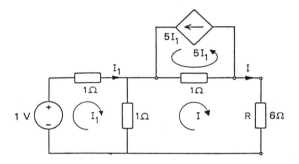

Figure 1.64

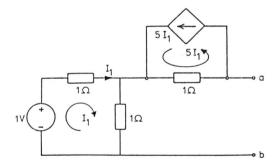

Figure 1.65

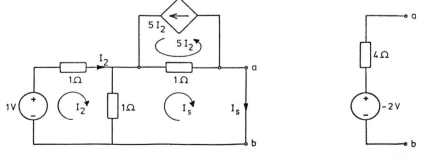

Figure 1.66

Figure 1.67

$$R_{ab} = \frac{V_B}{I_B} = R_T.$$

Note that I_1 of Figure 1.65 has been changed to I_3.
We find

$$0 = 2I_3 + I_B,$$

$$V_B = I_3 + 2I_B - 5I_3.$$

It follows that $V_B = 4I_B$, therefore $R_T = \frac{V_B}{I_B} = 4\ \Omega$, with which the correct value has been found.

If we finally set the external resistor at $6\ \Omega$, then with the Thévenin equivalence we find:

$$I = \frac{-2}{4+6} = -0.2\ \text{A}$$

in accordance with the value found with the mesh method.

1.20 The operational amplifier

The *operational amplifier* (*opamp*) is a voltage-voltage transactor (Figure 1.55) in which the amplification factor μ is very high. During the limit transition $\mu \to \infty$ the input voltage will be $V_1 = 0$, because

$$V_1 = \frac{V_2}{\mu}\ .$$

For this limit transition a new symbol has been introduced (Figure 1.69).
$I_1 = 0$; I_2 is also determined by the connected network. Further, the lower terminal of the output is connected to earth, as is usually done in practice. One of the input terminals can also be connected to earth, but this is not necessary. We now consider Figure 1.70.
For node a it holds that:

$$0 = (G_1 + G_2)V_a - G_1 V_1 - G_2 V_2.$$

Further $V_a = 0$.
So

$$V_2 = -\frac{G_1}{G_2} V_1$$

or

$$V_2 = -\frac{R_2}{R_1} V_1.$$

thus we find a (negative) *multiplier* (Figure 1.71).
For $R_1 = R_2$ we only have a change of sign.

In Section 8.7 we shall explain why the feedback is to the negative terminal here, instead of to the positive.

A circuit for adding two voltages is shown in Figure 1.72.
The node equations for the nodes a and b are:

(a) $0 = (G_1 + G_3)V_a \qquad -G_1V_1,$

(b) $0 = \qquad (G_2 + G_4)V_b \qquad -G_2V_2 - G_4V_3.$

It further holds that

$$V_a = V_b.$$

With this we find

$$V_3 = \frac{G_1(G_2 + G_4)}{G_4(G_1 + G_3)} V_1 - \frac{G_2}{G_4} V_2$$

or

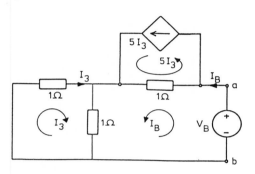

Figure 1.68

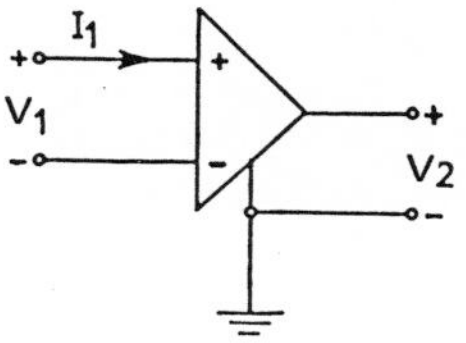

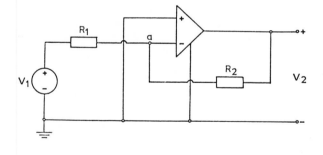

Figure 1.69

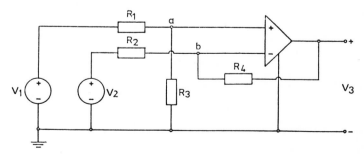

Figure 1.70

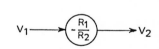

Figure 1.71

Figure 1.72

$$V_3 = \frac{1 + \dfrac{R_4}{R_2}}{1 + \dfrac{R_1}{R_3}} V_1 - \frac{R_4}{R_2} V_2.$$

An *adder* with coefficients α and β is shown in Figure 1.73.
Another circuit for adding two voltages is drawn in Figure 1.74.
We find

$$V_a = V_b = 0.$$

It further holds for node b that:

$$0 = (G + G_1 + G_2)V_b - G_1V_1 - G_2V_2 - GV_3.$$

So

$$V_3 = -\frac{G_1}{G} V_1 - \frac{G_2}{G} V_2$$

or

$$V_3 = -\frac{R}{R_1} V_1 - \frac{R}{R_2} V_2.$$

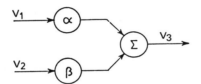

Figure 1.73

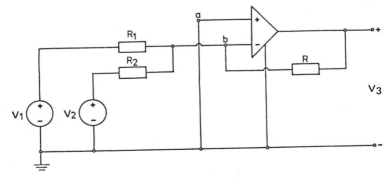

Figure 1.74

1.21 Problems

1.1 Calculate the current I.

1.2 a. Given that $V_{AB} = 5$ V,
$\qquad\qquad V_{CB} = 14$ V,
$\qquad\qquad V_{CD} = -2$ V,
$\quad$ find V_{DA}.

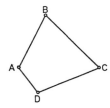

b. The current through a resistor R
is I and the voltage is V.
Prove that the dissipated power is
$P = I^2R = V^2G$.

1.3 Find the voltage across the current
source.

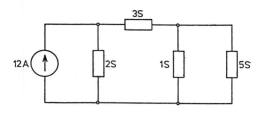

1.4 Find the current through the voltage
source.

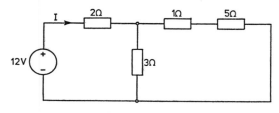

Find the power supplied by the
source and calculate the dissipated
power in each of the resistors.

1.5 Find R_{ab}.

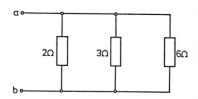

1.6 Find G_{ab}.

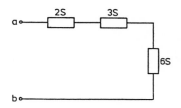

1.7 Calculate V.

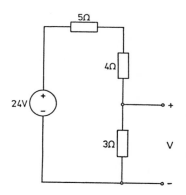

1.8 Calculate I.

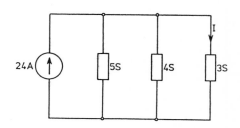

1.9   Give the solution of this network.
Find the power given by the 12 V-
source.

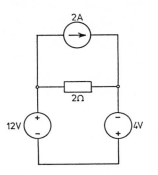

1.10   Calculate V_{ab}.

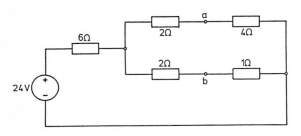

1.11   Find the power supplied by each
source and calculate the power
consumed by the resistor.

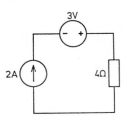

1.12   Find I.

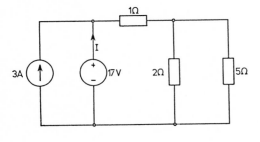

1.13   Find I_1 with
a. the branch method,
b. the mesh method.

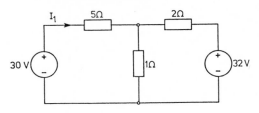

1.14   Find V_{ab}.
Calculate the substitute resistance
measured at the nodes a and b.

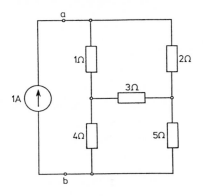

1.15   Calculate I_1, I_2 and I_3.

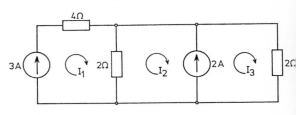

1.16   Find V.

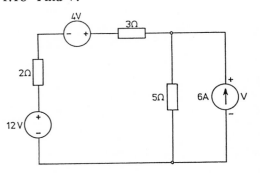

1.17 Find the voltages V_1 and V_2 with respect to earth by means of the node method.

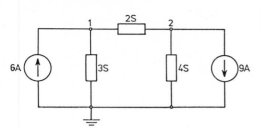

1.18 Give the solution with
a. the mesh method,
b. the node method.

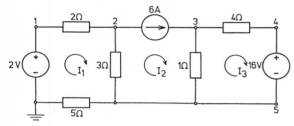

1.19 Find V_a, V_b and V_c using the node method.

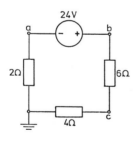

1.20 Find I_a, I_b and I_c using the mesh method.

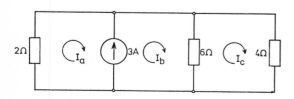

1.21 State if this graph is planar.

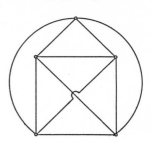

1.22 Find I.

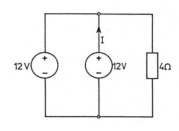

1.23 Find V.

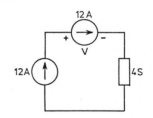

1.24 Find I.

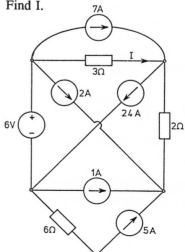

1.25 a. Find the node voltages using the
         node method.
      b. Draw the graph of the circuit and
         indicate the branch currents.

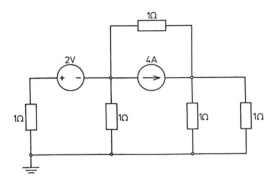

1.26 The bridge circuit in which the
      resistor R is variable is given. The
      other resistances are constant.

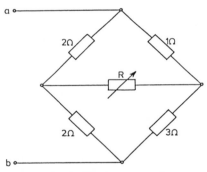

      a. Find the substitute resistance R_{ab}
         between the nodes a and b as a
         function of R.
      b. Calculate $\lim_{R \to 0} R_{ab}$ and $\lim_{R \to \infty} R_{ab}$
         and explain your answers by an
         examination of the network.

1.27 One wants to solve this network
      with the node method.

      a. Write the equations necessary to
         do this. (Solution is not
         necessary).

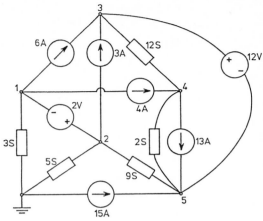

      b. Might we solve this network with
         the mesh method? Explain your
         answer.
      c. Calculate the current in the
         conductor of 9 S directly from
         the circuit.

1.28 Find the power that each of the
      sources delivers.
      Find the power that each of the
      resistors dissipates.

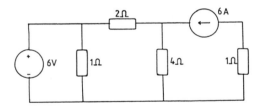

1.29 a. Find V_1 and V_2 with the node
         method.
      b. Give all branch currents in the
         graph of the network.

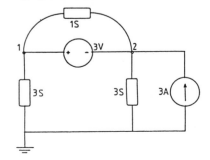

1.30 a. Use the node method to calculate the node voltages with respect to earth.

b. Give all branch currents in the graph of the network.

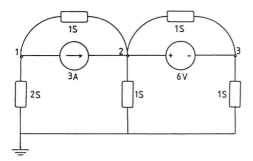

1.31 a. Use the node method to solve this network.

b. Give all branch currents in the graph of the network.

c. Find the power supplied by each of the sources.

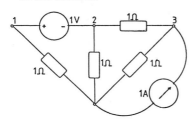

1.32 Use the node method to find the node voltages with respect to earth.

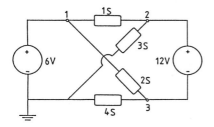

1.33 State whether I_x is larger than, smaller than or equal to I_y.

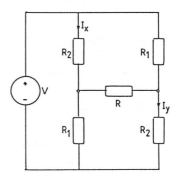

1.34 a. Use the node method to find the node voltages with respect to earth.

b. Give all branch currents in the graph of the network.

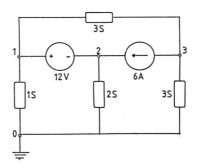

1.35 A network consists of two d.c. current sources, I_1 and I_2, and also of resistors. A voltage V is measured between two arbitrary nodes, using different values of the current source intensities.

The following result is found:

If one chooses the source intensities $I_1 = 2$ A and $I_2 = 1$ A, then V = 2 V, while for $I_1 = 1$ A and $I_2 = 2$ A, V = 3 V.

Find V if I_1 is 8 A and $I_2 = 1$ A.

1.36 A network consists of a voltage source V, a current source I and further of positive, constant resistors. One observes the current I_R through a certain resistor. V is given two different values in succession, but I is kept constant ($I \neq 0$).
One finds that $I_R = 5$ A if $V = 8$ V.
One finds that $I_R = 3$ A if $V = 14$ V.
Find I_R if $V = 12$ V and explain your answer.

1.37 Can one use the mesh method and the node method for a non-linear network?

1.38 A diode is characterised in the forward region by the voltage-current equation
$$I = 2(e^{U/4} - 1).$$
a. Give a plot of this equation.
b. Is this a linear element?

1.39 Solve Problem 1.13 with superposition.

1.40 Solve Problem 1.15 with superposition.

1.41 a. Use the node method to find the node voltages with respect to earth.
b. Use the superposition rule to find the voltage V_{30}.

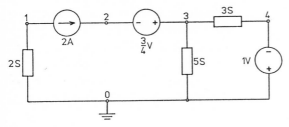

1.42 The value of R depends on the current I flowing through the resistor, according to the formula:
$$R = I.$$
Only positive values of I are considered.

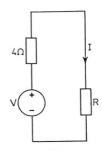

a. Express I in V.
b. Calculate I for $V = 5$ V and $V = 12$ V respectively.
c. Calculate I for $V = 17$ V ($= 5$ V + 12 V).
d. Is the superposition rule valid here?

1.43 Find the branch voltages of network a and the branch currents of network b. Verify Tellegen's theorem for this combination.

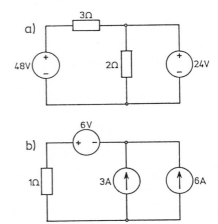

1.44   Prove that V″ = V′.

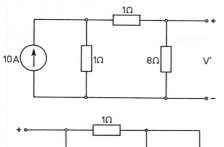

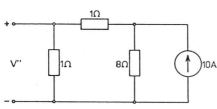

Give the equivalence of Thévenin and Norton of the following four networks.

1.45

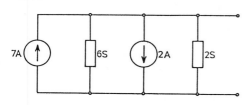

1.46

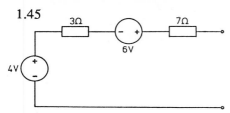

1.47

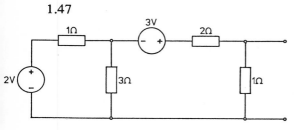

1.48

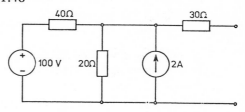

1.49

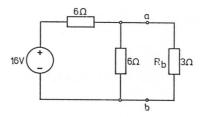

a. Find the power supplied by the source.

b. Find the power dissipated by R_b.

c. Find the Thévenin equivalence of the network to the left of the terminals a and b.

d. Find the power supplied by the Thévenin source.

e. Find the dissipated power in R_b in the Thévenin equivalence.

f. Compare your answers and explain them.

1.50

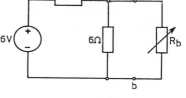

a. Find the Thévenin equivalence of the network to the left of the nodes a and b.

b. Calculate the dissipated power in R_b for

$R_b = 0\ \Omega,\ R_b = 2\ \Omega,\ R_b = 3\ \Omega,$
$R_b = 4\frac{1}{2}\ \Omega,\ R_b \to \infty.$

1.51

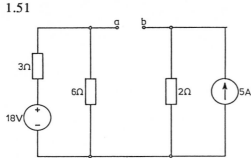

a. Calculate V_{ab}.
b. Calculate the current going from a to b if these terminals are short-circuited with Thévenin's theorem.
c. Which resistor must be connected to a and b so that a maximum power will be developed in that resistor? Calculate the maximum power.

1.52 Find I with Thévenin's theorem.

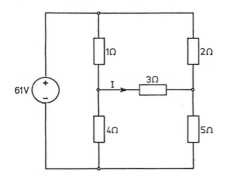

1.53 Find the value of R with which a maximum power is developed in R.

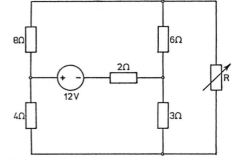

1.54 One first sets $R = 12\ \Omega.$
a. Find I.
b. Find the power P dissipated by R and examine for which other value of R the same power is dissipated by R.

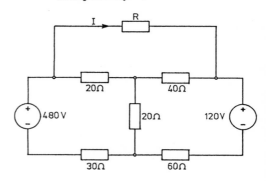

1.55 Find the substitute resistance between the terminals a and b.

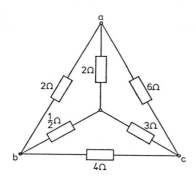

1.56 For the current source intensities I_a, I_b and I_c the values 0 A or 1 A are chosen. We want to know the voltage V for all eight possible combinations of I_a, I_b and I_c.

I_a	I_b	I_c	V
0	0	0	
0	0	1	
0	1	0	
0	1	1	
1	0	0	
1	0	1	
1	1	0	
1	1	1	

Find V for all eight cases.

1.57 Using superposition find the current I.

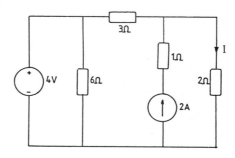

1.58 Give the Thévenin equivalence seen on the terminals a and b. Calculate the Thévenin resistance in two ways.

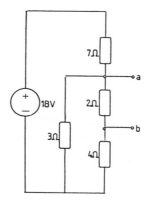

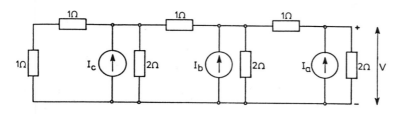

1.59 Using Thévenin's theorem find the current I.

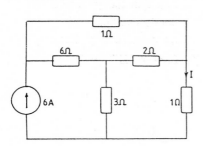

1.60 Find V_{ab} as a function of V and I.

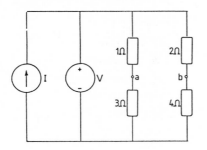

1.61 Which transactor is formed if we cascade a voltage-current transactor and a current-voltage transactor?

1.62 Find the two-port equations of two cascaded gyrators with equal gyration resistance R.

1.63 Find V_1.

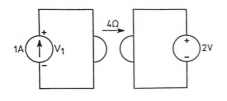

1.64 Given $V_1 = 0.01$ V, find V_2.

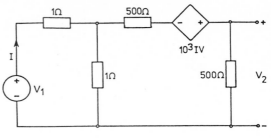

1.65 Find the voltage transfer function V_2/V_1.

See Figure at bottom of page.

1.66 Find V.

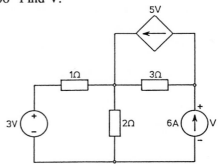

1.67 Find V_1, V_2 and V_3.

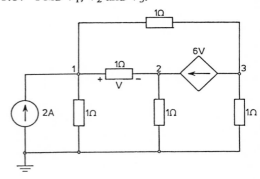

Figure of Problem 1.65

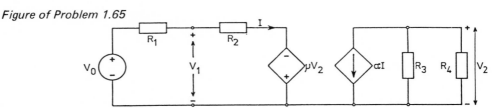

1.68 Find I.

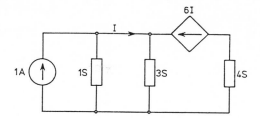

1.69 Find V.

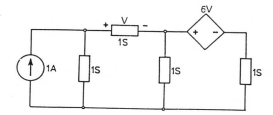

1.70 Find I.

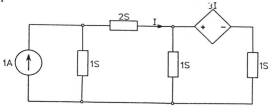

1.71 Find the Thévenin equivalence seen
at the terminals a and b.

1.72

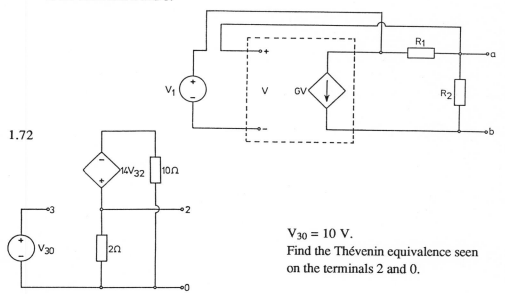

$V_{30} = 10$ V.
Find the Thévenin equivalence seen
on the terminals 2 and 0.

1.73 Find the Thévenin equivalence seen on the terminals a and b.

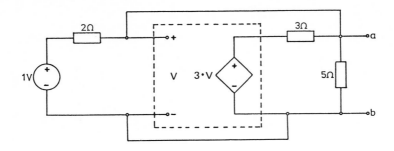

1.74 The network within the dotted rectangle is a voltage-voltage transactor for which it holds that:

$$V_{23} = AV_{12}.$$

The voltage source intensity V_{10} is 1 volt.

Compute the Thévenin equivalence at the terminals 2 and 0.

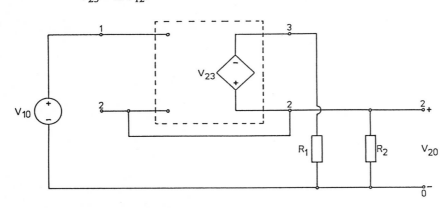

1.75 This is a transistor circuit and the equivalence of a transistor.

In order to make the calculation not too implicated all resistors are set to 1 Ω. We further choose $\alpha = 2$.

Find I_b.

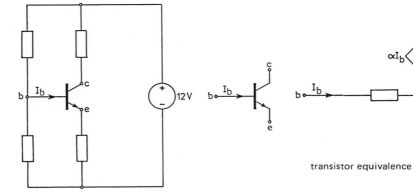

transistor equivalence

1.76 Given is a network with an opamp.
All conductances are G.

Find $\dfrac{V_2}{V_1}$.

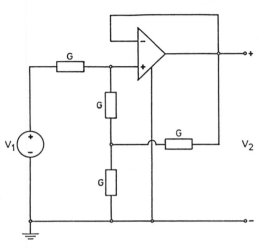

1.77 a. Give the node equations for the
nodes 4 and 5.
b. Express V_3 in V_1, V_2, the
conductances and K.
c. Find $\lim\limits_{K\to\infty} V_3$.

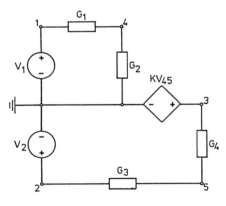

1.78 $\mu \to \infty$.
Use the node method to find the
voltage V_4 with respect to earth.

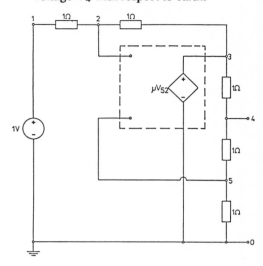

1.79 Use the node method to find the
Thévenin equivalence seen on the
nodes 3 and 0.

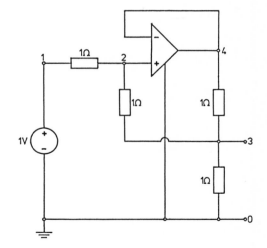

1.80 Find V_{20}.

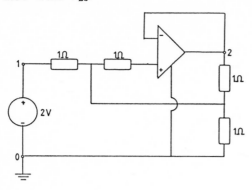

1.81 Find the power supplied by each of the sources and find the power dissipated by each of the resistors.

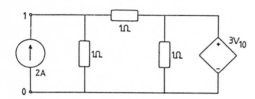

1.82 Find the power supplied by each of the sources and find the power dissipated by each of the resistors.

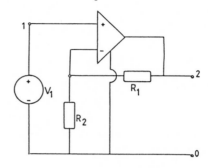

1.83 Find $H = \dfrac{V_2}{V_1}$.

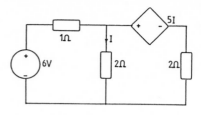

2

Varying currents and voltages

2.1 Introduction

In the previous chapter it was assumed that all voltages and currents in the considered network were constant as a function of time. We shall now examine networks in which the voltages and currents are functions of time. We shall use lower case letters; v and i.
We assume a current to be a varying current of which the intensity varies and of which the direction changes one or more times or not at all (see Figures 2.1, 2.2 and 2.3).

2.2 Direction and polarity

The positive current direction is a reference direction. This corresponds to the positive part of the plot. Consequently it is unmistakably clear which intensity and which direction the current has at any moment. For each plot the current direction must be indicated.
The same holds for the voltage. The plus and minus indicate the reference polarity. This corresponds to the positive part of the plot.
The positive polarity has to be indicated for each voltage with plus and minus. If a voltage is indicated by two indices (the terminals) we can omit the polarity (see Figure 2.4).

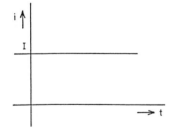

Figure 2.1 Direct current.

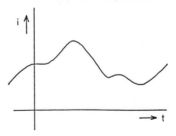

Figure 2.2   A current in which the intensity varies only.

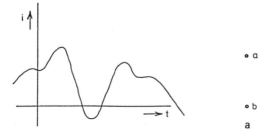

Figure 2.3 A current in which the sign changes.

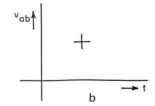

Figure 2.4

The positive part of the plot shows at which points of time (in which time intervals) the terminal a will have a higher potential than terminal b.

2.3 Kirchhoff's laws and Ohm's law

Kirchhoff's laws hold for any moment, so

$$\sum_{n=1}^{b} i_n = 0 \tag{2.1}$$

and

$$\sum_{m=1}^{l} v_m = 0. \tag{2.2}$$

Further, Ohm's law for a resistance holds for any moment, so

$$v = Ri, \tag{2.3}$$

$$i = Gv. \tag{2.4}$$

2.4 Periodical quantities

Often voltages and currents are periodical, after a certain time T the phenomenon is repeated. T is called the *period*.

So for a periodical current it holds that

$$i(t) = i(t + kT), \tag{2.5}$$

in which k is an integer.

Examples of periodical currents are shown in Figure 2.5.

2.5 Average value

The average value over a time interval T of a current that is variable or not is by definition

$$I_{av} = \frac{1}{T} \int_0^T i\,dt. \tag{2.6}$$

Usually T is the period, but that is not necessary. Because the average value is a constant it is indicated by a capital.

2.6 Power

The consumed immediate power is

$$p = v\,i. \tag{2.7}$$

As v and i are functions of time the product p is a function of time as well; that is why it is written with a lower case p. The power is consumed if v and i belong together (see Section 1.3).

It is also possible to find the average power over a certain interval:

$$P = \frac{1}{T} \int_0^T p \, dt.$$

This is the *average power*. Its unit is W.

2.7 Effective value

A resistor consumes power:

$$p = vi.$$

With $v = Ri$ it follows

$$p = i^2 R.$$

So the average power is

$$P = \frac{1}{T} \int_0^T i^2 R \, dt.$$

We assume i to be periodical with period T.
If there is a d.c. current I_d in resistor R the power is

$$P_d = I_d^2 R.$$

The d.c. current, which develops the same power in a resistor as a varying current, is called the *effective value* I_{eff}:

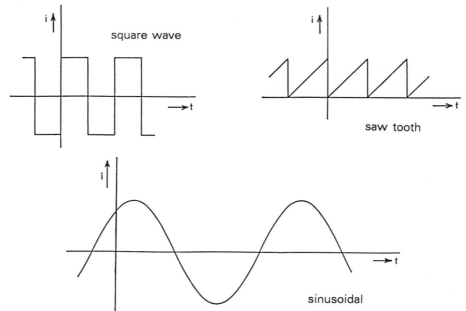

Figure 2.5

$$I_{eff} = \sqrt{\frac{1}{T} \int_0^T i^2 \, dt}. \tag{2.8}$$

The same formula also holds for voltages.

The effective value is often called *root mean square* (RMS).

Example

We calculate the effective value of a sawtooth wave current (Figure 2.6).

For $0 < t < T$ we have $i = \frac{I}{T} t$.

So

$$I_{eff}^2 = \frac{1}{T} \int_0^T i^2 \, dt = \frac{1}{T} \int_0^T \frac{I^2 t^2}{T^2} \, dt = \left[\frac{I^2}{T^3} \frac{I^2}{3}\right]_0^T = \frac{I^2}{3}.$$

So

$$I_{eff} = \frac{1}{3} I \sqrt{3} \text{ A}.$$

2.8 Sinusoidal quantities

The sinusoidal voltages and currents, also called *harmonic* voltages and currents are of fundamental importance. We often meet them in practice. For instance, the voltage we receive from the central power station in our homes is sinusoidal with an effective value of 240 V and a period of 20 ms.

This voltage comes from a *generator* in which a rotating magnetic field creates an a.c. voltage in the stationary wires (bars). The formula for a.c. voltage and current can be derived mathematically as follows: Imagine a pointer or vector rotating around its foot. The vertical projection of this vector gives a sinusoidal phenomenon as a function of the rotating angle (see Figure 2.7).

The length of the vector is $|I|$ (*modulus*), the vertical projection is i and the angle is α. We find

$$i = |I| \sin \alpha.$$

Now we rotate the vector with constant angle velocity ω rad/s anti-clockwise. Consequently

$$\alpha = \omega t.$$

So

$$i = |I| \sin \omega t. \tag{2.9}$$

In this formula $|I|$ is the *amplitude*, *top value* or *maximum value*. In Figure 2.8 this function is shown.

From $\alpha = \omega t$ follows

$$2\pi = \omega T. \tag{2.10}$$

The number of periods per second is called *frequency* with the unity 1 Hz (Hertz) = 1 s^{-1}.

$$f = \frac{1}{T} \tag{2.11}$$

so

$$\omega = 2\pi f, \tag{2.12}$$

ω is called the *angular frequency* (unit 1 rad/s). The frequency of the power supply in Europe is

$$f = 50 \text{ Hz.}$$

In practice one comes across frequencies of some Hz up to some GHz.

It often occurs that there are currents in a network with the same frequency but with different phases. In Figure 2.9 the currents i_1 and i_2 are drawn where i_2 leads the current i_1 by an angle φ.

We have

$$i_1 = |I_1| \sin \omega t,$$

$$i_2 = |I_2| \sin(\omega t + \varphi).$$

If $\varphi = 90°$ then $i_2 = |I_2| \sin(\omega t + 90°) = |I_2| \cos \omega t$.

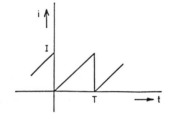

Figure 2.6

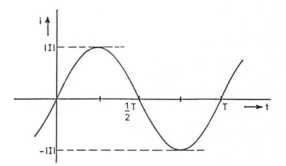

Figure 2.8

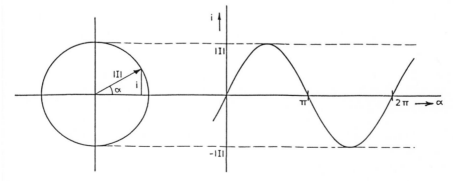

Figure 2.7

We shall call these expressions *sinusoidal currents*, also if the analytic expression contains the cosine function.

The above dissertation can, of course, also be held for voltages.

2.9 Power for sinusoidal voltages and currents

Suppose, for a two-terminal network (see Figure 2.10):

$$v = |V| \cos(\omega t + \alpha),$$

$$i = |I| \cos(\omega t + \beta).$$

The immediate power consumed is

$$p = vi = |V| |I| \cos(\omega t + \alpha) \cos(\omega t + \beta).$$

The average power over one period is

$$P = \frac{1}{T} \int_0^T p \, dt.$$

We find

$$P = \frac{1}{T} \int_0^T |V| |I| \cos(\omega t + \alpha) \cos(\omega t + \beta) \, dt.$$

With $\cos A \cos B = \frac{1}{2} \cos(A + B) + \frac{1}{2} \cos(A - B)$ we derive

$$P = \frac{|V| |I|}{2T} \int_0^T \{\cos(2\omega t + \alpha + \beta) + \cos(\alpha - \beta)\} \, dt.$$

The first term between the braces is zero because the integral is taken over two periods. So we get

$$P = \frac{|V| |I|}{2T} \int_0^T \cos \varphi \, dt \quad \text{with } \varphi = \alpha - \beta,$$

$$P = \frac{|V| |I| \cos \varphi \; T}{2T},$$

$$P = \frac{|V| |I| \cos \varphi}{2}. \tag{2.13}$$

In Figure 2.11 P is determined graphically. Note the creation of the double frequency in the immediate power.

If the two-terminal network is a resistor R we find ($\varphi = 0$ and $|V| = R |I|$):

$$P = \frac{|I|^2 R}{2}. \tag{2.14}$$

A d.c. current I_d in a resistor R means a power of $P_d = I_d^2 R$.

After equalling both powers we get:

$$I_d{}^2 = \frac{|I|^2}{2} \, .$$

I_d was called the effective value I_{eff}, so

$$I_{\text{eff}} = \frac{|I|}{\sqrt{2}} \, . \tag{2.15}$$

The effective value of the power supply of our homes is 240 V, so the amplitude is $240 \sqrt{2} = 339$ V.

We stress here that the factor $\sqrt{2}$ only holds for sinusoidal voltages and currents.

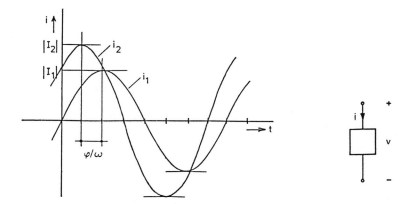

Figure 2.9   Figure 2.10

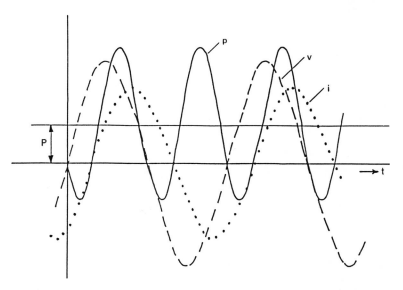

Figure 2.11

2.10 The sum of two sinusoidal quantities

Given that:

$$i_1 = |I_1| \sin(\omega t + \varphi_1),$$

$$i_2 = |I_2| \sin(\omega t + \varphi_2).$$

The sum is $i = i_1 + i_2$ and will also be a sinusoidal current with angular frequency ω. So $i = |I| \sin(\omega t + \varphi)$.

We shall now examine how we can derive $|I|$ and φ from $|I_1|$, $|I_2|$, φ_1 and φ_2. We develop the right-hand parts of i_1 and i_2, using the properties of sines and cosines goniometry:

$$i = i_1 + i_2 = |I_1| \sin \omega t \cos \varphi_1 + |I_1| \cos \omega t \sin \varphi_1 + |I_2| \sin \omega t \cos \varphi_2 + |I_2| \cos \omega t \sin \varphi_2$$

$$= \{|I_1| \cos \varphi_1 + |I_2|)\cos \varphi_2\} \sin \omega t + \{|I_1| \sin \varphi_1 + |I_2| \sin \varphi_2\} \cos \omega t.$$

So

$$i = i_1 + i_2 = A \sin \omega t + B \cos \omega t$$

with

$$A = |I_1| \cos \varphi_1 + |I_2| \cos \varphi_2$$
$$B = |I_1| \sin \varphi_1 + |I_2| \sin \varphi_2$$

$$\left.\vphantom{\begin{matrix}A\\B\end{matrix}}\right\} \quad \text{(See Figure 2.12)}$$

$$\frac{A}{\sqrt{A^2 + B^2}} = \cos \varphi \text{ and } \frac{B}{\sqrt{A^2 + B^2}} = \sin \varphi.$$

$$i = \sqrt{A^2 + B^2} \ (\sin \omega t \cos\varphi + \cos \omega t \sin \varphi),$$

$$i = \sqrt{A^2 + B^2} \ \sin(\omega t + \varphi).$$

So

$$|I| = \sqrt{A^2 + B^2}$$

and

$$\tan \varphi = \frac{\sin \varphi}{\cos \varphi} = \frac{B}{A}.$$

The conclusion is that the sum of two (and also of more than two) sinusoidal currents can be found by the vectorial adding of the vectors. We shall discuss this later, but suffice to mention that vectorial adding can be simply realised by placing the vectors 'head to tail'. (see Figure 2.13 in which we add three vectors). In Figure 2.13 the vectors are indicated by their amplitudes; we shall later introduce another indication.

2.11 The capacitor

The capacitor consists of two conductors (plates) separated by an insulator (dielectric). The symbol is shown in Figure 2.14.

The charge on the positively marked plate is +q, on the other plate it is –q. We define the charge of the capacitor as the charge on the positive plate. Between the plates there is an electric field. The charge q is a function of the voltage v.

We assume the relation between q and v to be linear:

$$q = Cv,$$ (2.16)

in which C is the capacitance in Farad (F).

The positive current i in Figure 2.14 increases the positive charge on the upper plate while at the same time the charge on the lower plate decreases at the same rate.
With $i = \dfrac{dq}{dt}$ and C = constant we find

$$i = C\frac{dv}{dt} .$$ (2.17)

We may regard this formula as the definition formula for the capacitor. Such a capacitor is an ideal one, i.e. a capacitor without a 'leak current' through the dielectric. Such a leak current can be interpreted as a (large) resistance in parallel with the capacitor (*parasitic resistance*). Note that (2.17) is only valid if voltage and current belong to each other. There are capacitors with a constant capacity and with a varying capacity. A varying capacity is obtained by varying the distance between the plates. Theoretically one can vary a capacitance between zero and infinity.

2.12 The inductor

An inductor consists of one or more wire turns, either coiled on a core or not. The symbol is shown in Figure 2.15.

Inside (and also outside) the coil there is a magnetic field. The magnetic field lines are closed. The coupled magnetic flux φ is a function of the current i, which we assume to be linear):

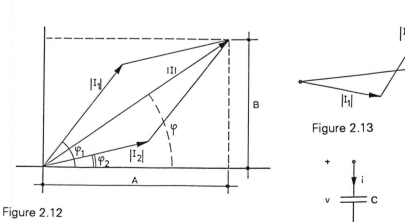

Figure 2.12

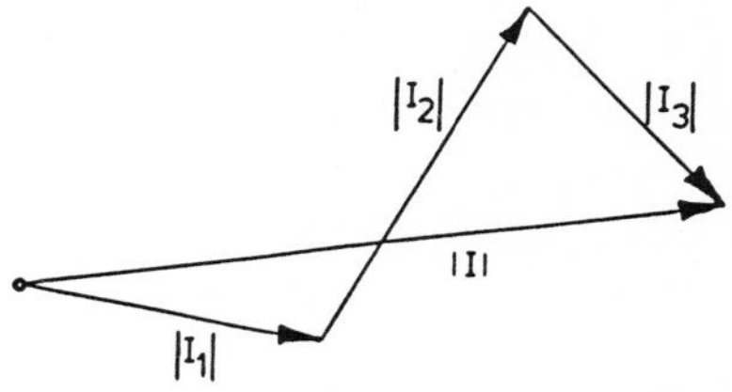

Figure 2.13

Figure 2.14

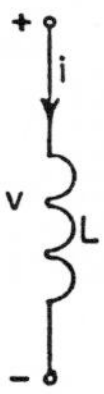

Figure 2.15

$$\phi = Li, \tag{2.18}$$

in which the self-inductance L with the unit henry (H) is constant.
With Faraday-Maxwell's induction law

$$e = -\frac{d\phi}{dt}$$

and Kirchhoff's voltage law

$$v + e = 0$$

we find

$$v = L\frac{di}{dt}. \tag{2.19}$$

This is the definition formula for the (ideal) inductor in networks. In practice there are losses which can be interpreted by a (small) series resistance (parasitic resistance). Note the duality of the capacitor and the inductor. Most inductors have a constant inductance. Variable inductances can be made by moving the magnetic material within the coil. One can also obtain a discontinuous change of the inductance by switching parts of the coil on and off. Such inductors then have branch-terminals. Finally, one can also use the varying magnetic coupling between two inductors to make a variable inductance (see Chapter 4). Theoretically one can vary an inductance between zero and infinity.

2.13 Energy stored in capacitor and inductor

A capacitor can be charged by connecting it to a current source and to make the source intensity zero after some time (Figure 2.16).
The power supplied by the source is

$$p = vi.$$

We further have $i = C\frac{dv}{dt}$.
If the source strength is made zero after t_1 seconds, the energy supplied is

$$W = \int_0^{t_1} p\, dt = \int_0^{t_1} C\, v\, \frac{dv}{dt}\, dt.$$

Suppose the voltage is V after t_1 seconds, then

$$W = \int_0^V C\, v\, dv = \tfrac{1}{2} C\, V^2.$$

(We assume the starting energy to be zero).
This energy is stored in the capacitor and is in fact in the electric field. So, in general the energy of a charged capacitor is

$$W_C = \tfrac{1}{2} C\, V^2, \tag{2.20}$$

in which V is the voltage of the capacitor.

Note that after t_1 seconds a capacitor with open terminals develops. Theoretically speaking, the capacitor will remain charged indefinitely. In practice the voltage will become zero after some time (sometimes only after some days). This is due to the fact that in practice there always is a parallel parasitic resistance which converts the stored energy into heat.

Using the dual way we can deduce the energy stored in an inductor (see Figure 2.17). We find

$$p = vi,$$

$$v = L \frac{di}{dt}.$$

After t_1 seconds the source strength is made zero. If the current at that moment is called I, then the energy supplied is

$$W = \int_0^{t_1} p \, dt = \int_0^{t_1} Li \frac{di}{dt} \, dt = \int_0^{I} Li \, di = \tfrac{1}{2} L I^2.$$

The energy stored in an inductor is

$$W_L = \tfrac{1}{2} L I^2. \tag{2.21}$$

This energy is in the magnetic field.

Note that after t_1 seconds the inductor is short-circuited. We have the ideal situation for a short-circuited inductor, in which the total resistance is zero. Theoretically speaking, the current will keep the same value infinitely long. In practice the current will become zero after some seconds. The energy is then converted into heat in the loss resistance.

So, the capacitor and the inductor are elements which can store energy. The resistor, however, can only *dissipate* energy, i.e. convert it into heat. All three are *passive* elements, elements that cannot supply more electrical energy than was previously given. The voltage source and the current source are *active* elements. The energy stored in capacitors and inductors can be supplied to resistors which will dissipate this energy.

2.14 The passive elements in sinusoidal excitation

We connect a resistor R to a voltage source with strength

$$v = |V| \sin(\omega t + \varphi).$$

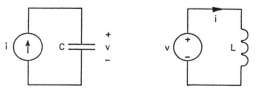

Figure 2.16   Figure 2.17

Therefore the current is (Figure 2.18)

$$i = \frac{v}{R} = \frac{|V|}{R}\, \sin(\omega t + \varphi).$$

Voltage and current are in phase.

Now take the same source and connect a capacitor C to it. We find

$$i = C\frac{dv}{dt} = \omega C\,|V|\cos((\omega t + \varphi) = \omega C\,|V|\sin(\omega t + \varphi + \frac{\pi}{2})$$

$$= |I|\sin(\omega t + \varphi + \frac{\pi}{2}).$$

(Figure 2.19). The current leads the voltage by 90° (i.e. $\frac{\pi}{2}$ rad).

For the amplitudes it holds that $|I| = \omega C\,|V|$.
This formula looks like Ohm's law ($I = GV$). So ωC can be regarded as a sort of conductance.
Finally we take a current source with strength

$$i = |I|\sin(\omega t + \varphi)$$

and connect an inductor L to it.
The voltage becomes (Figure 2.20)

$$v = L\frac{di}{dt} = \omega L\,|I|\cos(\omega t + \varphi) = \omega L\,|I|\sin(\omega t + \varphi + \frac{\pi}{2})$$

$$= |V|\sin(\omega t + \varphi + \frac{\pi}{2}).$$

We find Figure 2.20. The voltage leads the current by 90° ($\frac{\pi}{2}$ rad).

For the amplitudes holds $|V| = \omega L\,|I|$. If we compare this with $V = RI$ we see that ωL is a sort of resistance.

We summarise the results and only draw the vector diagrams. The position of the diagrams is not important, but we are interested in the mutual angle and the relation of the amplitudes (see Figure 2.21).

2.15 A larger network in sinusoidal excitation

Example (Figure 2.22).
Given is $v = |V|\cos\omega t$. Find i.

Solution
We have

$$v = L\frac{di}{dt} + Ri + v_C$$

and

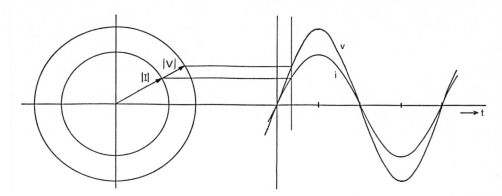

Figure 2.18

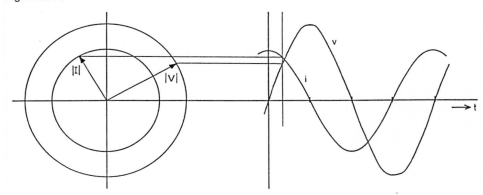

Figure 2.19

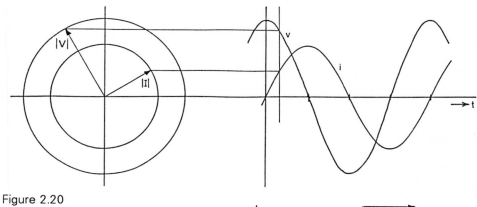

Figure 2.20

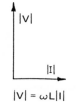

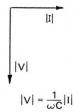

Figure 2.21

$$i = C \frac{dv_C}{dt} .$$

So

$$\frac{dv}{dt} = L \frac{d^2i}{dt^2} + R \frac{di}{dt} + \frac{i}{C} .$$

We find a linear, second order differential equation. We assume the current to be sinusoidal and to have an angle α behind that of the voltage.
So

$$i = |I| \cos(\omega t - \alpha).$$

We must find $|I|$ and α. The first derivative of i is

$$\frac{di}{dt} = -\omega \, |I| \sin(\omega t - \alpha).$$

The second derivative is

$$\frac{d^2i}{dt^2} = -\omega^2 \, |I| \cos(\omega t - \alpha).$$

We find

$$-\omega \, |V| \sin \omega t = -\omega^2 L \, |I| \cos(\omega t - \alpha) - \omega R \, |I| \sin(\omega t - \alpha) + \frac{|I|}{C} \cos(\omega t - \alpha).$$

So

$$\omega \, |V| \sin \omega t = \omega R \, |I| \sin(\omega t - \alpha) + |I| \, (\omega^2 L - \frac{1}{C}) \cos(\omega t - \alpha),$$

hence

$$\omega \, |V| \sin \omega t = \omega R \, |I| \, (\sin \omega t \cos \alpha - \cos \omega t \sin \alpha) +$$

$$+ |I| \, (\omega^2 L - \frac{1}{C}) \, (\cos \omega t \cos \alpha + \sin \omega t \sin \alpha).$$

This must hold for all t, so also for t = 0:

$$0 = -\omega R \, |I| \sin \alpha + |I| \, (\omega^2 L - \frac{1}{C}) \cos \alpha$$

and also for $t = \frac{\pi}{2\omega}$, so that $\omega t = \frac{\pi}{2}$:

$$\omega \, |V| = \omega R \, |I| \cos \alpha + |I| \, (\omega^2 L - \frac{1}{C}) \sin \alpha.$$

From the first equation it follows that:

$$\tan \alpha = \frac{\sin \alpha}{\cos \alpha} = \frac{\omega^2 L - \frac{1}{C}}{\omega R} = \frac{\omega L - \frac{1}{\omega C}}{R} .$$

If we substitute this in the second equation we get

$$\omega |V| = |I| \cos \alpha \left\{ \omega R + \omega \frac{(\omega L - \frac{1}{\omega C})^2}{R} \right\}$$

or

$$|V| = |I| \cos \alpha \left\{ R + \frac{(\omega L - \frac{1}{\omega C})^2}{R} \right\}.$$

From $\tan \alpha = \dfrac{\omega L - \dfrac{1}{\omega C}}{R}$ it follows that $\cos \alpha = \dfrac{R}{\sqrt{R^2 + (\omega L - \dfrac{1}{\omega C})^2}}.$ so

$$\cos^2 \alpha = \frac{R^2}{R^2 + (\omega L - \frac{1}{\omega C})^2}, \text{ so that}$$

$$\frac{|V|}{R} = \frac{|I|}{\cos \alpha},$$

with which we find

$$|I| = \frac{|V|}{\sqrt{R^2 + (\omega L - \frac{1}{\omega C})^2}}.$$

So the current as a function of time becomes

$$i = \frac{|V|}{\sqrt{R^2 + (\omega L - \frac{1}{\omega C})^2}} \cdot \cos(\omega t - \arctan \frac{\omega L - \frac{1}{\omega C}}{R}).$$

The vector diagram gives a faster solution (see Figure 2.23).
The inductor voltage leads the current by 90°, the amplitude is $\omega L \, |I|$.
The capacitor voltage is 90° behind the current, the amplitude is $\frac{1}{\omega C} \, |I|$.

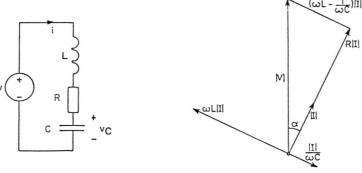

Figure 2.22   Figure 2.23

Vectorially adding the inductor voltage, the capacitor voltage and the resistor voltage gives the source voltage.

From the diagram we see:

$$\tan \alpha = \frac{\omega L - \dfrac{1}{\omega C}}{R} \quad \text{and} \quad |V|^2 = \{(\omega L - \frac{1}{\omega C})^2 + R^2\}\, |I|^2,$$

with which the above result has been found.

Notes

− The solution found only gives the response in the so-called *steady state*, i.e. if previous switch operation phenomena have been damped.

− For more complicated circuits the above-mentioned method is too digressive. The vector diagrams do give rapid results, but they cannot always be constructed directly, at least not in first instance.

− A two-terminal network in which the current is an angle φ behind the voltage ($0 < \varphi \le \frac{\pi}{2}$) is called *inductory*. If the current leads the voltage by an angle φ, we speak of *capacitory* behaviour.

Next we give a short and elegant procedure, also suitable for more complicated networks. Then we shall also give a systematic procedure for drawing vector diagrams. But first complex numbers have to be introduced.

2.16 Complex numbers

In mathematics the imaginary unit is indicated by i. In electrotechnology the letter j is usually taken, due to the fact that i is already used for the current. The imaginary unit is a number for which

$$j^2 = -1.$$

A complex number is

$$Z = A + jB \qquad \text{with A and B real.} \tag{1}$$

In Figure 2.24 Z is drawn as a vector in the complex plane.

Im is the imaginary axis and Re the real axis. The real part of Z is Re $Z = A$, the imaginary part of Z is Im $Z = B$. The magnitude of Z is called the *modulus* (plural *moduli*):

$$|Z| = \sqrt{A^2 + B^2}.$$

The angle α is called the *argument* of Z:

$$\arg Z = \alpha = \arctan \frac{B}{A}.$$

So we can write Z as follows:

$$Z = |Z| (\cos \alpha + j \sin \alpha). \tag{2}$$

According to Euler

$$e^{j\alpha} = \cos \alpha + j \sin \alpha.$$

With this a third way to write Z has been found:

$$Z = |Z| e^{j\alpha}. \tag{3}$$

Examples

1. The number $Z = 1 + 2j$ can be written in form (3) as follows. We have $|Z| = \sqrt{5}$ and arg Z = arctan 2. So

$$Z = \sqrt{5} \cdot e^{j \text{ arctan } 2}.$$

2. The number j can therefore be written as

$$j = e^{j\frac{\pi}{2}}.$$

2.17 Operations with complex numbers

Adding two complex numbers can be done most easily if they are written in form (1).
Suppose $Z_1 = A_1 + jB_1$ and $Z_2 = A_2 + jB_2$, then the sum is

$$Z = Z_1 + Z_2 = A_1 + A_2 + j(B_1 + B_2).$$

Subtraction can also be done in this way.

Multiplication is simplest if the numbers are written in form (3).
Suppose $Z_1 = |Z_1| e^{j\alpha_1}$ and $Z_2 = |Z_2| e^{j\alpha_2}$, then the product is:

$$Z = Z_1 Z_2 = |Z_1| |Z_2| e^{j(\alpha_1 + \alpha_2)}.$$

So the modulus of a product is the product of the moduli and the argument of a product is the sum of the arguments of both factors.
So $|Z_1 Z_2| = |Z_1| |Z_2|$ and arg $Z_1 Z_2$ = arg Z_1 + arg Z_2.
Multiplication of two complex numbers in form (1) is as follows:

$$Z = (A_1 + jB_1)(A_2 + jB_2) = A_1 A_2 - B_1 B_2 + j(A_1 B_2 + A_2 B_1).$$

For a quotient we find the quotient of the moduli and the difference of the arguments.

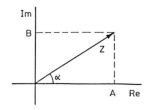

Figure 2.24

So $|\frac{Z_1}{Z_2}| = \frac{|Z_1|}{|Z_2|}$ and arg $\frac{Z_1}{Z_2}$ = arg Z_1 – arg Z_2.

Multiplying a complex number with j means a rotation over 90° in the complex plane:

Given that $Z_2 = jZ_1$,

$$|Z_2| = |Z_1|$$

$$\arg Z_2 = 90° + \arg Z_1.$$

(See Figure 2.25).

In the case of dividing by j (which is the same as multiplying by –j) we find a rotation over –90°. The *conjugate complex* of a number $Z = A + jB$ is

$$Z^* = A - jB.$$

(See Figure 2.26). So it also holds that

$$\{|Z| e^{j\alpha}\}^* = |Z| e^{-j\alpha}.$$

Further

$$Z + Z^* = 2\text{Re } Z \text{ and } Z - Z^* = 2j \text{ Im } Z.$$

Subsequently with $Z = |Z| e^{j\alpha}$ we have :

$$ZZ^* = |Z| e^{j\alpha} |Z| e^{-j\alpha} = |Z|^2.$$

If the complex number is a fraction with a complex numerator and a complex denominator, that number can be made into the form $A + jB$ by multiplying the numerator and the denominator with the conjugate complex of the denominator:

$$Z = \frac{C + jD}{E + jF} = \frac{(C + jD)(E - jF)}{E^2 + F^2} = \frac{CE + DF + j(DE - CF)}{E^2 + F^2}.$$

The complex number $\tilde{v} = |V| e^{j(\omega t + \varphi)}$ is a function of time t.
We have

$$\text{Re } \tilde{v} = |V| \cos(\omega t + \varphi),$$

which is a form equal to the sinusoidal voltage discussed before. It is indeed possible to describe sinusoidal phenomena with the aid of complex expressions.

2.18 Complex voltages and currents

A sinusoidal voltage may be written as the real part of a *complex voltage*:

$$v = |V| \cos(\omega t + \varphi) = \text{Re } |V| e^{j(\omega t + \varphi)}.$$

We write

$$\tilde{v} = |V| e^{j(\omega t + \varphi)}, \tag{2.22}$$

which is a complex quantity as a function of time. We use a lower case letter with a zigzag line (tilde) over it. In the complex plane that is a rotating vector turning anti-clockwise with angle velocity ω.

We can divide $\tilde{v}$ by $e^{j\omega t}$, with which the rotation disappears. The result is a complex number V, that is independent of time:

$$V = |V|\, e^{j\varphi}. \tag{2.23}$$

One can visualise the voltage v, i.e. it is physical; the voltage $\tilde{v}$ is not physical. The complex voltage is not physical either. We have formulated a mathematical complex voltage from the physical voltage. In other words we have performed a *transformation* from the (real) region to the (mathematical) complex plane.

Formulated in still another way; We have 'added' a complex voltage V to the physical voltage v.

With the following transformation we can once more obtain the correct physical voltage from the complex voltage:

$$v = \mathrm{Re}\,(V\, e^{j\omega t}) = \mathrm{Re}\,\tilde{v}. \tag{2.24}$$

So we first multiply the complex voltage with $e^{j\omega t}$, so that the time-dependent complex voltage $\tilde{v}$ is created. Subsequently we take the real part.

Written in full:

$$v = \mathrm{Re}\,(V\, e^{j\omega t}) = \mathrm{Re}\,(\tilde{v}) = \mathrm{Re}\,(|V|\, e^{j\varphi} \cdot e^{j\omega t})$$

$$= \mathrm{Re}\,\{|V|\, e^{j(\omega t + \varphi)}\} = \mathrm{Re}\,\{|V| \cos(\omega t + \varphi) + j \sin(\omega t + \varphi)\}$$

$$= |V| \cos(\omega t + \varphi).$$

The transformation can be symbolically written as follows:

$$v \leftrightarrow V$$

or more extensively

$$v = |V| \cos(\omega t + \varphi) \leftrightarrow |V|\, e^{j\varphi} = V.$$

On the left-hand side are the physical voltage with amplitude $|V|$ and phase φ (as well as time t and the angular frequency ω), on the right-hand side we have the complex voltage V with the modulus $|V|$ and the argument φ.

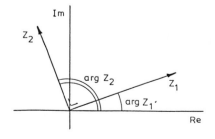

Figure 2.25

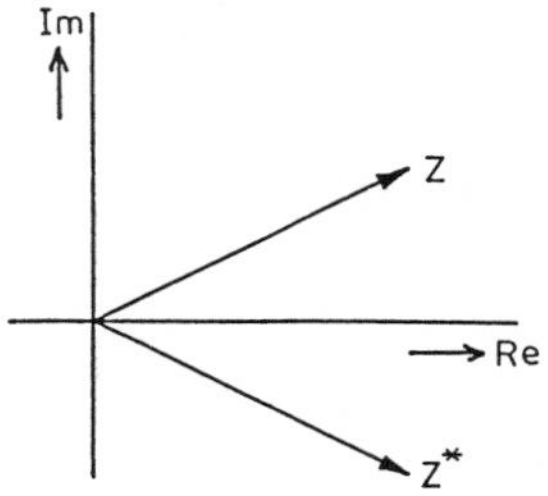

Figure 2.26

The concepts *amplitude* and *phase* occur in physics, the concepts *modulus* and *argument* are found in mathematics.

We can summarise the above symbolically with

- physical voltage $\leftrightarrow$ complex voltage
- amplitude $\leftrightarrow$ modulus
- phase $\leftrightarrow$ argument

Example

As an addition $v = 2\sqrt{2}\cos(\omega t + \frac{\pi}{4})$ V gets the complex voltage

$$V = 2\sqrt{2}\, e^{j\frac{\pi}{4}} = 2(1 + j)\text{ V}.$$

The above holds for currents as well.

So if one has

$$I = 8 + 3j\text{ A}$$

the additional physical current is

$$i = |I|\cos(\omega t + \arg I),$$

$$i = \sqrt{73}\cos(\omega t + \arctan\tfrac{3}{8})\text{ A}.$$

The transformation complex $\rightarrow$ time according to (2.24) can also be performed as follows:

$$v = \tfrac{1}{2}(\tilde{v} + \tilde{v}*). \tag{2.25}$$

After all

$$\tilde{v} = |V|\, e^{j(\omega t + \varphi)},$$

so

$$\tilde{v}* = |V|\, e^{-j(\omega t + \varphi)},$$

hence

$$\tilde{v} = |V|\,\{\cos(\omega t + \varphi) + j\sin(\omega t + \varphi)\}$$

and

$$\tilde{v}* = |V|\,\{\cos(\omega t + \varphi) - j\sin(\omega t + \varphi)\}.$$

Half the sum gives

$$v = |V|\cos(\omega t + \varphi).$$

The transformations given are unambiguous: For every time function there is only one complex quantity and for every complex quantity there is just one time function.

We now consider once more the network of Figure 2.22 with the excitation

$$v = |V| \cos(\omega t + \varphi) \tag{a}$$

and the response

$$i = |I| \cos(\omega t + \varphi - \alpha). \tag{b}$$

We shall now calculate $|I|$ and α with the complex quantities introduced above. The differential equations equation found is

$$\frac{dv}{dt} = L \frac{d^2 i}{dt^2} + R \frac{di}{dt} + \frac{1}{C} i. \tag{c}$$

This equation is linear (see Appendix I). So we are allowed to apply the principle of superposition. Choose two voltage sources in series:

$$\tfrac{1}{2} \tilde{v} = \tfrac{1}{2} |V| e^{j(\omega t + \varphi)} \tag{d}$$

and

$$\tfrac{1}{2} \tilde{v}* = \tfrac{1}{2} |V| e^{-j(\omega t + \varphi)}. \tag{e}$$

The sum of these is the excitation given:

$$\tfrac{1}{2} (\tilde{v} + \tilde{v}*) = |V| \cos(\omega t + \varphi) \tag{f}$$

So the response is the sum of the responses

$$\tfrac{1}{2} \tilde{\imath} = \tfrac{1}{2} |I| e^{j(\omega t + \varphi - \alpha)} \tag{g}$$

and

$$\tfrac{1}{2} \tilde{\imath}* = \tfrac{1}{2} |I| e^{-j(\omega t + \varphi - \alpha)} \tag{h}$$

because

$$\tfrac{1}{2} \tilde{\imath} + \tfrac{1}{2} \tilde{\imath}* = |I| \cos(\omega t + \varphi - \alpha), \tag{i}$$

while $\tfrac{1}{2} \tilde{\imath}$ results from $\tfrac{1}{2} \tilde{v}$ and $\tfrac{1}{2} \tilde{\imath}*$ from $\tfrac{1}{2} \tilde{v}$.
So

$$\frac{d \tfrac{1}{2} \tilde{v}}{dt} = L \frac{d^2 \tfrac{1}{2} \tilde{\imath}}{dt} + R \frac{d \tfrac{1}{2} \tilde{\imath}}{dt} + \frac{1}{C} \tfrac{1}{2} \tilde{\imath} \tag{j}$$

or

$$\frac{d \tilde{v}}{dt} = L \frac{d^2 \tilde{\imath}}{dt^2} + R \frac{d \tilde{\imath}}{dt} + \frac{1}{C} \tilde{\imath}, \tag{k}$$

consequently

$$j\omega |V| e^{j(\omega t + \varphi)} = (j\omega)^2 L |I| e^{j(\omega t + \varphi - \alpha)} + j\omega R |I| e^{(j\omega t + \varphi - \alpha)} + \frac{1}{C} |I| e^{(j\omega t + \varphi - \alpha)}$$

or

$$j\omega |V| e^{j\varphi} = \left\{ (j\omega)^2 L + j\omega R + \frac{1}{C} \right\} |I| e^{j(\varphi - \alpha)}. \tag{l}$$

With the complex voltage

$$V = |V| \, e^{j\varphi} \tag{m}$$

and the complex current

$$I = |I| \, e^{j(\varphi - \alpha)} \tag{n}$$

we find

$$j\omega V = \left\{ (j\omega)^2 L + j\omega R + \frac{1}{C} \right\} I, \tag{o}$$

which *algebraic* equation follows directly from the differential equation by replacing d/dt by a factor $j\omega$ and d^2/dt^2 by $(j\omega)^2$ (!).

From the last equation it follows that

$$I = \frac{V}{Z} \tag{p}$$

with

$$Z = j\omega L + R + \frac{1}{j\omega C} \, . \tag{q}$$

From (d) and (m) it follows that

$$\tilde{v} = V \, e^{j\omega t}, \tag{r}$$

and from (g) and (n)

$$\tilde{\imath} = I \, e^{j\omega t}. \tag{s}$$

With this we find

$$\tilde{\imath} = \frac{V}{Z} \, e^{j\omega t} \tag{t}$$

and thus

$$i = \tfrac{1}{2} (\tilde{\imath} + \tilde{\imath}^*) = \operatorname{Re} \tilde{\imath} = \operatorname{Re} \left(\frac{V}{Z} \, e^{j\omega t} \right). \tag{u}$$

Now it holds that (see (m) and (q))

$$\frac{V}{Z} = \frac{|V| \, e^{j\varphi}}{|Z| \, e^{j\beta}} = \frac{|V|}{|Z|} \, e^{j(\varphi - \beta)} \tag{v}$$

with

$$\beta = \arg Z = \arctan \frac{\omega L - \dfrac{1}{\omega C}}{R} \, , \tag{w}$$

so that

$$i = \frac{|V|}{\sqrt{R^2 + (\omega L - \frac{1}{\omega C})^2}} \cos(\omega t + \varphi - \arctan \frac{\omega L - \frac{1}{\omega C}}{R}), \qquad (x)$$

with which the response requested ($|I|$ and α) has been found (for $\varphi = 0$ we have exactly the same excitation and response as in Section 2.15).

We shall now summarise the essential points:

α) The system is linear, because the network elements L, R and C are constant;

β) Instead of solving a differential equation (c) we solve a (complex) algebraic equation (o);

γ) A formula (q) arises, which reminds us of the series connection of elements.

Generally the above principle can be used. In the solution simultaneous differential equations arise, which are linear. Transformation to the complex plane gives algebraic equations. It turns out to be possible to avoid writing down the differential equations.

Next we shall individually investigate the behaviour of an inductor, a capacitor and a resistor and subsequently prove that Kirchhoff's laws remain valid for complex voltages and currents. This opens the road to analyse each network with the aid of complex quantities.

Let us examine the behaviour of the inductor, using complex quantities.

If the current is

$$i = |I| \cos(\omega t + \varphi)$$

the voltage is

$$v = L \frac{di}{dt} = -\omega L\, |I| \sin(\omega t + \varphi).$$

So

$$v = \omega L\, |I| \cos(\omega t + \varphi + \frac{\pi}{2}).$$

Transformation to the complex plane gives

$$I = |I|\, e^{j\varphi}$$

and

$$V = \omega L\, |I|\, e^{j(\varphi + \frac{\pi}{2})} = \omega L\, |I|\, e^{j\varphi} \cdot e^{j\frac{\pi}{2}} = j\omega L\, |I|\, e^{j\varphi}$$

$$V = j\omega L I. \qquad (2.26)$$

The result is an unexpectedly elegant formula, and gives the relation between the complex voltage and current for the inductor. It resembles Ohm's law.

For the moduli,

$$|V| = \omega L\, |I|,$$

a result we also found in Section 2.14 for the amplitudes.

For the arguments,

$$\arg V = \arg I + \frac{\pi}{2}.$$

Transformed to the time domain this has a result that the voltage leads the current by 90°, which we also found in Section 2.14.

We now write (2.26) as follows

$$V = ZI,$$

where

$$Z = j\omega L \tag{2.27}$$

is the *impedance*. The dimension is that of resistance, so the unit is Ω.

In Figure 2.27 the complex current and voltage for an inductor are drawn.

Note the conformity with the vector part of Figure 2.20. From now on we shall write the complex quantities, instead of the moduli, near the vectors.

ωL is called the *reactance* X of the inductor:

$$X = \omega L. \tag{2.28}$$

The reactance, too, has the dimension of resistance.

Example

Through an inductor L of 4 H flows the complex current $I = 1 - 2j$ A. The frequency is $\omega = 2$ rad/s. The complex inductor voltage is

$$V = j\omega LI = 8j(1 - 2j) = 8(2 + j) \text{ V.}$$

For the time function we find

$$i = \sqrt{5} \cos(2t - \arctan 2) \text{ A}$$

and

$$v = 8\sqrt{5} \cos(2t + \arctan \tfrac{1}{2}) \text{ V.}$$

We shall now discuss the capacitor:

Suppose $v = |V| \cos(\omega t + \varphi),$

hence $i = C \dfrac{dv}{dt} = \omega C |V| \cos(\omega t + \varphi + \frac{\pi}{2}).$

So $V = |V| e^{j\varphi}$

$I = j\omega C |V| e^{j\varphi}$

$I = j\omega C V. \tag{2.29}$

$Y = j\omega C \tag{2.30}$

is called the *admittance* of the capacitor; it has the dimension of conductance.

In general

$$Z = \frac{1}{Y}.$$ (2.31)

The impedance of a capacitor is

$$Z = \frac{1}{j\omega C}.$$ (2.32)

The reactance is

$$X = \frac{-1}{\omega C}.$$ (2.33)

In Figure 2.28 the complex quantities of a capacitor are drawn.

For a resistor $v = Ri$, so voltage and current are in phase. In the complex region this means the same argument:

$$V = RI.$$ (2.34)

(See Figure 2.29).

An important characteristic of the above complex relations is that they are linear. Together with the theorem to be discussed in the next section this results in important consequences.

We note that the impedance of an inductor or a capacitor in a network is often written in Ohm, for instance $4j\ \Omega$ for an inductor or $-3j\ \Omega$ for a capacitor.

2.19 Kirchhoff's laws for complex quantities

We shall prove that Kirchhoff's current law remains valid if all branch currents are transformed to the complex plane.

We have

$$\sum_{n=1}^{b} i_n = 0$$

for all time t.

We set

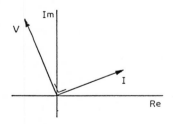

Figure 2.27

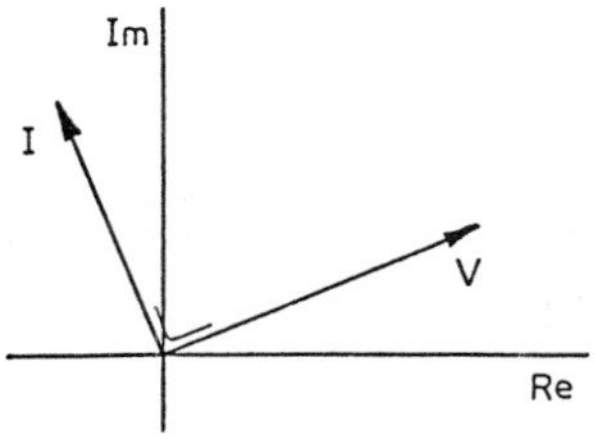

Figure 2.28

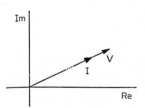

Figure 2.29

$$i_1 = |I_1| \cos(\omega t + \varphi_1)$$

$$i_2 = |I_2| \cos(\omega t + \varphi_2)$$

.

.

$$i_b = |I_b| \cos(\omega t + \varphi_b).$$

For $t = 0$ we get

$$|I_1| \cos\varphi_1 + |I_2| \cos\varphi_2 + \ldots + |I_b| \cos\varphi_b = 0. \qquad (a)$$

For $t = \dfrac{\pi/2}{\omega}$, i.e. $\omega t = \dfrac{\pi}{2}$, we find

$$-|I_1| \sin\varphi_1 - |I_2| \sin\varphi_2 + \ldots - |I_b| \sin\varphi_b = 0. \qquad (b)$$

Multiply (b) by $-j$ and add this to (a):

$$|I_1| (\cos\varphi_1 + j \sin\varphi_1) + |I_2| (\cos\varphi_2 + j \sin\varphi_2) + \ldots + |I_b| (\cos\varphi_b + j \sin\varphi_b) = 0.$$

All terms turn out to be those complex currents added to the sinusoidal branch currents. So

$$I_1 + I_2 + \ldots + I_b = 0$$

or

$$\sum_{n=1}^{b} I_n = 0.$$

We find a formula with exactly the same form as (1.9). In (1.9) we had d.c. currents, now we have complex currents.

In a similar way we can derive Kirchhoff's voltage law for complex voltages:

$$\sum_{m=1}^{l} V_m = 0.$$

The validity of both Kirchhoff's laws, together with the linearity of the complex element relations has as the important consequence that all previously derived theorems, formulas and rules for d.c. circuits keep their validity! An exception is power which, as we have seen, was a non-linear expression in voltage and current. We shall further discuss this in Section 2.22.

The following do remain valid: The formulas for series and parallel connection, voltage and current partition, the mesh and node methods, Tellegen's theorem, the reciprocity theorem, Thévenin's and Norton's theorems, the star-delta transformation and the various properties of transactors. Below are some examples.

Example 1 (Figure 2.30).

The total impedance of this one-port is

$$Z = j\omega L + R_1 + \frac{R_2 \cdot \dfrac{1}{j\omega C}}{R_2 + \dfrac{1}{j\omega C}} = \frac{V}{I},$$

where V is the port voltage and I the port current. It follows that

$$Z = j\omega L + R_1 + \frac{R_2}{1 + j\omega C R_2} = R_1 + j\omega L + \frac{R_2(1 - j\omega C R_2)}{1 + \omega^2 C^2 R_2^2},$$

hence

$$\operatorname{Re} Z = R_1 + \frac{R_2}{1 + \omega^2 C^2 R_2^2}$$

and

$$\operatorname{Im} Z = \omega L - \frac{\omega C R_2^2}{1 + \omega^2 C^2 R_2^2}.$$

Note that both the real part R and the imaginary part X of the impedance $Z = R + jX$ depend on the frequency!

This is also the case with admittances. If we set $Y = G + jB$ then, in general, G and B are functions of ω. B is called the *susceptance*.

Example 2 (Figure 2.31).

The angular frequency is $\omega = 3$ rad/s. The complex voltage source strength is 2 V. We calculate the voltage V_2 with voltage partition.

$$V_2 = \frac{2}{2 + 3j} \cdot 2 = \frac{4}{13}\,(2 - 3j)\ \text{V}.$$

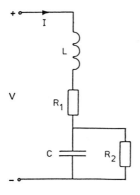

Figure 2.30

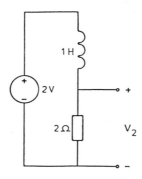

Figure 2.31

Example 3 (Figure 2.32).
The complex voltage source intensities are 2 V and $1 + j$ V. From these values we conclude that the second source strength as a function of time leads the first one by 45°. With the mesh method we find

$$2 = 2jI_1 - jI_2$$

$$1 + j = -jI_1 + (2 - j)I_2.$$

After some calculation we find $I_1 = \frac{1}{5}(1 - 3j)$ A and $I_2 = \frac{2}{5}(1 + 2j)$ A.

Example 4 (Figure 2.33).
Calculate the Thévenin equivalence on terminals a and b.
The open gate voltage is

$$V_T = \frac{1-j}{4-j}(4-j) = 1 - j \text{ V}.$$

The Thévenin impedance is

$$Z_T = Z_{ab} = \frac{(1-j)\cdot 3}{1-j+3} = \frac{3}{17}(5-3j) \ \Omega.$$

The Thévenin equivalence is shown in Figure 2.34.

Example 5 (Figure 2.35).
Check the validity of Tellegen's theorem for the network shown in Figure 2.35(a). Its graph is shown in Figure 2.35(b). After some calculation we find the branch currents $I_2 = I_3 = 6 - 2j$ A and $I_4 = 4 + 2j$ A. From this the branch voltages follow which enable us to draft the following table:

branch	voltage	current	product
1	−8 − 4j	+10	−80 − 40j
2	2 + 6j	6 − 2j	24 + 32j
3	6 − 2j	6 − 2j	32 − 24j
4	8 + 4j	4 + 2j	24 + 32j
			0 + 0j +

Addition of the last column gives zero. We note that, in contrast to the d.c. case, the product of complex branch voltage and complex branch current is not power. The phenomenon of *complex power* will be discussed in Section 2.22.

Example 6 (Figure 2.36).
We examine the reciprocity: We interchange the voltage source with the short circuit. In Figure 2.36(a) we find

$$I_1 = \frac{5}{2j + \frac{-j}{1-j}} = \frac{5(1-j)}{2+j} \text{ A}.$$

With current partition

$$I_2 = I_1 \frac{1}{1-j} .$$

It follows that

$$I_2 = \frac{5}{2+j} \text{ A.}$$

In Figure 2.36(b) we find

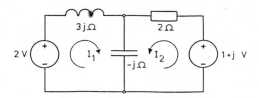

$3j\,\Omega$ $2\,\Omega$

2 V I_1 I_2 $1+j$ V

$-j\,\Omega$

Figure 2.32

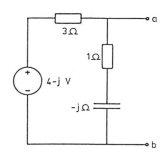

$3\,\Omega$ a

$1\,\Omega$

$4-j$ V

$-j\,\Omega$

b

Figure 2.33

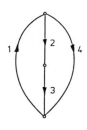

$\frac{3}{17}(5-3j)\,\Omega$ a

$1-j$ V

b

Figure 2.34

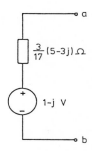

$j\,\Omega$

$\uparrow$ 10 A $2\,\Omega$

$1\,\Omega$

Figure 2.35   a

1   2   4   3

b

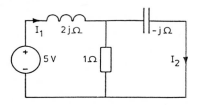

I_1 $2j\,\Omega$ $-j\,\Omega$

5 V $1\,\Omega$ I_2

Figure 2.36   a

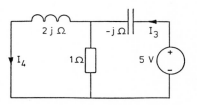

$2j\,\Omega$ $-j\,\Omega$ I_3

I_4 $1\,\Omega$ 5 V

b

$$I_3 = \frac{5}{-j + \frac{2j}{1 + 2j}} = \frac{5(1 + 2j)}{2 + j} \text{ A}$$

and

$$I_4 = I_3 \frac{1}{1 + 2j} .$$

We find

$$I_4 = \frac{5}{2 + j} \text{ A}.$$

The result is indeed $I_4 = I_2$.

Example 7 (Figure 2.37).
We calculate the current I with the superposition law (see Figure 2.38(a) and (b)).
We find

$$I' = \frac{2}{3 + 2j} = \frac{2}{13} (3 - 2j) \text{ A}$$

and

$$I'' = \frac{2j}{3 + 2j} (1 + j) = \frac{2}{13} (-1 + 5j) \text{ A}.$$

It follows that

$$I = I' + I'' = \frac{2}{13} (2 + 3j) \text{ A}.$$

2.20 The complex → time transformation

In Section 2.18 we discussed the transformation

$$v = \text{Re}(V \, e^{j\omega t}) = |V| \cos(\omega t + \varphi).$$

In this transformation the cosine function occurs. With the sine function

$$v = |V| \sin(\omega t + \varphi)$$

it can be said that

$$v = \text{Im}(V \, e^{j\omega t}) = \text{Im} \, \tilde{v}.$$

Instead of the real part the imaginary part of the complex time function $\tilde{v} = V \, e^{j\omega t}$ is taken. It is also possible to divide the difference of v and v* by 2j instead of taking half the sum as in (2.25):

$$v = \frac{1}{2j} (\tilde{v} - \tilde{v}*).$$

After all

$$\frac{1}{2j} (\tilde{v} - \tilde{v}*) = \frac{1}{2j} (V \, e^{j\omega t} - V* \, e^{-j\omega t}) = \frac{1}{2j} |V| (e^{j\varphi} \, e^{j\omega t} - e^{-j\varphi} \, e^{-j\omega t})$$

$$= \frac{1}{2j} \, |V| \, \{\cos(\omega t + \varphi) + j \sin(\omega t + \varphi) - \cos(\omega t + \varphi) - j \sin(-\omega t - \varphi)\}$$

$$= |V| \sin(\omega t + \varphi).$$

One can, of course, also first convert the sine function to a cosine function:

$$\sin(\omega t + \varphi) = \cos(\omega t + \varphi - \frac{\pi}{2}).$$

Example

In a network the voltage source intensity is $v = 6 \sin 8t$ V. So $v = 6 \cos(8t - \frac{\pi}{2})$ V. The additional (complex) voltage is $V = 6 \, e^{-j(\pi/2)} = -6j$ V. If we find the complex current $I = 2 - j$ A in a branch then the time function belonging to this complex current is

$$i = \sqrt{5} \, \cos(8t - \arctan \tfrac{1}{2}) \text{ A},$$

after which we can change back to the sine function:

$$i = \sqrt{5} \, \sin(8t - \arctan \tfrac{1}{2} + \tfrac{\pi}{2}) \text{ A}.$$

It is simpler to convert the sine function to a real number:

$$v = 6 \sin 8t \text{ V} \quad \rightarrow \quad V = 6 \text{ V}.$$

Hence we now use the transformation

$$v = \mathrm{Im}(V \, e^{j\omega t}).$$

We now find

$$I = 1 + 2j \text{ A}$$

(viz. the current in the last case multiplied by j) so

$$i = \sqrt{5} \, \sin(8t + \arctan 2) \text{ A}.$$

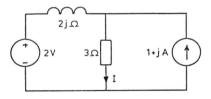

Figure 2.37

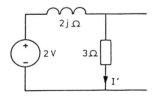

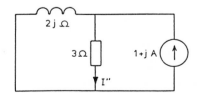

Figure 2.38   a   b

We find the same function because

$$\arctan 2 + \arctan \tfrac{1}{2} = \frac{\pi}{2}.$$

This can be summarised as follows:

$V = |V| \sin \omega t$. Add to this the real voltage $V = |V|$. Calculation results in the current $I = |I| e^{j\alpha}$. Transformation gives $i = |I| \sin(\omega t + \alpha)$.

Note that we do not realise in this summary that we are using the imaginary part.

We can even go a step further and add a real number to a cosine function with an arbitrary phase φ. This is possible because we are discussing the stationary state here, i.e. non-damped harmonic voltages and currents. If that is the case it does not matter which moment of time we choose as the origin.

Example.

$v = 6 \cos(8t + \varphi)$ V. Formally we must add

$$V = 6 \, e^{j\varphi} \text{ V},$$

after which this power of e is present in the remainder of the calculation. However, we can shift the time axis a little so that $v_s = 6 \cos 8t$ V results. v_s is the shifted v.

We now add $V_s = 6$ V. So in the result we have to perform the shift in the opposite direction.

Example.

The voltage source intensity is $v = 24 \cos(15t + \varphi)$ V. The phase is φ rad. We add the complex voltage $V = 24$ V (real).

Suppose we find $I = 2(1 - j)$ A in a branch of the network.

So

$$|I| = 2\sqrt{2} \text{ A}$$

and

$$\arg I = -\frac{\pi}{4} \text{ rad.}$$

Adding the phase φ we find:

$$i = 2\sqrt{2} \cos(15t + \varphi - \frac{\pi}{4}) \text{ A.}$$

If a network contains several sources of which the intensities have different phases, only one of the complex source intensities can be set to real.

As an example consider Figure 2.37. The voltage source strength as a function of time is, for instance, $2 \sin\omega t$ V, while the phase of the current source strength leads by 45° so $\sqrt{2} \sin(\omega t + \frac{\pi}{4})$ A. Only the complex voltage source is set real, the complex current source strength must be necessarily complex.

So far we have performed the transformation from complex to time with the rules

$$\text{modulus} \quad \rightarrow \quad \text{amplitude}$$

$$\text{argument} \quad \rightarrow \quad \text{phase.}$$

However, it is also possible to transform the real and the imaginary parts separately. We then make use of the superposition principle.

Example.

Suppose the voltage source intensity is $v_1 = 12 \cos 6t$ V. We add $V_1 = 12$ V. Suppose that for a voltage elsewhere in the network we find $V_2 = 2 + 5j$ V.

We transform that back to the time region by considering the real and the imaginary parts separately and by adding the results:

$$v_2 = 2 \cos 6t - 5 \sin 6t \text{ V}.$$

The modulus of the real part is 2 and the argument is zero, whereas the modulus of the imaginary part is 5 and the argument is +90°. Thus we find
$5 \cos(6t + \frac{\pi}{2}) = -5 \sin 6t$ for the imaginary part. Direct back-transformation results in

$$v_2 = \sqrt{29} \cos(6t + \arctan \tfrac{5}{2}) \text{ V}.$$

With the aid of goniometry it can simply be proved that both expressions are equal.

To this end we consider the tangent function (see Figure 2.39). In one period this function is dimorphic. In order to be able to write an unambiguous expression the region $-\frac{\pi}{2} < \varphi < \frac{\pi}{2}$ has been chosen.

In this interval the function is singular.

The complex current $I = -2 + 5j$ A should not be transformed to

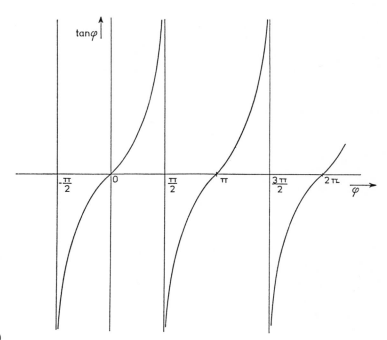

Figure 2.39

$$i = \sqrt{29} \cos(\omega t - \arctan \tfrac{5}{2}) \, A$$

but through $-I = 2 - 5j \, A$
and

$$-i = \sqrt{29} \cos(\omega t - \arctan \tfrac{5}{2}) \, A$$

to

$$i = -\sqrt{29} \cos(\omega t - \arctan \tfrac{5}{2}) \, A.$$

In the complex plane the angles are limited to the region where the real part is positive (see Figure 2.40).

2.21 Vector diagrams

The node potentials in the networks considered in this chapter are nearly all sinusoidal. The potentials are measured with respect to a point with potential zero (e.g. earth), which may be a node of the network. To this node potential complex potentials can be added. In Figure 2.41 we have drawn the complex potentials of the nodes 1 and 2 of a network.
However, in networks we are more interested in mutual voltages instead of the potentials themselves.
With the construction of

$$V_{12} = V_1 - V_2$$

we see a vector V_{12} resulting, which represents the voltage V_{12} and for which the first index is the arrowhead. Because the position of the point with potential zero is not important, one can leave out the co-ordinate system without problem. If we do this for all n nodes in the network this results in a polygon with n vertices, in which the sides represent the voltages.
One cannot use this method with the currents, because there may be two branches in parallel between the nodes 1 and 2 in which two different currents may occur. We shall therefore indicate all branch currents with one index: I_1, I_2, I_3, etc. possibly $I_a, I_b, I_c, \ldots$

Example 1 (Figure 2.42).
Given $I = 2 \, A$, find the vector diagram of the voltages and the currents.

Solution
$V_{12} = 4j \, V$, $V_{23} = 6 \, V$, with these voltages we find Figure 2.43, in which we start with the horizontally drawn current vector:
In Figure 2.43(a) the current vector diagram is drawn, in Figure 2.43(b) the voltage vector diagram is drawn.

Example 2 (Figure 2.44).
In order to draw the vector diagrams we once more start with the horizontal position of

the current I (this choice is arbitrary). From this the branch voltages result (see Figure 2.45).

Starting with I the voltages $V_{13} = 3jI$, $V_{34} = 2I$ and $V_{42} = -jI$ follow, after which V_{12} is found. If we draw the vectors to scale then $|V_{12}|$ and φ can be measured. If the voltage V_{12} is given in the network then the current and all branch voltages, their magnitudes and their phases follow with this construction.

A great advantage of vector diagrams is that one is able to survey the whole situation in the network at a glance. For instance, the voltage V_{32} in Figure 2.45 can be found directly by measuring it. One may further leave out the voltage notation and the arrowheads in the voltage diagram without any objection.

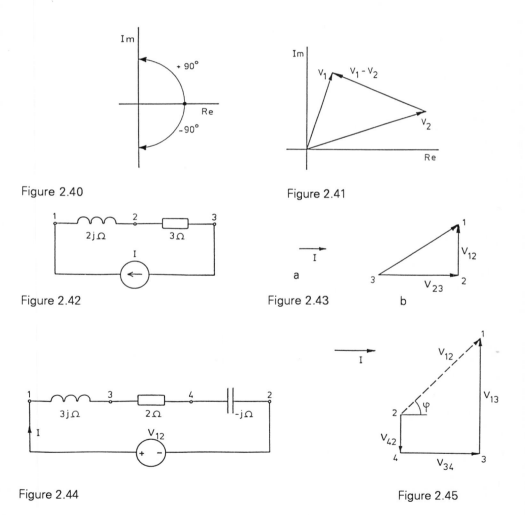

Figure 2.40

Figure 2.41

Figure 2.42

Figure 2.43

Figure 2.44

Figure 2.45

Example 3 (Figure 2.46).

We start with I_2. V_{14} and V_{42} follow (see Figure 2.47).

Next we continue with I_1, ignoring the above. Then V_{13} and V_{32} follow (see Figure 2.48).

By rotating and changing the magnitudes both diagrams can be combined. This has been done in Figure 2.49.

Note that the points 3 and 4 are situated on a circle of which V_{12} is the diameter. The branch currents I_1 and I_2 give the total current I after vectorial adding. For this reason they are placed 'head to tail'.

In the above examples it was always possible to find the vector diagrams directly by drawing them. For larger networks this is often not possible. One must then do a calculation with complex voltages and currents first before being able to make a drawing.

Example (Figure 2.50).

The mesh method gives:

$$0 = (1 - j)I_1 + 2jI_2 - 12j,$$

$$0 = 2jI_1 + (2 - 2j)I_2 - 12.$$

It follows that $I_1 = 3 + 3j$ A and $I_2 = 6 + 3j$ A, after which the branch voltages and the branch currents can be found:

$V_{14} = 3(1 + j)$ V	$I_3 = -3$ A
$V_{42} = 3(2 + j)$ V	$I_4 = 3(3 - j)$ A
$V_{13} = 3(1 + 3j)$ V	$I_5 = 3(2 - j)$ A
$V_{32} = 3(2 - j)$ V	
$V_{12} = 3(3 + 2j)$ V	
$V_{43} = 6j$ V.	

Choosing point 2 as the origin the vector diagram of Figure 2.51 results.

The place of the co-ordinate system is not important, any other node at the origin would have been possible. In Figure 2.52 the current vector diagram is drawn.

For the sake of clearness the arrows of some vectors have not been drawn at the end but a little before it. We may also leave out the co-ordinate system in Figure 2.52. The current vectors are drawn in such a way that Kirchhoff's current law is an easy 'head to tail' construction. For instance $I_2 + I_3 = I_1$.

The voltage vector diagram can only be drawn in one unambiguous way, however, this is not the case with the current vector diagram. A diagram that is also correct is shown in Figure 2.53.

We conclude this section with some general rules for drawing vector diagrams.:

a. Number all nodes and name all branch currents.

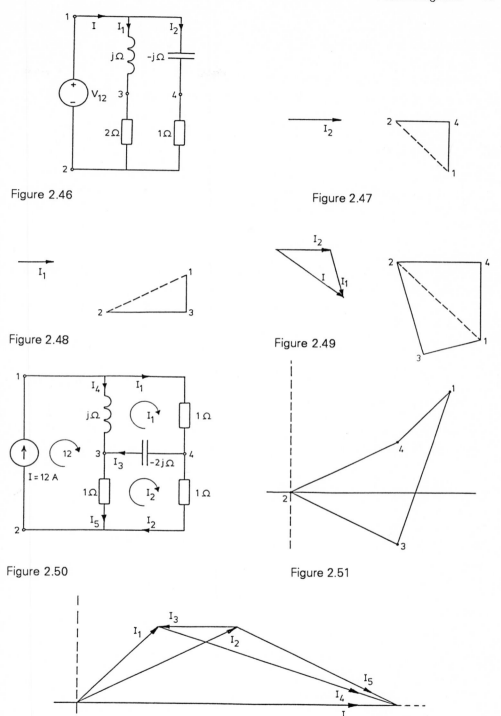

Figure 2.46

Figure 2.47

Figure 2.48

Figure 2.49

Figure 2.50

Figure 2.51

Figure 2.52

b. A positive branch voltage belongs to each branch current, thus the current in a branch flows from the node mentioned in the first index to the node mentioned in the second index of the branch voltage.

c. The first voltage index is the arrowhead.

d. Place the currents 'head to tail' in order to enable vectorial adding without a parallellogram.

2.22 Complex power

If one knows the complex voltage and the complex current of a two-terminal network it is possible to compute the power consumed by transforming these quantities to the time region first and subsequently use the following formula:

$$P = \frac{|V|\,|I|\cos\varphi}{2}.$$

However, it is possible to compute this power directly with the complex quantities themselves.

If $V = |V|\,e^{j\alpha}$ and $I = |I|\,e^{j\beta}$ then the angle between v and i is $\varphi = \alpha - \beta$ (see Figure 2.54). If we choose the conjugate of I then a sign change results in the angle β:

$$I^* = |I|\,e^{-j\beta}.$$

So then $P = \mathrm{Re}(\tfrac{1}{2}VI^*)$ turns out to be the power.

For

$$VI^* = |V|\,e^{j\alpha}\,|I|\,e^{-j\beta} = |V|\,|I|\,e^{j\varphi},$$

of which the real part is $|V|\,|I|\cos\varphi$. Evidently P is the real part of a complex number. This complex number is called the *complex power* S:

$$S = \tfrac{1}{2}VI^*. \tag{2.35}$$

So

$$P = \mathrm{Re}\ S. \tag{2.36}$$

P is called the *real (average) power*. The imaginary part of S is called the *reactive power* Q, so

$$Q = \mathrm{Im}\ S. \tag{2.37}$$

Therefore

$$S = P + jQ. \tag{2.38}$$

See Figure 2.55, in which Q is positive real. Q can also be negative real. For a passive two-port $P \geq 0$ always holds.

Finally we know the *apparent power*

$$|S| = \tfrac{1}{2}|V|\,|I|. \tag{2.39}$$

The unit of the real power P is W. It is exclusively reserved for the real power. The unit of all other powers introduced here is VA, while the reactive power gets an extra index r, VA$_r$. (In high power technology one often uses kVA or MVA).

In high power technology it is also usual not to use the amplitudes of the voltages and currents but the effective values. This results in, for example,

$$S = V_{eff} I_{eff}^*$$

and

$$P = |V_{eff}| |I_{eff}| \cos \varphi$$

and

$$P = |I_{eff}|^2 R.$$

Because the index 'eff' is often omitted in this field, one should be well aware of the notation convention.

According to Tellegen's theorem we have

$$\sum_{n=1}^{b} V_n I_n = 0,$$

in which b is the number of branches.

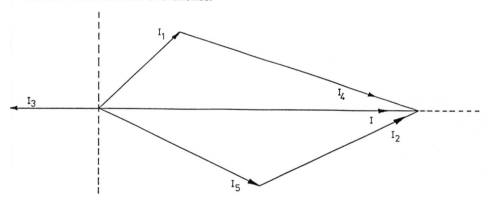

Figure 2.53

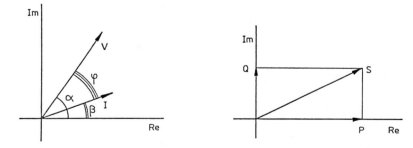

Figure 2.54   Figure 2.55

It also holds that

$$\sum_{n=1}^{b} V_n I_n^* = 0.$$

So

$$\sum_{n=1}^{b} S_n = 0. \tag{2.40}$$

S_n is the complex power consumed by branch n. This formula is called the rule of *power balance*.

$$\sum_{n=1}^{b} P_n = 0$$

and

$$\sum_{n=1}^{b} Q_n = 0.$$

Note that

$$\sum_{n=1}^{b} |S_n| = 0$$

does *not* follow from (2.40)

Example 1 (Figure 2.56).
Suppose V = 50 V. We find $I = \dfrac{50}{3 + 4j}$ A, so $I^* = \dfrac{50}{3 - 4j}$ A.

The complex power supplied by the source is

$$S = \tfrac{1}{2} VI^* = \tfrac{1}{2} \cdot 50 \cdot \frac{50}{3 - 4j} = 150 + 200 j \text{ VA}.$$

So P = 150 W and Q = 200 VA$_r$.

The resistor consumes a real power

$$P = \tfrac{1}{2} |I|^2 R = \tfrac{1}{2} \left| \frac{50}{3 + 4j} \right|^2 \cdot 3 = 150 \text{ W}.$$

The inductor voltage is

$$V_L = 4jI = 4j \frac{50}{3 + 4j} \text{ V}.$$

The complex power consumed by the inductor is thus

$$S_L = \tfrac{1}{2} V_L I^* = \tfrac{1}{2} \cdot 4j \cdot \frac{50}{3 + 4j} \cdot \frac{50}{3 - 4j} = 200j \text{ VA}.$$

So the inductor consumes a reactive power

$$Q = 200 \text{ VA}_r.$$

Example 2 (Figure 2.57).
Given are V = 5 V and I = 1 + j A.
The mesh method results in

$$5 = (1 + 2j)I_1 + 2j(1 + j).$$

From this it follows that

$$I_1 = \frac{3 - 16j}{5} \text{ A}.$$

So the voltage source supplies

$$S_V = \tfrac{1}{2} V I_1^* = \tfrac{1}{2} \cdot 5 \cdot \frac{3 + 16j}{5} = 1.5 + 8j \text{ VA}.$$

The current source voltage is

$$V_1 = V - I_1 - j(1 + j) = \frac{27 + 11j}{5} \text{ V}.$$

So the current source supplies

$$S_I = \tfrac{1}{2} \frac{27 + 11j}{5} \cdot (1 - j) = 3.8 - 1.6j \text{ VA}.$$

So both sources supply the complex power S = 5.3 + 6.4j VA.
The resistor consumes

$$S_R = P = \tfrac{1}{2} |I_1|^2 R = \tfrac{1}{2} \frac{|3 - 16j|^2}{25} = 5.3 \text{ W},$$

which is the real part of S.
If we call the inductor voltage V_L and the inductor current I_L then the complex power consumed by the inductor is:

$$S_L = \tfrac{1}{2} V_L I_L^* = \tfrac{1}{2} j\omega L I_L I_L^* = \tfrac{1}{2} j\omega L |I_L|^2 = \tfrac{1}{2} \cdot 2j \cdot |I_1 + I|^2$$

$$= j \left| \frac{8 - 11j}{5} \right|^2 = 7.4 \text{ VA}.$$

For the capacitor we find

$$S_C = \tfrac{1}{2} V_C I_C^* = \tfrac{1}{2} \frac{1}{j\omega C} I_C I_C^* = \tfrac{1}{2} \frac{1}{j\omega C} |I_C|^2 = -\tfrac{1}{2} j \cdot 2 = -j \text{ VA}.$$

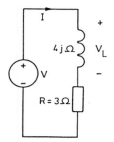

Figure 2.56

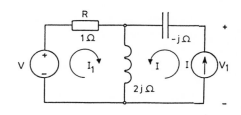

Figure 2.57

$S_L + S_C = 6.4j$ VA and this is the imaginary part of S.
We have

$$S_V + S_I = S_R + S_L + S_C \qquad \text{(power balance)}.$$

The apparent powers supplied by the sources are:

$$|S_V| = \sqrt{66.25} = 8.14 \text{ VA}; \qquad |S_I| = \sqrt{17} = 4.12 \text{ VA}.$$

The apparent powers consumed are:

$$|S_R| = 5.3 \text{ VA } (=W); \qquad |S_L| = 7.4 \text{ VA} \qquad \text{and} \qquad |S_C| = 1 \text{ VA}.$$

2.23 Maximum power transfer for complex quantities

Consider the network of Figure 2.58.
The part to the left of nodes a and b is a non-ideal voltage source, but it may also be the
Thévenin equivalence of another network.
We have $Z_T = R_T + jX_T$.
The part right of nodes a and b is the load $Z = R + jX$.
We now calculate Z, given V_T and Z_T, so that a maximum power will be dissipated in the
impedance Z. The current is

$$I = \frac{V_T}{Z_T + Z} = \frac{V_T}{R_T + R + j(X_T + X)} \ .$$

So the real power in the load Z is

$$P = \tfrac{1}{2} |I|^2 R = \frac{|V_T|^2}{2} \frac{R}{(R_T + R)^2 + (X_T + X)^2} = f(R,X)$$

with $R > 0$ and $-\infty < X < \infty$.
For each value of R the power P is maximal if

$$X_T = -X.$$

For this value

$$P = \frac{|V_T|^2}{2} \frac{R}{(R_T + R)^2}$$

is a function of R only.
We find the same formula as in Section 1.14 (besides the coefficient $\tfrac{1}{2}$): for the value
$R = R_T$ the power P has a maximum.
The result is that for

$$Z = Z_T^* \qquad\qquad (2.41)$$

in the load Z a maximum power will be developed.
Later on we will see how it is possible to *match* a given load with the aid of a transformer.

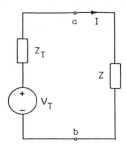

Figure 2.58

2.24 Problems

2.1 Given is a half-wave-rectified sinusoidal current with amplitude $|I|$.

    a. Calculate the average value,
    b. Calculate the effective value.

2.2 Given is a full-wave-rectified sinusoidal voltage with amplitude $|V|$.
    a. Calculate the average value,
    b. Calculate the effective value.

2.3 Find V_{eff} and V_{av}.

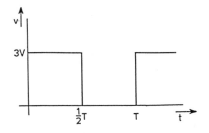

2.4 Find V_{eff} and V_{av}.

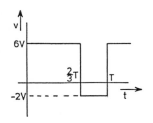

2.5 Find I_{eff}.

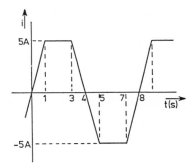

2.6 Compute $p(t)$ and P if
    a. $v = 4$ V
        $i = 3$ A.
    b. $v = 4 \cos 2t$ V
        $i = 6 \cos 2t$ A.
    c. $v = 4 \cos(3t + \frac{\pi}{8})$ V
        $i = 6 \cos 3t$ A.

2.7 Find the effective value of
$$i = 5 + 3 \cos(7t + \tfrac{\pi}{4}) \text{ A.}$$

2.8 Given
$$i = |I_1| \cos \omega t + |I_2| \cos 2\omega t \text{ A.}$$

    a. Find the period.
    b. Find the effective value.

2.9   The current i is the sum of two harmonic terms with frequencies of 100 Hz and of 700 Hz. Find the period of i.

2.10   Write

$$v = 7 \cos(3t + \frac{\pi}{8}) \text{ V} \qquad \text{in the form}$$

$$v = A \cos 3t + B \sin 3t \text{ V}.$$

2.11   Write

$$i = 5 \cos 4t + 2 \sin 4t \text{ A} \quad \text{in the form}$$

a. $i = |I| \sin(4t + \varphi)$ A.
b. $i = |I| \cos(4t + \varphi)$ A.

2.12   a. The current in an inductor of 0.1 H is in accordance with Figure A. Draw $v_L(t)$ to scale.
   b. Now draw $v_L(t)$ to scale for a current in accordance with Figure B.
   c. Regard Figure C as a limit case of Figure B. Find $v_L(t)$.

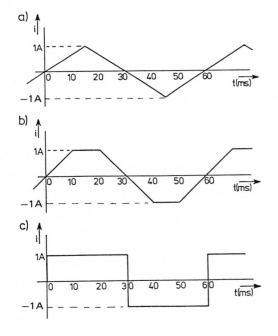

2.13   The voltage on a two-terminal network is v = –100 sin 377t V. The current is i = 10 cos 377t A. With which network element are we concerned here? Find the value.

2.14   i = |I| cos ωt A. Find v, if at t = 0 the voltage v = 0 V.

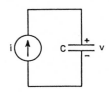

2.15   v = |V| sin ωt.

Find i, if at t = 0 the current is

$$i = -\frac{|V|}{\omega L} \text{ A}.$$

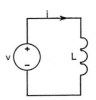

2.16   The current as shown in the plot goes through a capacitor. Draw the capacitor voltage $v_C(t)$ if $v_C(0) = 0$ V.

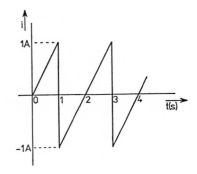

2.17 Give a plot of $v_{ab}(t)$ to scale.

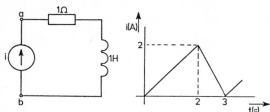

2.18 Given he current i of Problem 2.17, and both voltage source intensities are constant (V = 2 V).
a. Draw $i_v(t)$ to scale.
b. Draw $i_v(t)$ to scale, but now for V = 6 V.

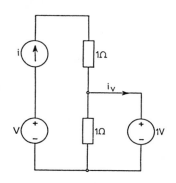

2.19 v = −10 sin 2t V; i = 5 sin 2t A. Find v_{ab}.

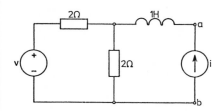

2.20 This network exists infinitely. At t = 0 S is closed.
Find the dissipated energy between t = 0 and t → ∞.

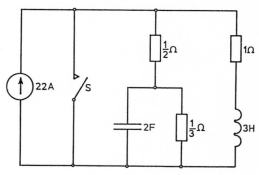

2.21 v = 25 sin 20t V.
Find i.

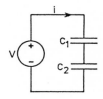

2.22 What is the substitute self-inductance of two inductors L_1 and L_2 in parallel?

2.23 Given the network with a d.c. voltage source and an a.c. voltage source, we are concerned with the steady state here.
v_1 = 200 sin 5000t V, v_2 = 300 V.
a. Find i(t).
b. Find the effective value of i(t).

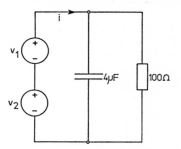

2.24   Given a network with a voltage
source, of which the intensity is
sinusoidal.
The current i_L is according to the
formula

$$i_L = 8 \sin 3t \text{ A.}$$

a.   Find $v_{ab}(t)$.
b.   Find $i(t)$.
c.   Give a simpler substitute circuit
for the network to the right of the
terminals a and b.

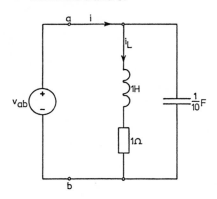

2.25   Given a network with a current
source, of which the intensity is
sinusoidal, the capacitor voltage is
$v_C = 8 \sin 3t \text{ V.}$
a.   Find $i(t)$.
b.   Find $v_{ab}(t)$.
c.   Give a simpler substitute circuit
for the network to the right of the
terminals a and b.

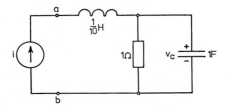

2.26 $i_1 = 5 \cos \omega t \text{ A}$, $i_2 = 12 \sin \omega t \text{ A.}$
a.   Find i_3.
b.   Calculate v so that $i_v = 0$ for
all t.

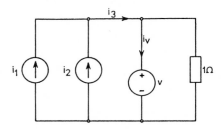

2.27   The voltage v is trapezoidal
according to the plot given. Draw
the current i as a function of time to
scale.

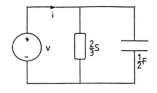

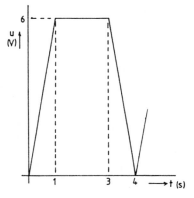

2.28 Given $i_L = 3 \sin 2t$ A.
   a. Find v and i.

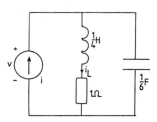

The network can be replaced by the circuit below, while v and i remain the same.
   b. Find L and R.

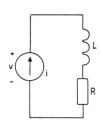

2.29 Write in the form $A + jB$:
   a. $(7 - 3j) + (8 + 2j)$.
   b. $(2 + 3j) \cdot (1 + j)$.
   c. $\dfrac{2 + j}{-1 + 3j}$.
   d. $(6 - 3j)^*$.

2.30 Write in the form $|Z| e^{j\varphi}$:
   a. $2 + 7j$.
   b. j.
   c. $j(3 - j)$.
   d. $-2 - j$.

2.31 Find
   a. $\mathrm{Re}\{(2 - j)(2 - 2j)\}$.
   b. $\mathrm{Im}\,(5e^{j\frac{\pi}{6}})$.
   c. $\mathrm{Re}\left(\dfrac{\sqrt{2}\ e^{j\frac{\pi}{4}}}{1 + 2j}\right)$.

2.32 It is given that A and B are complex numbers. Write without parentheses:
   a. $(A + B)^*$.
   b. $(A^*)^*$.
   c. $(A^*B)^*$.
   d. $\left(\dfrac{A}{B}\right)^*$.

2.33 Calculate
   a. $|5 + j|$.
   b. $|3e^{j\frac{\pi}{8}}|$.
   c. $\left|\dfrac{5 + 2j}{5 - 2j}\right|$.

2.34 Calculate
   a. $\arg(5 + j)$.
   b. $\arg\{3e^{j\frac{\pi}{8}}\}$.
   c. $\arg\dfrac{5 + 2j}{5 - 2j}$.

2.35 Calculate
   a. j^5 b. 5^j c. 2^{1+j}
   d. j^j e. $(5 + j)^{1+j}$.

2.36 If it is given that Z is a complex number:
   a. Prove that $ZZ^* = |Z|^2$.
   b. $Z + Z^* = 2\mathrm{Re}Z$.
   c. $Z - Z^* = 2\mathrm{Im}Z$.

2.37 Examine whether the following relations are valid:
   a. $A^* + B = B^* + A$.
   b. $(|A|)^* = |A|$.
   c. $\arg(A^*) = -\arg(A)$.
   d. $(A^B)^* = (A^*)^{B^*}$.

2.38 Find
   a. $\mathrm{Re}(4e^{j\omega t})$.
   b. $\mathrm{Re}(4e^{j(\omega t+\frac{\pi}{4})})$.
   c. $\mathrm{Re}(4e^{j\omega t+\frac{\pi}{4}})$.

2.39 Of which exponential functions are the following functions the real part?
   a. $4 \cos 5t$.
   b. $4 \cos(5t + \dfrac{\pi}{4})$.

c. $4 \sin 5t$.

d. $4 \sin(5t + \frac{\pi}{4})$.

e. $4e^{-t} \cos 5t$.

f. $4e^{-t} \sin 5t$.

2.40

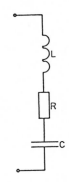

Find $Z(\omega)$.
Find $Y(\omega)$.

2.41

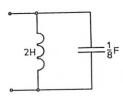

Find $Z(\omega)$.
Give a simpler circuit for

a. $\omega = 1$ rad/s.

b. $\omega = 2$ rad/s.

c. $\omega = 3$ rad/s.

2.42

For which angular frequency does
this two-terminal network behave as
a short circuit?

2.43 a. Which relation is there between
       the four impedances if $V_2 = 0$?

     b. Find two relations for the case

$$Z_1 = R_1$$

$$Z_4 = R_4$$

$$Z_3 = R_3 + j\omega L$$

$$Y_2 = \frac{1}{Z_2} = G_2 + j\omega C.$$

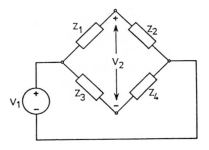

2.44 Calculate R so that Z is real. Express
      R in L, C and ω. $\omega > 0$.

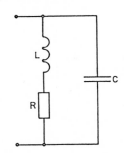

2.45 Calculate ω so that Z is real. $\omega > 0$.

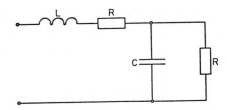

2.46 Find the conditions for which Z is real.

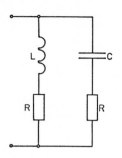

2.47 Prove that |Z| is independent of R if $2\omega^2 LC = 1$.

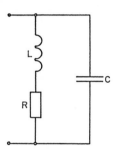

2.48 Prove that $|V_1| = |V_2|$ if $\omega = 10^4$ rad/s.

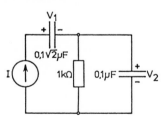

2.49 Find Z.

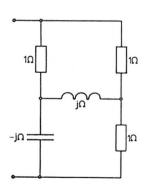

2.50 Find the capacitor current
   a. With Thévenin's theorem
   b. With superposition.

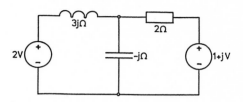

2.51 Check Tellegen's theorem by combining the branch voltages with the conjugates of the branch currents in the network of 2.50.

2.52 Find the time function i(t) if the complex voltage V = 5 V is added to v = 5 cos 4t V and
   a. I = 7 A.
   b. I = 3 + 8j A.

2.53 If the complex voltage V = 5 V is added to v = 5 sin 4t V, find the time function i(t) if
   a. I = 7 A.
   b. I = 3 + 8j A.

2.54 If the complex voltage V = 5 V is added to v = 5 sin 3t V find the complex current I if
   a. i = 4 sin 3t A.
   b. i = 4 sin(3t + $\frac{\pi}{4}$) A.
   c. i = 2 cos 3t A.
   d. i = 9 cos(3t − $\frac{\pi}{6}$) A.
   e. i = 14 cos 3t + 15 sin 3t A.

2.55 Find i(t) if

    a. v = 15 cos 3t V.

    b. v = 15 sin 3t V.

    c. v = 15 sin(3t + φ) V.

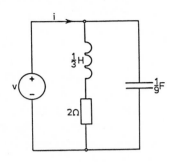

2.56 Given

$$v_{34}(t) = 10\sqrt{2}\,\sin(\omega t - \frac{\pi}{4})\ \text{V},$$

$$\omega = 10^3\ \text{rad/s}.$$

a. Find $v_{24}(t)$.

Next it is given that

$$v_{14}(t) = 20\cos\omega t\ \text{V}.$$

b. Find L.

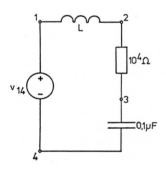

2.57 Find the phase difference between the time functions i and i_5.

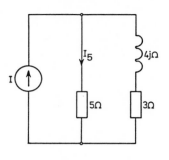

Find the voltage and current vector diagram of the following four networks

2.58

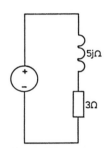

2.59

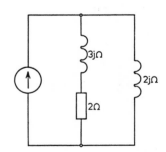

2.60

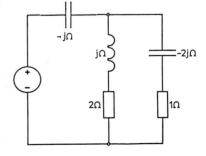

2.61

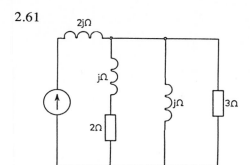

2.62 Given the network with a current source of which the intensity is a sinusoidal function of time, the amplitude is 13 A. The complex current is set to 13 A (real).
a. Find the complex currents I_1, I_2 and I_3.
b. Draw the vector diagram of the voltages to scale. Choose point 5 in the centre of your sheet. Set 1 V $\triangleq$ 1 cm.

2.63 Given $v_{10} = 9 \sin 2t$ V,
$v_{20} = 27 \cos 2t$ V,
the complex voltage V_{10} is set to $V_{10} = 9$ V.
a. Find the complex currents I_1 and I_2 (added to i_1 and i_2).
b. Find I_3 and I_4 and all branch voltages and subsequently draw the voltage vector diagram.
c. Find $i_1(t)$.

See Figure at bottom of page.

2.64 a. Draw the vector diagrams of all voltages and currents to scale.
In the figure the angle between the voltages V_{12} and V_{45} appears to be 90°.
b. Calculate and check if this angle is exactly 90°.

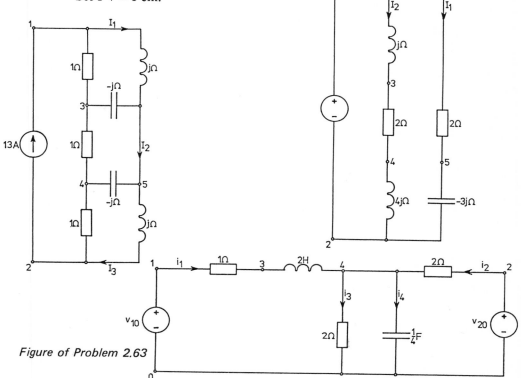

Figure of Problem 2.63

2.65 The current source intensity is
i(t) = sin(t) A. Set I = 1 A.
  a. Draw the vector diagram of the
     voltages.
  b. Find the dissipated power in the
     circuit.
  c. Give $v_{12}(t)$.

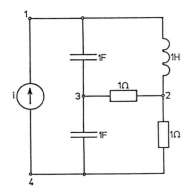

2.66 $I_a = 13$ A (real).
  a. Find the mesh currents I_1 and I_2.
  b. Find all branch currents and
     branch voltages.
  c. Draw the vector diagrams of all
     voltages and currents to scale.
  d. If $i_a = 13 \cos 300t$, give $v_{23}(t)$.

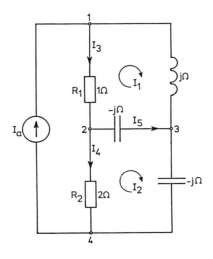

2.67 The source intensity is sinusoidal
with an amplitude of 5 V and an
angular frequency of 1 rad/s.
Set the complex source intensity to
real: 5 V.
  a. Prove (by means of the mesh
     method) that: $I_1 = 10$ A.
  b. Draw all mesh and branch
     currents in the complex plane.
  c. Construct the voltage vector
     diagram using the current vector
     diagram found under b.
  d. If $v_{10} = 5 \cos t$, find $v_{32}(t)$.

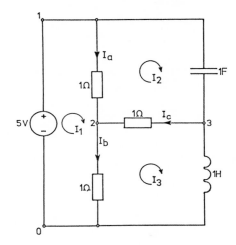

2.68 $V_{12} = 2$ V, $V_{34} = (-1 + j)$ V,
$\omega = 1$ rad/s.
  a. Find I.

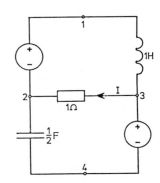

b. Find the complex power each of the sources supplies.
c. Find the reactive power that will be consumed by the inductor and the capacitor.

2.69

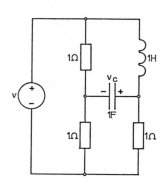

a. Find the angular frequency ω for which the capacitor voltage v_C is in phase with the source voltage v.
$\omega = 1$ rad/s is chosen and the amplitude $|V|$ of the voltage source intensity is set to 1 V.
b. Find the reactive power consumed by the capacitor.

2.70

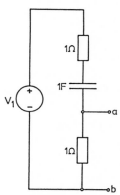

The complex voltage source intensity V is 1 V. The complex current source intensity I is $1 + j$ A. The voltage source intensity as a function of time is $v = \cos \omega t$.
a. Find A and B in the expression

of the current source intensity
$$i = A \cos \omega t + B \sin \omega t.$$
b. Find the complex current I_v.
c. Find the complex power S_V supplied by the voltage source and the complex power S_I supplied by the current source.
d. Calculate the total real power dissipated by both resistors and compare that with your answer to c.

2.71

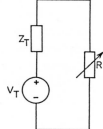

a. Find $H = \dfrac{V_{ab}}{V_1}$ as a function of ω.
Next $V_1 = 4$ V and $\omega = \frac{1}{2}$ rad/s.
b. Which impedance is to be connected to a and b so that a maximum power is developed in that impedance?
c. Calculate the power.

2.72 $Z_T = R_T + jX_T$, R is positive real. For which value of R is the power in R maximal?

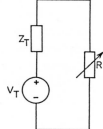

2.73

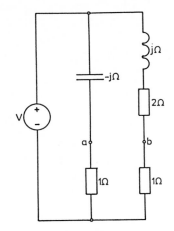

The complex voltage V is 5 V.
a. Find V_{ab}.
Next one short-circuits a and b.
b. Find the short circuit current.
The short circuit is taken away
again.
c. Which impedance is to be
connected to the terminals a and b
so that the power in that
impedance is maximal? Construct
the impedance with two elements
in parallel if the angular
frequency is 3 rad/s. Give the
values of those elements.
The impedance between a and b is
taken away again and a positive real
resistor R is connected to a and b.
d. How large should R be for the
power in R to be maximal?

2.74 Given a linear network with voltage
and current sources and with
resistors, inductors and capacitors.
All source intensities are sinusoidal
with the same angular frequency. To
all source intensities complex values
have been added.
One considers one of the voltage
sources with the terminals a and b.

Call that source intensity V_{ab}.
One further considers the current I
through the source V_{ab}, directed
from a to b. All complex source
intensities are kept constant, except
V_{ab}.
One measures the following results:

if $V_{ab} = 2$ V then $I = 1 + 3j$ A

if $V_{ab} = 5$ V then $I = -2 + 6j$ A.

a. What will I be if $V_{ab} = 10$ V?
Next one takes away the voltage
source V_{ab}.
b. Find the Thévenin equivalence on
the terminals a and b of the
remaining network.

2.75 Given are $V_{12} = 2$ V,
$V_{34} = (-1 + j)$ V, $\omega = 1$ rad/s.
a. Find the complex current I.
Next $v_{12} = 2 \cos 5t$ V.
b. Find i.

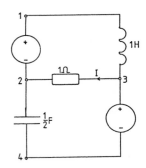

2.76 Given $I = 2j$ A.
a. Find I_Z as a function of Z using

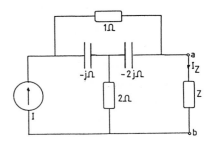

Thévenin's theorem.

b. Choose Z so that a maximum real power will be developed and subsequently find the complex power that Z consumes.

2.77 Both resistances are equal. Examine for which value of R the voltage V_{ab} is zero.

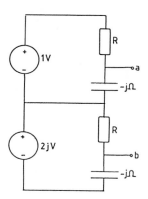

3

Some properties of networks

3.1 Polar diagrams

If we vary a real quantity in a network, e.g. the angular frequency ω, then the complex quantities in that network will also vary. We shall direct our attention to such a complex quantity. For instance, the impedance of a one-port will result in different complex values if we vary the angular frequency ω. Although variation of ω occurs at the most, one sometimes finds the variation of a resistance, a capacitor or an inductor.

The complex quantity may be an impedance or an admittance of a one port or a voltage to voltage relation, a current to current relation or a voltage to current relation of a two-port. It results in a function of the form

$$H = f(\rho). \tag{3.1}$$

Here ρ is the real variable (parameter) and H the complex function value. It is not possible to draw such a function in the way we draw real functions, $y = f(x)$, because H as a complex number already has two independent values. H can be drawn in the complex plane for different values of ρ, so that the complex numbers H describe a curve.

In this way a so-called *polar diagram* is created (see Figure 3.1).

The real parameter passes through all values from zero to infinity here. For any other value of ρ, H takes another complex value. The curve is the set of points of all complex values H.

Example (Figure 3.2).
Compute Z as a function of ω.
We find

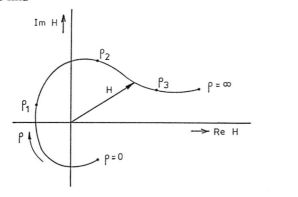

Figure 3.1   Figure 3.2

$$Z = 3 + 2j\omega \ \Omega.$$

The real part of Z is a constant 3 while the imaginary part is not negative and increases if ω becomes larger (see Figure 3.3).

Here the polar diagram is a straight line. By measuring one can derive both $|Z|$ and $\arg Z$ as a function of ω from the polar diagram.

In many cases one finds a function of the following form:

$$H = \frac{A + B\rho}{C + D\rho} . \tag{3.2}$$

In this equation A, B, C and D are constant complex numbers and ρ is the real parameter with $-\infty < \rho < \infty$. This function is called *bilinear* because both numerator and denominator are linear.

We note that one or more of the constants may also be zero or real.

We shall show that the polar diagram corresponding to this function is a circle. For the proof we start with

$$H_1 = \frac{1}{E + \rho} \qquad (H_1 = 0 \text{ for } \rho = \pm \infty)$$

in which E is a constant complex number and ρ is real. So $E = F + jG$ with F and G real:

$$H_1 = \frac{1}{F + \rho + jG} = \frac{F + \rho - jG}{(F + \rho)^2 + G^2} .$$

Thus

$$x = \text{Re } H_1 = \frac{F + \rho}{(F + \rho)^2 + G^2} \qquad \text{and}$$

$$y = \text{Im } H_1 = \frac{-G}{(F + \rho)^2 + G^2} .$$

We eliminate ρ, so that $y = f(x)$ results:

$$\frac{x}{y} = \frac{F + \rho}{-G} , \text{ so } \frac{x^2}{y^2} = \frac{(F + \rho)^2}{G^2} .$$

$$x = \frac{-Gx/y}{G^2 \frac{x^2}{y^2} + G^2} = \frac{-Gxy}{G^2 (x^2 + y^2)} .$$

Hence

$$G^2(x^2 + y^2) = -Gy$$

$$x^2 + y^2 = -\frac{y}{G}$$

$$x^2 + (y + \frac{1}{2G})^2 = \frac{1}{4G^2} .$$

This is the equation of a circle with radius $\frac{1}{2G}$ and co-ordinates of the centre: $M\left(0, -\frac{1}{2G}\right)$ (see Figure 3.4).

We find that $H_1 = \dfrac{1}{E + \rho}$ is a circle in the complex plane. If we multiply H_1 by a constant real number the figure remains a circle. Only the magnitude alters.

We also keep the shape of a circle if we rotate all vectors H_1 over a certain angle. This means that we keep a circle if we multiply H_1 by F/D, in which F and D are constant complex numbers.

So

$$H_2 = \frac{F}{D} H_1 = \frac{F}{D}\frac{1}{E + \rho}$$

represents a circle (see Figure 3.5).

If we add the constant complex number G to H_2 we once more have a circle: $H = G + H_2$ (see Figure 3.6).

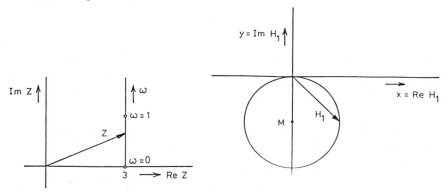

Figure 3.3   Figure 3.4

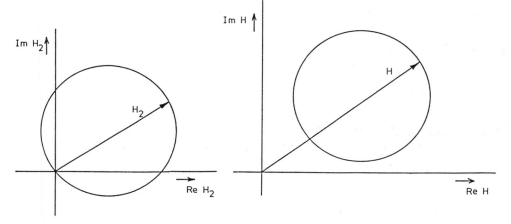

Figure 3.5   Figure 3.6

The result is that

$$H = G + \frac{F}{DE + D\rho} = \frac{DEG + DG\rho + F}{DE + D\rho}$$

represents a circle.

If we set

$$DEG + F = B$$
$$DG = B$$
$$DE = C$$

it follows that

$$H = \frac{A + B\rho}{C + D\rho}$$

represents a circle, with which the proof is completed.

Notes
- In some cases a straight line is found, i.e. a circle with an infinitely large radius.
- It is possible to derive expressions for the radius and the centre. We shall not do this because it does not increase our understanding and because it is hardly used in practice.
- If the real parameter is the angular frequency ω one often finds that part of a circle is mapped, because $\omega \geq 0$. The same holds for R, C or L as parameters.
- In some cases one can get a bilinear shape from a non-bilinear one by altering the real parameter. This will, of course, only succeed if the polar diagram (after calculation) turns out to be a circle (see Example 4 below).
- If the polar diagram has turned out to be a circle the calculation of three different points is sufficient to be able to construct the circle or the circle arc: the perpendicular bisectors cut each other in the centre M (see Figure 3.7).

Example 1 (see Figure 3.8).

Calculate the admittance $Y(\omega)$:

$$Y = \frac{1}{2 + 3j\omega} \cdot$$

This form is bilinear because comparison with (3.2) results in $A = 1$, $B = 0$, $C = 2$, $D = 3j$, and these numbers are all constant complex. So the polar diagram is (part of) a circle. Substitution gives

$$\omega = 0 \qquad Y = \tfrac{1}{2} \text{ S}$$

$$\omega = \infty \qquad Y = 0 \text{ S}$$

$$\omega = \tfrac{2}{3} \text{ rad/s} \quad Y = \frac{1}{2 + 2j} = \tfrac{1}{4}(1 - j) \text{ S}. \qquad \text{(See Figure 3.9).}$$

Note

We write $\omega = \infty$. Strictly speaking we should write $\omega \to \infty$.

We find half a circle with its centre at the real axis.

Example 2 (see Figure 3.10).
Set $\omega = 1$ rad/s while C is variable. Consider $H = \dfrac{V_2}{V_1}$.
We find

$$H = \frac{2}{2 + \dfrac{1}{jC}} = \frac{2jC}{1 + 2jC} \cdot$$

This form is bilinear, so the polar diagram is (part of) a circle.

$$
\begin{array}{ll}
C = 0 & H = 0 \\
C = \infty & H = 1 \\
C = \tfrac{1}{2}\,F & H = \dfrac{j}{1 + j} = \tfrac{1}{2}(1 + j).
\end{array}
$$

This results in Figure 3.11.

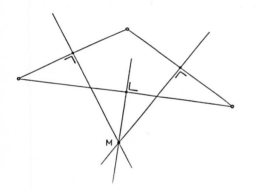

Figure 3.7

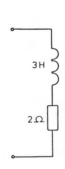

Figure 3.8

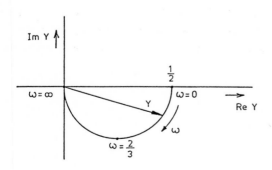

Figure 3.9

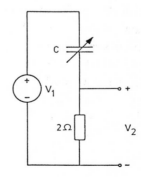

Figure 3.10

Example 3 (see Figure 3.12).

Determine $H(R) = \dfrac{V_2}{V_1}$. After some calculations we find

$$H = \frac{-1 + j + R(1 + j)}{1 - j + R}.$$

This form is also bilinear. So the polar diagram once more is a circle arc.

$$R = 0 \qquad H = -1$$

$$R = \infty \qquad H = 1 + j$$

$$R = 1\Omega \qquad H = -0.4 + 0.8j.$$

(See Figure 3.13).

Example 4 (see Figure 3.14).

The admittance is $Y(\omega) = \dfrac{1}{R + j(\omega L - \dfrac{1}{\omega C})}$.

The real variable is ω. This form is not bilinear. We shall now introduce another variable. First we note that Y is real for $\omega = (1/\sqrt{LC}) = \omega_0$.

ω_0 is called the *resonance frequency*. We shall discuss these resonance phenomena more detailed in Section 3.3.

We now introduce *detuning*

$$d = \frac{\omega}{\omega_0} - \frac{\omega_0}{\omega} \tag{3.3}$$

and *quality*

$$Q_0 = \frac{\omega_0 L}{R}. \tag{3.4}$$

Detuning and quality are dimensionless.

Quality is the ratio of the reactance and the loss resistance (drawn here as a series resistor) of an inductor. In a wire the current for high frequencies is larger at the outside than in the middle (skin effect), so that the resistance for higher frequencies is greater than that for lower frequencies or d.c. currents. If the inductor is wound on a ferromagnetic core, then losses occur in that core which can be interpreted as a contribution to the series resistance. For instance, a high-quality inductance (e.g. $Q_0 = 100$) requires careful construction and choice of material for say 1 MHz.

For $\omega = \infty \qquad d = \infty.$
For $\omega = 0 \qquad d = -\infty.$
For $\omega = \omega_0 \qquad d = 0.$

Further

$$\frac{\mathrm{d}d}{\mathrm{d}\omega} = \frac{1}{\omega_0} + \frac{\omega_0}{\omega^2}.$$

The derivative is positive for positive ω and decreases for increasing ω. The function is as shown in Figure 3.15.

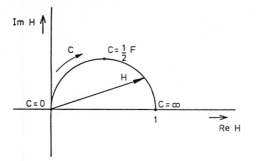

Figure 3.11

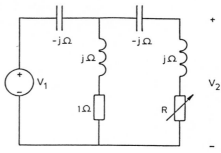

Figure 3.12

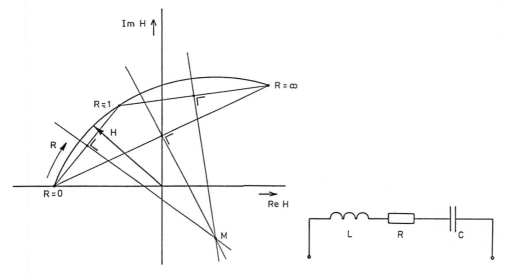

Figure 3.13

Figure 3.14

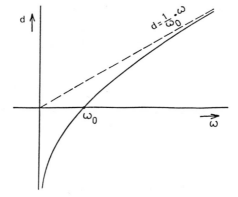

Figure 3.15

For large ω the slope is $\frac{1}{\omega_0}$ and is therefore constant (asymptote, see dotted line).

So (3.3) has the important characteristic that for every value of $\omega > 0$ there is only one d and for every value of d there is only one $\omega > 0$. The denominator of the admittance (i.e. the impedance) now becomes

$$Z = R + j(\omega L - \frac{1}{\omega C}) = R + j\omega_0 L(\frac{\omega}{\omega_0} - \frac{1}{\omega C \cdot \omega_0 L})$$

$$= R + j\omega_0 L \cdot d = R(1 + jdQ_0).$$

So

$$Y(d) = \frac{1}{R(1 + jdQ_0)}$$

which is bilinear.
Subsequently

$$d = \pm \infty \qquad Y = 0 \text{ S}$$

$$d = 0 \qquad Y = \frac{1}{R} \text{ S}$$

$$d = \frac{1}{Q_0} \qquad Y = \frac{1}{R(1 + j)} = \frac{1}{2R}(1 - j) \text{ S}$$

$$d = -\frac{1}{Q_0} \qquad Y = \frac{1}{R(1 - j)} = \frac{1}{2R}(1 + j) \text{ S}.$$

See Figure 3.16. We find a complete circle.
Applying (3.3) we can add an ω-scale to the plot besides the d-scale (which we have not done).

Example 5 (see Figure 3.17).
We examine $I(\omega)$. We set V real. We find

$$I = YV = (j\omega C + \frac{1}{R + j\omega L}) \text{ V}.$$

This form is not bilinear and the polar diagram also turns out not to be to be a circle.
This can be seen by drawing the polar plots $I_1(\omega)$ and $I_2(\omega)$ separately (see Figure 3.18).
$I_1 = \dfrac{V}{R + j\omega L}$ is half a circle whereas $I_2 = j\omega C V$ is a straight line along the positive imaginary axis.
Adding both figures for all values of $\omega \geq 0$ gives the polar plot of Figure 3.19.
We finally note that the current I is real for $\omega = \omega_0$ and that it can further be seen from the figure that $|I|$ is minimal for a frequency larger than ω_0.

3.2 Bode diagrams

From the polar plot $H = f(\omega)$ one can find both $|H|$ and arg H as functions of the frequency. These plots are called *frequency characteristics*.

Example (see Figure 3.20).
We find

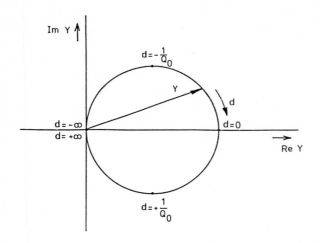

Figure 3.16

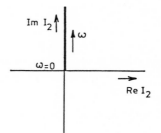

Figure 3.17

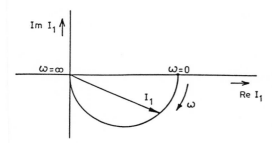

Figure 3.18

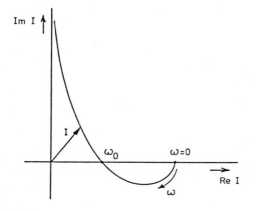

Figure 3.19

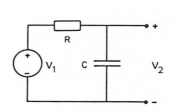

Figure 3.20

$$H = \frac{V_2}{V_1} = \frac{\frac{1}{j\omega C}}{R + \frac{1}{j\omega C}} = \frac{1}{1 + j\omega RC} = \frac{1}{1 + j\omega\tau}$$

with

$$\tau = RC. \tag{3.5}$$

τ is called the *time constant* or, in this case, the *RC-time*, because we are concerned with a resistor and a capacitor here. For the dimension of RC is that of time.

If we choose $R = 1\ \Omega$ and $C = 2\ F$ then $\tau = 2\ s$ and therefore $H = 1/(1 + 2j\omega)$. This form is bilinear, the polar diagram is therefore the arc of a circle.

For $\omega = 0$ $H = 1.$

For $\omega = \infty$ $H = 0.$

For $\omega = \frac{1}{2}\ rad/s$ $H = \dfrac{1}{1 + j} = \frac{1}{2}(1 - j).$

The polar diagram is drawn in Figure 3.21.

From this we find the frequency characteristics of Figure 3.22.

Figure 3.22(a) is called the *amplitude characteristic* and Figure 3.22(b) the *phase characteristic.*

Instead of a linear scale of angular frequency a *logarithmic* scale is often chosen. We further set

$$G = 20 \log |H|, \tag{3.6}$$

where G is called the *voltage ratio* in dB (decibel).

Note

The dB has its origin in the logarithm of the power ratio

$$G_P = \log \frac{P_2}{P_1}$$

with the unit B (bel).

If the powers P_1 and P_2 are dissipated by the resistors R_1 and R_2 respectively we obtain

$$P_1 = \frac{|V_1|^2}{R_1} \quad \text{and} \quad P_2 = \frac{|V_2|^2}{R_2} \quad \text{respectively, so that}$$

$$G_P = 2 \log \left|\frac{V_2}{V_1}\right| \quad \text{provided that } R_1 = R_2.$$

In dB (decibel) we get

$$G = 20 \log \left|\frac{V_2}{V_1}\right|.$$

Although it is not mathematically correct, it is usual to regard the expression as valid if $R_1 \neq R_2$.

For our example we find

$$G = 20 \log \left| \frac{1}{1 + j\omega\tau} \right| = -20 \log |1 + j\omega\tau|.$$

An approximation is used to draw this function.

For small ω: $G \approx -20 \log 1 = 0 = G_1$.

For large ω: $G \approx -20 \log \omega\tau = -G_2$.

We examine $G_2 = 20 \log \omega\tau$. If the frequency is chosen two times larger (*octave*) we have

$$G_2 = 20 \log 2\omega\tau = 20 \log 2 + 20 \log \omega\tau = 6 + 20 \log \omega\tau,$$

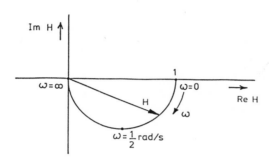

Figure 3.21

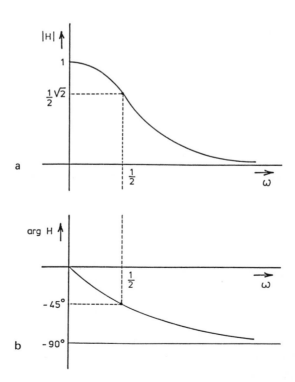

Figure 3.22

this means that G_2 has increased by 6 dB. So G_2 increases by 6 dB over one octave. A frequency $10 \times$ as large is called a *decade*. It follows that G_2 increases by 20 dB per decade. The *corner frequency* is the intersection of G_1 and G_2:

$$20 \log \omega\tau = 0.$$

So the corner frequency is

$$\omega_c = \frac{1}{\tau}. \tag{3.7}$$

This construction is called *asymptotic approximation*.

For $\tau = 2$ s we find the plot of Figure 3.23.

The correct plot has G_1 and G_2 as asymptotes, see the dotted line.

In the corner frequency we have $G = -20 \log |1 + j| = -20 \log \sqrt{2} = -3$ dB.

The behaviour of a network with one capacitor or with one inductor and one or more resistors can be described with a first order differential equation. For that reason these networks are called *first order networks*. In the network of Figure 3.20 the numerator of H was constant. In general the numerator can also be a function of ω:

$$H = K \frac{j\omega - z_1}{j\omega - p_1} \qquad (z_1 \neq p_1). \tag{3.8}$$

We find

$$G = 20 \log |H| = 20 \log K + 20 \log \left| \frac{z_1 - j\omega}{p_1 - j\omega} \right|$$

$$= 20 \log K + \tilde{G}$$

with

$$\tilde{G} = 20 \log |z_1 - j\omega| - 20 \log |p_1 - j\omega| = G_1 - G_2$$

with

$$G_1 = 20 \log |z_1 - j\omega| \quad \text{and} \quad G_2 = 20 \log |p_1 - j\omega|.$$

For small ω

$$G_1 = 20 \log |z_1| \quad \text{and} \quad G_2 = 20 \log |p_1|$$

and for large ω

$$G_1 = 20 \log \omega \quad \text{and} \quad G_2 = 20 \log \omega.$$

Both these latter straight lines with positive slope cut the horizontal axis at $\omega = 1$ rad/s. The corner frequency for G_1 is $\omega = |z_1|$ and for G_2 $\omega = |p_1|$.

Assume that $z_1 < p_1$ and also that $z_1 > 1$ and $p_1 > 1$.

We then find the plots of Figure 3.24(a) in which $K = 1$ is set.

The resulting plot is shown in Figure 3.24(b).

Example (see figure 3.25).
We find

$$H = \frac{V_2}{V_1} = \cfrac{1}{2 + \cfrac{1/j\omega}{1/j\omega + 1}} = \cfrac{1}{2 + \cfrac{1}{1 + j\omega}} = \frac{1 + j\omega}{3 + 2j\omega}$$

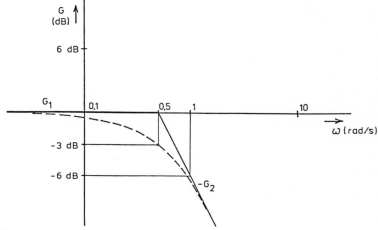

Figure 3.23

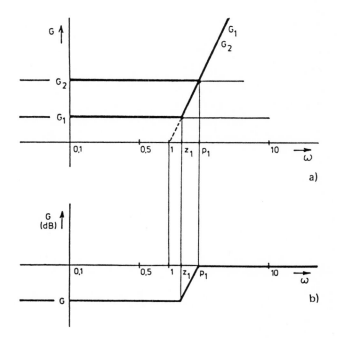

Figure 3.24

$$H = \frac{1}{2} \frac{1 + j\omega}{\frac{3}{2} + j\omega}.$$

$$G = 20 \log |H| = -6 + 20 \log |1 + j\omega| - 20 \log |\tfrac{3}{2} + j\omega|$$

$$= -6 + G_1 - G_2$$

$$G_1 = 20 \log |1 + j\omega|.$$

For small ω: $G_1 = 20 \log 1 = 0$ dB.

For large ω: $G_1 = 20 \log \omega$ and corner frequency: $\omega = 1$ rad/s.

$$G_2 = 20 \log |\tfrac{3}{2} + j\omega|.$$

For small ω: $G_1 = 20 \log \frac{3}{2} = 3.5$ dB.

For large ω: $G_2 = 20 \log \omega$ and corner frequency: $\omega = \frac{3}{2}$ rad/s.

In Figure 3.26 the separate plots are shown, after which the total plot of $G = f(\omega)$ follows by adding (Figure 3.27).

3.3 Duality

In Section 1.2 we met the phenomenon of duality. We are now able to compile a more detailed list of dual elements, formulas and concepts.

The substitute impedance of n impedances in series is $Z = \sum_{k=1}^{n} Z_k$, while the substitute admittance of n admittances in parallel is $Y = \sum_{k=1}^{n} Y_k$.

Consequently impedance and admittance are dual concepts, but also series and parallel connections. A switch, too, can be added to the list: a closed switch has zero resistance, an open switch has zero conductance.

In advanced network theory the *graph theory* is used, which enables us to formalise the general concept of dualism. It then turns out that the node and the internal surface of a mesh are fundamental dual concepts; this will not be discussed in this book.

The list is as follows:

- voltage   current
- voltage source   current source
- voltage polarity   current direction
- Kirchhoff's voltage law   Kirchhoff's current law
- resistance   conductance
- short circuits   open terminals
- series   parallel
- voltage division   current division
- mesh method   node method
  (*only for planar networks*)
- Thévenin's theorem   Norton's theorem

- star configuration
- impedance
- inductor
- flux
- magnetic energy
- closed switch
- current-current transactor
- voltage-current transactor

delta configuration
admittance
capacitor
charge
electric energy
open switch
voltage-voltage transactor
current-voltage transactor

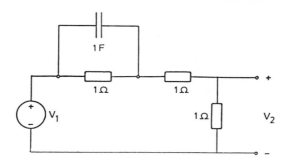

Figure 3.25

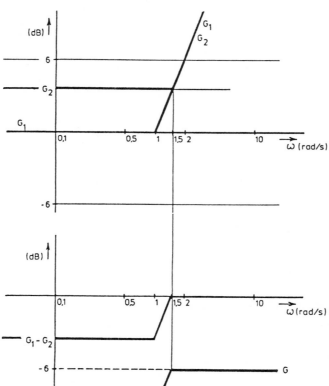

Figure 3.26

Figure 3.27

3.4 Resonance

Resonance is an important phenomenon in physics and in technology. In network theory one distinguishes

- *Phase resonance.* Here the phase difference between two voltages, between two currents or between a voltage and a current is zero at a certain angular frequency $\omega = \omega_0$ (in some rare cases a 90°-angle is used instead of zero).
- *Amplitude resonance.* The amplitude of a voltage or a current has an extreme value at a certain angular frequency ω_a, i.e. a maximum or a minimum.

Consider the series circuit of an inductor, a resistor and a capacitor (Figure 3.28). The source intensity is sinusoidal with amplitude |V| and with a variable angular frequency ω. The complex source voltage is V.

The complex current is $I = \dfrac{V}{R + j(\omega L - \frac{1}{\omega C})}$.

For $\omega L = \dfrac{1}{\omega C}$ is $I = \dfrac{V}{R}$. i.e. The phase between the time functions v and i is zero.

The angular frequency $\omega_0 = \dfrac{1}{\sqrt{LC}}$ is thus the phase resonance frequency (only the positive root suffices).

Further $|I| = \dfrac{|V|}{\sqrt{R^2 + (\omega L - \frac{1}{\omega C})^2}}$. In the denominator we have $(\omega L - \frac{1}{\omega C})^2$. This term

is not negative. If this term is zero the denominator is minimal, so |I| is maximal: $|I|_{max} = \dfrac{|V|}{R}$. The condition is $\omega L - \dfrac{1}{\omega C} = 0$. The frequency for amplitude resonance is therefore

$$\omega_a = \frac{1}{\sqrt{LC}} .$$

For this circuit ω_0 and ω_a coincide. In Figure 3.29 |I| as a function of ω is shown (amplitude characteristic).

This plot can also be found from the polar diagram (Figure 3.16). Both frequencies for which $|I| = \frac{1}{2} \sqrt{2} \cdot |I|_{max}$ are indicated in Figure 3.29.

From the formula for the complex current we find that these frequencies satisfy

$$\omega L - \frac{1}{\omega C} = \pm R.$$

From this follow 4 angular frequencies, of which two are positive:

$$\omega_1 = -\frac{R}{2L} + \sqrt{\frac{1}{LC} + \frac{R^2}{4L^2}}$$

and

$$\omega_2 = \frac{R}{2L} + \sqrt{\frac{1}{LC} + \frac{R^2}{4L^2}}$$

(3.9)

The interval form ω_1 to ω_2 is called the *bandwidth* B, so

$$B = \frac{R}{L}.$$ (3.10)

We can also express the band width in the resonance frequency and the quality. If we multiply the numerator and the denominator of the right-hand part of (3.10) by ω_0, then with (3.4) we find:

$$B = \frac{\omega_0}{Q_0}.$$ (3.11)

This formula says that the bandwidth is small if the quality is high.

With $\omega_0 = \dfrac{1}{\sqrt{LC}}$ we can also write the quality $Q_0 = \dfrac{\omega_0 L}{R}$ as

$$Q_0 = \frac{1}{R}\sqrt{\frac{L}{C}}.$$ (3.12)

From the section on polar diagrams it follows that $I = \dfrac{V}{R(1 + jdQ_0)}$, so the bandwidth is determined by

$$dQ_0 = \pm 1.$$ (3.13)

The detuning d_1 corresponding to ω_1 is thus

$$d_1 = -\frac{1}{Q_0}$$

and the detuning corresponding to ω_2 is

$$d_2 = +\frac{1}{Q_0},$$

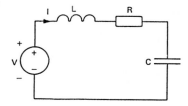

Figure 3.28

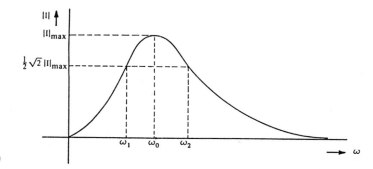

Figure 3.29

so that the band width in the polar diagram of Figure 3.16 is determined by half the circle going from $-\dfrac{1}{Q_0}$ over $d = 0$ to $+\dfrac{1}{Q_0}$.

From (3.9) it follows that

$$\frac{\omega_1 + \omega_2}{2} = \sqrt{\frac{1}{LC} + \frac{R^2}{4L^2}} = \sqrt{\frac{1}{LC}} \sqrt{1 + \frac{R^2C}{4L}}.$$

So

$$\frac{\omega_1 + \omega_2}{2} = \omega_0 \sqrt{1 + \frac{1}{4Q_0^2}}.$$

If $Q_0 \gg 1$ the right-hand part becomes ω_0. In that case the resonance frequency is halfway the band width (see Figure 3.30).

Finally we construct the argument of the current I as a function of the angular frequency from the polar diagram (see Figure 3.31).

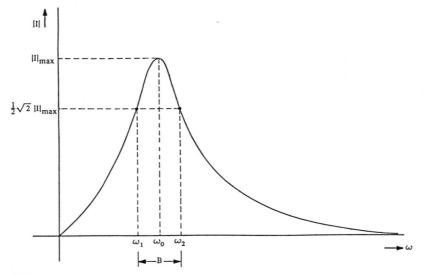

Figure 3.30

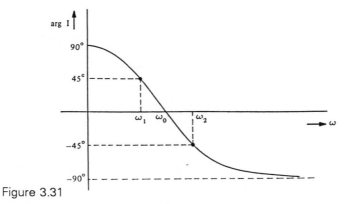

Figure 3.31

This is called the *phase characteristic*. The limits of the bandwidth are determined by +45°
and –45°, as follows from the polar plot.

We can summarise the above with an example.
Choose V = 1 V, L = 1 H, C = 1 F and first take R = 1 Ω so that ω_0 = 1 rad/s and
Q_0' = 1. Subsequently we choose R = 0.2 Ω so that Q_0'' = 5.
We find

$$I' = \frac{V}{R(1 + jd' Q_0)} = \frac{1}{1 + jd'} \quad \text{and} \quad I'' = \frac{5}{1 + 5jd''} \quad \text{respectively.}$$

This is shown by the polar plots of Figure 3.32.
From (3.9) we find the band limit frequencies: $\omega_1' \approx 0.62$ rad/s and $\omega_2' \approx 1.62$ rad/s,
$\omega_1'' \approx 0.9$ rad/s and $\omega_2'' \approx 1.1$ rad/s respectively. The bandwidths are B' = 1 rad/s and
B'' = 0.2 rad/s respectively. In Figure 3.33 both |I| and arg I are drawn as a function of d
and ω.
We see that for a large value of Q_0 the plots are more pronounced.

Next investigate the capacitor voltage (see Figure 3.34).
We find

$$H = \frac{V_C}{V} = \frac{\dfrac{1}{j\omega C}}{R + j\omega L + \dfrac{1}{j\omega C}} = \frac{1}{1 - \omega^2 LC + j\omega R C} . \tag{3.14}$$

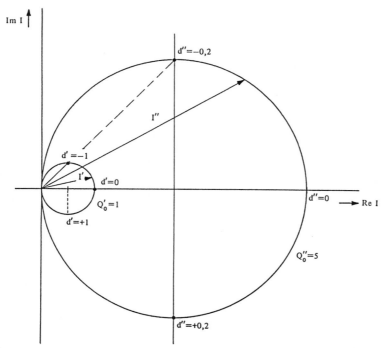

Figure 3.32

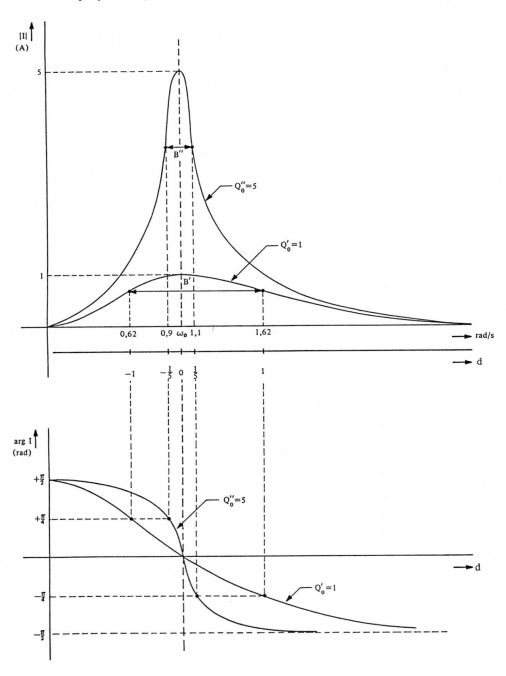

Figure 3.33

It follows that

$$|H|^2 = \frac{1}{(1 - \omega^2 LC)^2 + \omega^2 R^2 C^2} = \frac{1}{D(\omega)}$$

with

$$D(\omega) = (1 - \omega^2 LC)^2 + \omega^2 R^2 C^2 = 1 - 2\omega^2 LC + \omega^4 L^2 C^2 + \omega^2 R^2 C^2.$$

The frequency $\omega_a > 0$ for which $|H|$ is maximal means a minimum for $D(\omega)$.
We find

$$\frac{dD}{d\omega} = -4\omega LC + 4\omega^3 L^2 C^2 + 2\omega R^2 C^2.$$

$\dfrac{dD}{d\omega} = 0$ for $\omega = 0$ and for $4\omega^2 L^2 C^2 = 4LC - 2R^2 C^2$.
So

$$\omega_a = \sqrt{\frac{4\,LC - 2R^2 C^2}{4L^2 C^2}},$$

hence

$$\omega_a = \sqrt{\frac{1}{LC} - \frac{R^2}{2L^2}} \tag{3.15}$$

or

$$\omega_a^2 = \frac{1}{LC}\left(1 - \frac{R^2 C}{2L}\right) = \omega_0^2\left(1 - \frac{1}{2Q_0^2}\right).$$

So

$$\omega_a = \omega_0 \sqrt{1 - \frac{1}{2Q_0^2}}. \tag{3.16}$$

Consequently

$$\omega_a < \omega_0. \tag{3.17}$$

The frequency at which the capacitor voltage is a maximum is smaller than the frequency at which the current becomes maximal. At first sight this is a little surprising, but it will be understood if we consider the relation between the capacitor voltage and the current:

$$|V_C| = \frac{1}{\omega C}\,|I|.$$

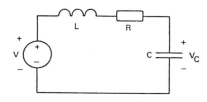

Figure 3.34

Both $\dfrac{1}{\omega C}$ and $|I|$ depend on the frequency. For the angular frequency ω_0, $|I|$ is maximal. Because the function $|I| = f(\omega)$ is a flat curve in the neighbourhood of $\omega = \omega_0$ (Figure 3.29) one can increase the product $\dfrac{1}{\omega C} \times |I|$ still a little further by decreasing ω a little. For the first factor $\dfrac{1}{\omega C}$ varies much more than $|I|$. The result is that the capacitor voltage becomes maximal at a somewhat lower frequency than ω_0.

$D(\omega)$ does indeed reach a minimum for $\omega = \omega_a$, which is shown by examining the second derivative:

$$\frac{d^2D}{d\omega^2} = -4LC + 12\omega^2 L^2 C^2 + 2R^2 C^2.$$

For $\omega^2 = \omega_a^2 = \dfrac{1}{LC} - \dfrac{R^2}{2L^2}$ this becomes:

$$(\frac{d^2D}{d\omega^2})_{\omega=\omega_a} = -4LC + 12LC - 6R^2 C^2 + 2R^2 C^2$$

$$= 8LC - 4R^2 C^2 = 4C(2L - R^2 C) > 0,$$

if $2L > R^2 C$, so if $\dfrac{L}{R^2 C} > \dfrac{1}{2}$, so if

$$Q_0 > \frac{1}{2}\sqrt{2}. \tag{3.18}$$

This is also the condition for the existence of a real value of ω_a according to (3.16). The magnitude of $|H|_{max}$ for the frequency ω_a can be found by substituting $\omega = \omega_a$ in the formula for $|H|$. We get

$$D(\omega_a) = (1 - \omega_a^2 LC)^2 + \omega_a^2 LC.$$

Now according to (3.15)

$$\omega_a^2 LC = 1 - \frac{R^2 C}{2L}, \quad \text{so} \quad 1 - \omega_a^2 LC = \frac{R^2 C}{2L} = \frac{1}{2Q_0^2}.$$

Further

$$\omega_a^2 R^2 C^2 = \frac{R^2 C}{L} - \frac{R^4 C^2}{2L^2} = \frac{1}{Q_0^2} - \frac{1}{2} \cdot \frac{1}{Q_0^4}.$$

$$D(\omega_a) = \frac{1}{4Q_0^4} + \frac{1}{Q_0^2} - \frac{1}{2Q_0^4} = \frac{1}{Q_0^2} - \frac{1}{4Q_0^4} = \frac{1 - \dfrac{1}{4Q_0^2}}{Q_0^2}.$$

So

$$|H|_{max} = \frac{Q_0}{\sqrt{1 - 1/4Q_0^2}}. \tag{3.19}$$

If $Q_0 > 1$ it follows that

$$|H|_{max} = Q_0. \tag{3.20}$$

So the capacitor voltage is Q_0 times as large as the source voltage. This is why Q_0 is sometimes called the *amplifying factor*.

In Figure 3.35 $|H|$ is drawn for different values of Q_0.

The larger Q_0 the higher the maximum according to (3.19) and the closer the position of the maximum approaches ω_0 according to (3.16).

For $\omega = \omega_0$, $I = V/R$ so in that case the capacitor voltage is

$$(V_C)_{\omega_0} = \frac{1}{j\omega_0 C} \cdot I = \frac{1}{j\omega_0 C} \frac{V}{R} = -jQ_0 V. \tag{3.21}$$

So

$$|V_C|_{\omega_0} = Q_0 |V|. \tag{3.22}$$

Equation (3.19) gives the amplification at ω_a, (3.22) is the amplification at ω_0. For large Q_0 the values are the same.

In practice the quality is determined by measuring the capacitor voltage. Theoretically, measuring the inductor voltage is also suitable:

$$(V_L)_{\omega_0} = \omega_0 L I, \text{ so } |V_L|_{\omega_0} = \omega_0 L \frac{|V|}{R} = Q_0 |V|,$$

but the skin effect and loss in iron prevent the exclusive measuring of the inductor voltage.

Note

In practice a capacitor also has losses, which can be indicated with a resistance in parallel (or with a resistance in series). These losses, however, are extremely small in comparison with the losses of an inductor.

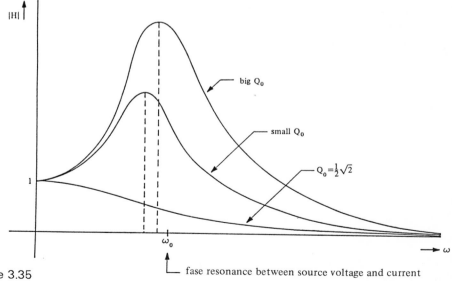

Figure 3.35

We shall now give a sketch of the polar plot H(ω). Its function is not bi-linear and the polar plot is not a circle.

We find:

for $\omega = 0$, H = 1. See (3.14)

for $\omega = \omega_0$, H = $-jQ_0$. See (3.21).

The maximum of $|H|$ is reached at $\omega = \omega_a$, so that $|H|_{\omega_a} > |H|_{\omega_0}$.

Finally from (3.14) we derive: arg H = $-\pi$ for $\omega \to \infty$. Consequently we arrive at Figure 3.36.

The preceding theory can be applied to a parallel network with no problems at all (see Figure 3.37).

This is the dual of the network of Figure 3.28. So we also find the dual formulas. We shall not write down these formulas, but only mention the dual formulas for the quality:

$$Q_0 = \frac{\omega_0 C}{G} = \frac{R}{\omega_0 L} = \frac{1}{G} \sqrt{\frac{C}{L}}. \tag{3.23}$$

We note that the formulas (3.12) and (3.23) only hold for the series and parallel circuits respectively. It is not allowed to use these formulas for other networks.

We finally give a general formula for the quality of a system (also a non-electric one) without discussing it further.

$$Q = 2\pi \frac{\text{accumulated energy}}{\text{energy losses per period}}. \tag{3.24}$$

3.5 Loss-less one-ports

Consider one-ports (two-terminal networks) which only contain inductors and capacitors. Because resistors are not present the real power consumed will be zero if a source is connected to the terminals.

So P = Re S = Re $\frac{1}{2}$ VI* = 0. With V = ZI it follows that P = Re $\frac{1}{2}$ ZII* = $\frac{1}{2}$ |I|² Re Z = 0. So Re Z = 0. This means that

$$Z = jX. \tag{3.25}$$

We consider the reactance X as as function of the angular frequency ω. The most simple one-port is one single inductor or one single capacitor. We start with the inductor (Figure 3.38).

We have Z = $j\omega L$, so X = ωL.

X = X(ω) is thus a straight line with slope L (see Figure 3.39).

The single capacitor is shown in Figure 3.40.

We find

$$Z = \frac{1}{j\omega C}, \text{ so } X = -\frac{1}{\omega C}.$$

The plot of this function is a hyperbola (Figure 3.41).

Now consider the series circuit of an inductor and a capacitor (Figure 3.42).

$Z = j\omega L + \dfrac{1}{j\omega C}$, so $X = \omega L - \dfrac{1}{\omega C} = \dfrac{\omega^2 LC - 1}{\omega C}$.

This function is zero for $\omega = \dfrac{1}{\sqrt{LC}} = z_1$. It is called a *zero*. In plots the zero is indicated by a circle.

If there are several zeros these are written as z_1, z_2, z_3, etc.

The function is $-\infty$ for $\omega = 0 = p_1$. This is called a *pole*. In a pole the function has the value $+\infty$ or $-\infty$ and is indicated by a cross. For $0 < \omega < z_1$ we have $X < 0$ and for $z_1 < \omega < \infty$ is $X > 0$.

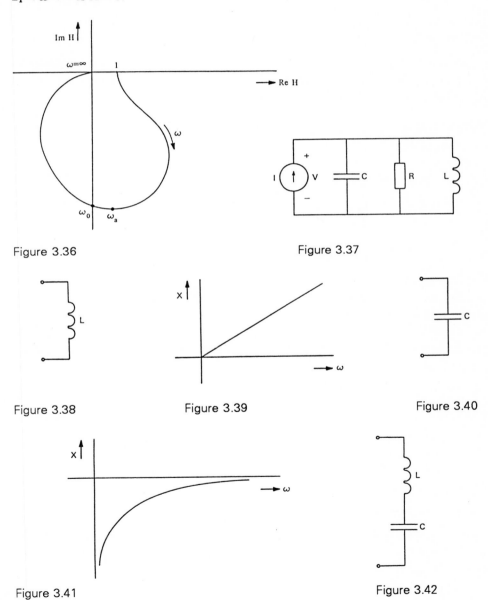

Figure 3.36

Figure 3.37

Figure 3.38

Figure 3.39

Figure 3.40

Figure 3.41

Figure 3.42

For large values of ω X approaches the function $X = \omega L$, a straight line. Consequently we find Figure 3.43.

The dotted line is the *asymptote* with the equation $X = \omega L$. For $\omega < z_1$ X is negative, i.e. the circuit behaves capacitively. For $\omega > z_1$ X is positive, the network behaves inductively.

The plots of these examples meet

$$\frac{dX}{d\omega} \geq 0. \tag{3.26}$$

This is *Foster's theorem*.

We shall prove this theorem in Appendix VI.

The equal sign in (3.26) only holds in some cases when $\omega \to \infty$. We can thus say that $X(\omega)$ is always increasing for finite ω.

We shall now use Foster's theorem in the following example.

Example (Figure 3.44).

We find

$$Z = \frac{j\omega L \cdot \dfrac{1}{j\omega C}}{j\omega L + \dfrac{1}{j\omega C}} = \frac{j\omega L}{1 - \omega^2 L C}, \text{ so } X = \frac{\omega L}{1 - \omega^2 L C}.$$

Zero: $z_1 = 0$,

Pole: $p_1 = \dfrac{1}{\sqrt{LC}}$ (we only take the positive frequencies).

The plot is shown in Figure 3.45.

After having determined the poles and the zeros a rough sketch of the plot can be made. Starting with z_1, where the function is zero, the function must rise to $+\infty$ in the pole p_1. The function is $+\infty$ for low frequencies approaching p_1 and $-\infty$ for high frequencies approaching p_1. In the pole the function leaps from $+\infty$ to $-\infty$. By increasing ω the function increases to the value zero.

Only for $\omega \to \infty$ do we have $\dfrac{dX}{d\omega} = 0$.

The tangent in the origin can be determined in three ways:

a. By differentiating:

$$\frac{dX}{d\omega} = \frac{(1 - \omega^2 LC)\, L + 2\omega^2 L^2 C}{(1 - \omega^2 LC)^2}, \text{ so } \left(\frac{dX}{d\omega}\right)_{\omega=0} = L.$$

The equation of the tangent thus is $X_r = \omega L$.

b. By only writing the term with the smallest power of ω of the numerator and the denominator of $X(\omega)$. The higher power terms can be neglected compared to the lower power terms for small ω.

We find $X(\omega) \rightarrow \dfrac{\omega L}{1} = \omega L$, which is the tangent: $X_r = \omega L$.

Both of the above methods are mathematical, while method (a) requires a lot of calculation if the network is large. There is still a third, physical method:

c. Consider the network at low frequencies. The capacitors may be removed then, because the impedance of a capacitor is very large in that case. In our example we are left with an inductor only, of which we already know the reactance function. Again we find the tangent: $X_r = \omega L$.

A consequence of Foster's theorem is that the poles and zeros *alternate* at the x-axis. It is not possible to have two zeros next to one another, without a pole between the two. After all, the function $X(\omega)$ must go through the first zero while it increases and will have to go down to reach the following zero (see figure 3.46).

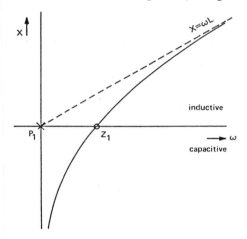

Figure 3.43

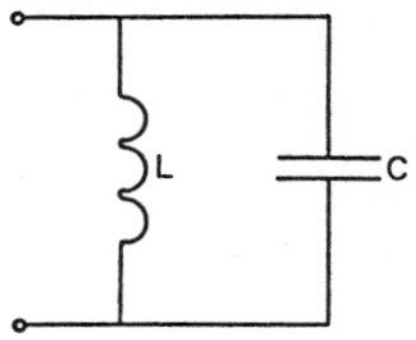

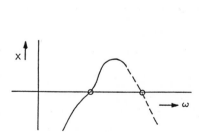

Figure 3.44

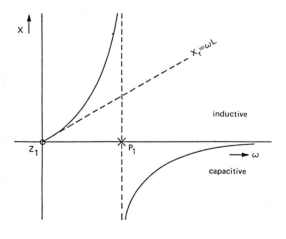

Figure 3.45

Figure 3.46

This is in contradiction with Foster's theorem. A similar view can be given for two successive poles.

The network of Figure 3.44 behaves as a short circuit for high frequencies. For those frequencies the impedance of a capacitor is very small. This means: $X(\omega)$ is zero for $\omega \to \infty$.

A network that behaves differently for $\omega \to \infty$ is shown in Figure 3.47.

$$\text{We find } Z = \frac{3j\omega(6j\omega + \frac{1}{j\omega})}{9j\omega + \frac{1}{j\omega}} = \frac{3j\omega(1 - 6\omega^2)}{1 - 9\omega^2}.$$

$$\text{So } X = \frac{3\omega(1 - 6\omega^2)}{1 - 9\omega^2}.$$

The zeros are $z_1 = 0$ and $z_2 = \sqrt{1/6} = 0.41$ rad/s. There is one pole $p = 0.33$ rad/s. These zeros and pole do indeed alternate: $z_1 < p_1 < z_2$.

For the tangent in the origin we find $X_r = 3\omega$. For the tangent in $\omega = \infty$, i.e. the crooked asymptote, we again have three methods at our disposal. We shall not do method a, the differentiating. In method (b) we reason as follows: for large values of ω the lower powers in the numerator and denominator can be neglected with respect to the higher powers, with which we find $X \to \dfrac{-18\omega^3}{-9\omega^2} = 2\omega$, so the equation of the asymptote is $X_a = 2\omega$.

On *examining* the network we also find this result, because the capacitor is a short circuit for high frequencies, so that both inductors are connected in parallel with the substitute self-induction $\dfrac{3 \times 6}{3 + 6} = 2$ H.

Consequently we have the plot of Figure 3.48.

In the zeros the reactance is zero. Because the impedance of an inductor and a capacitor in series is zero at the resonance frequency a zero is often called a *series resonance frequency*. For dual reasons it is said that there is *parallel resonance* in poles.

We shall now solve a *synthesis* problem. Synthesis means constructing a network from a number of data, while *analysis* (which we have been concerned with so far) means, starting with a given network, finding certain voltages, currents, transfer functions, etc.

Given are the poles $p_1 = 0$ rad/s and $p_2 = 3$ rad/s and the zero $z_1 = 1$ rad/s of a reactance function $X(\omega)$. Asked is to construct a network with this reactance.

Solution

We have $\lim\limits_{\omega \to 0} = -\infty$ because there is a pole in the origin. This means that there must be a path going from one terminal to the other, that is not allowed to contain inductors only.

There further is one series resonance frequency and one parallel resonance frequency. The simplest networks that meet these requirements are drawn in Figure 3.49.

In Figure 3.49.a the resonance frequency of L_2 and C_2 in parallel is the pole $p_2 = 3$ rad/s, while this parallel circuit behaves inductively at the frequency 1 rad/s (see Figure 3.45), so that together with C_1 it causes series resonance.

This can also be argued for Figure 3.49(b). We shall now continue analysing the network of Figure 3.49(a).

We find $\dfrac{1}{\sqrt{L_2C_2}} = 3$ so $L_2C_2 = \dfrac{1}{9}$.

The impedance is $Z = \dfrac{1}{j\omega C_1} + \dfrac{j\omega L_2 \cdot \dfrac{1}{j\omega C_2}}{j\omega L_2 + \dfrac{1}{j\omega C_2}}$.

It follows that $X = \dfrac{\omega^2 L_2(C_1 + C_2) - 1}{\omega C_1(1 - \omega^2 L_2 C_2)}$.

The zero of this is $z_1 = \sqrt{\dfrac{1}{L_2(C_1 + C_2)}}$ and this equals 1 according to the values given.

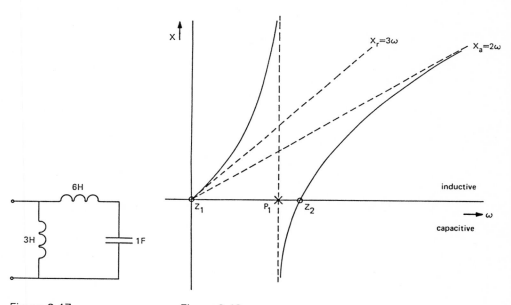

Figure 3.47   Figure 3.48

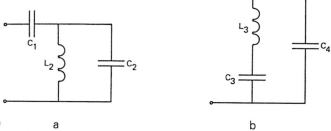

Figure 3.49   a   b

We note here that the frequency z_1 found is the parallel resonance frequency of L_2, C_1 and C_2 *in parallel*. The underlying principle will be discussed in Chapter 7.

So we find $L_2(C_1 + C_2) = 1$. Evidently the values given are not sufficient to find the values of the elements. If we also require that $X = 5\Omega$ for $\omega = 2$ rad/s we find

$$X = \left\{ \frac{\omega^2 - 1}{\omega C_1(1 - \omega^2/9)} \right\}_{\omega=2} = 5\Omega$$

from which $C_1 = 0.54$ F follows. Subsequently $L_2 = 1.65$ H and $C_2 = 0.07$ F. The plot is shown in Figure 3.50.

As said above, Foster's theorem will be proved in Appendix VI.

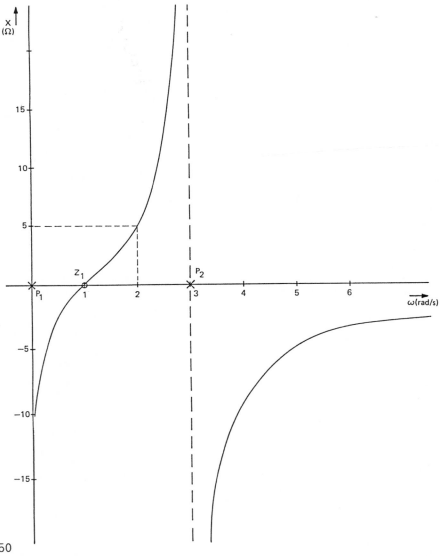

Figure 3.50

3.6 Zobel networks

An interesting one-port is the *Zobel network* (see Figure 3.51).

Z_1 and Z_2 are impedances. The impedance of this one-port is

$$Z_{ab} = \frac{(R + Z_1)(R + Z_2)}{2R + Z_1 + Z_2} = \frac{R^2 + R(Z_1 + Z_2) + Z_1 Z_2}{2R + Z_1 + Z_2}.$$

If $Z_1 Z_2 = R^2$, i.e. if

$$Z_1 = R^2 Y_2 \ (\text{or } Y_1 = \frac{1}{R^2} Z_2), \tag{3.27}$$

$Z_{ab} = R$ and is therefore real for all ω (!).

In this case the impedances Z_1 and Z_2 are called *inverse*. If $Z_1 = j\omega L$ then
$Z_2 = \dfrac{R^2}{j\omega L} = \dfrac{1}{j\omega L/R^2}$ which is the impedance of a capacitor with the capacity

$$C = \frac{L}{R^2}.$$

Example

For $R = 2\ \Omega$ and $L = 4$ H the one-port of Figure 3.52 is found.

The impedance is $Z_{ab} = 2\ \Omega$. This means that the network behaves as a resistance of 2 Ω for all frequencies.

We now consider the one-ports of Figure 3.53.

For Figure 3.53(a) we find $Z_1 = R_1 + j\omega L$ and for Figure 3.53(b) $Y_2 = \dfrac{1}{R_2} + j\omega C$.
Thus Z_1 and Z_2 are inverse if

$$R_1 + j\omega L = \frac{R^2}{R_2} + jR^2\omega C.$$

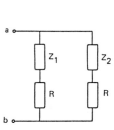

Figure 3.51

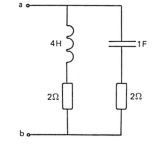

Figure 3.52

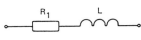

Figure 3.53   a

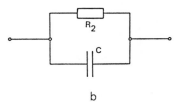

b

So

$$R_1R_2 = R^2 \quad \text{and} \quad L = R^2C.$$

With this we can also construct a Zobel network (see Figure 3.54). We choose R = 4 Ω, L = 1 H, R_1 = 2 Ω, so that R_2 = 8 Ω and C = $\frac{1}{16}$ F. The impedance is Z_{ab} = 4 Ω for all ω ≥ 0.

Another Zobel network is shown in Figure 3.55. It is the dual of Figure 3.51. We find

$$Z_{ab} = \frac{1}{G + Y_1} + \frac{1}{G + Y_2} = \frac{2G + Y_1 + Y_2}{(G + Y_1)(G + Y_2)} .$$

So

$$Y_{ab} = \frac{G^2 + G(Y_1 + Y_2) + Y_1Y_2}{2G + Y_1 + Y_2} .$$

If $Y_1Y_2 = G^2$ (i.e. if $Z_1Z_2 = R^2$) or

$$Y_1 = G^2Z_2$$

the input admittance is $Y_{ab} = G$.

These are the same conditions as (3.27). Y_1 and Y_2 are inverse admittances, or, which is the same, Z_1 and Z_2 are inverse impedances.

Finally two somewhat more intricate networks with inverse impedances have been drawn in Figure 3.56.

Similar to the above, one can find the relation between the element values of Figure 3.56(a) and those of Figure 3.56(b), after which the Zobel network can be constructed.

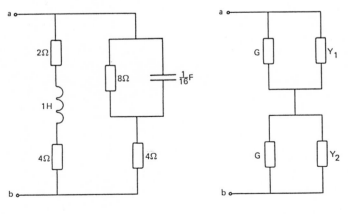

Figure 3.54   Figure 3.55

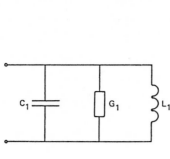

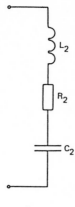

Figure 3.56   a   b

3.7 Problems

3.1 Draw the polar plots $Z(\omega)$ of these networks.

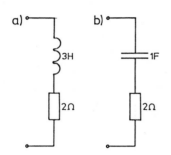

3.2 Draw the polar plots $Y(\omega)$ of the networks of problem 3.1.

3.3 $\omega = 1$ rad/s. C is variable.
$I_b = 18$ A.

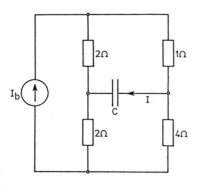

Find I as a function of C and draw this polar plot.

3.4 a. Give an expression for
$$H = \frac{V_{34}}{V_{12}}.$$

b. Draw the polar plot $H(\omega)$.
c. Sketch $|H|$ and arg H as functions of ω.

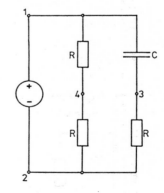

3.5   a. Give the polar plot I(R).
      b. Check for which value of R the
         current I is real and for which
         value of R the current I is
         imaginary.

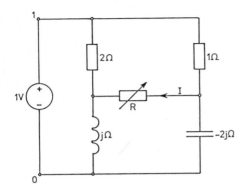

3.6   a. Find $H = \dfrac{V_2}{V_1}$ as a function of C
         and draw this polar plot.
      b. Sketch |H| = f(C) and check for
         which value of C this function is
         maximal.
      c. Sketch arg H as a function of C.

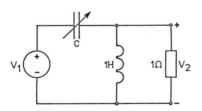

3.7   V = 2 V, $I_0 = 1 + j$ A. R is variable.
      a. Find the current I in the resistor

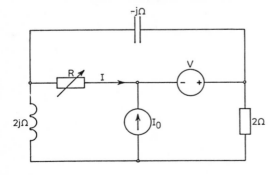

R as a function of R with
   Thévenin's theorem.
b. Draw the polar plot of I as a
   function of R with $0 \leq R < \infty$ and
   examine for which value of R the
   current I is imaginary.

3.8   V = 1 V, I = 1 + 2j A, $\omega = 1$ rad/s.
      a. Find I_1 if L = 4 H, using
         Thévenin's theorem.
      Next one makes the self-inductance
      L of the inductor variable.
      b. Find $I_1 = f(L)$ and sketch this
         function.
      c. Sketch |I_1| and arg I_1 as functions
         of L, using your answer to b.
      d. Examine for which value of L the
         voltage source supplies a reactive
         power only.

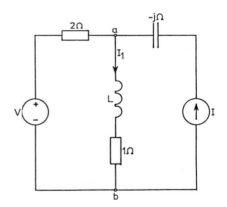

3.9   a. Find $H = \dfrac{V_2}{V_1}$ as a function of the
         resistance R.

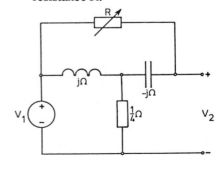

b. Draw the polar plot H(R) to scale.

Next we set R = 1 Ω and the voltage source intensity
$v_1 = 3 \cos \omega t$ V.

c. Find the output voltage v_2 as a function of time.

3.10

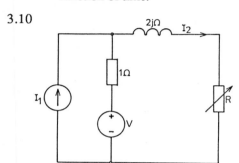

The complex source intensities are:
$I_1 = j$ A and V = 1 V. The resistance R is variable.

a. Express I_2 in R.

b. For which value of R is I_2 real?

c. Draw the polar plot $I_2 = f(R)$.

d. Sketch arg I_2 as a function of R.

Draw the asymptotic approach
$G = 20 \log |H|$ *in which* $H = V_1/V_2$, *of the following seven two-ports. Also find the corner frequencies and the real value of G in those points.*

3.11

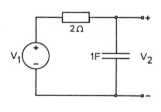

3.12

3.13

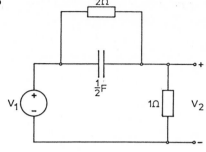

3.14

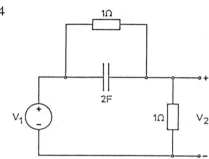

3.15

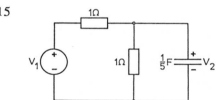

3.16

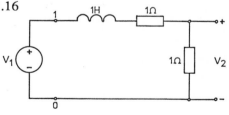

3.17   For the transactor μ = 1.

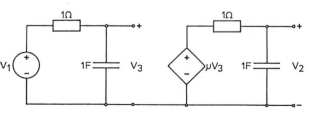

Give the dual of the following three networks.

3.18

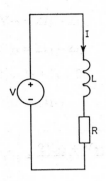

3.19

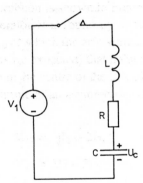

3.20

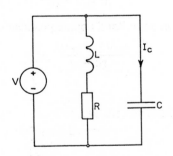

3.21 For which angular frequency is $|V|$ minimal?

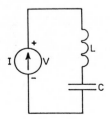

3.22 For which angular frequency is $|I|$ minimal?

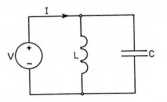

3.23 Give a rough sketch of

$$|H| = \left| \frac{V_2}{V_1} \right| \text{ as a function of } \omega$$

(a calculation is not necessary).

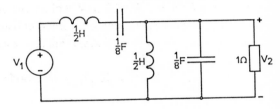

3.24 a. Find the resonance frequency. Next it is given that the voltage source supplies 200 W.
  b. Find both possible frequencies.
  c. Find the quality and the band width.

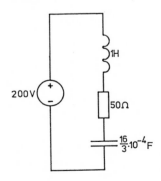

3.25 For which angular frequency does this network behave as a resistance?

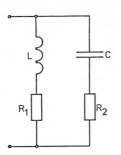

3.28 For which angular frequency is the impedance real?

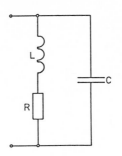

3.26 For which frequency is Z real?

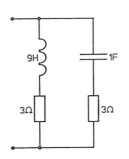

3.27 For which frequency is Z imaginary?

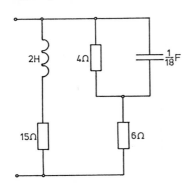

Sketch the reactance as a function of the angular frequency for the following eight two-terminal networks. Give the tangents.

3.29

3.30

3.31

3.32

3.33

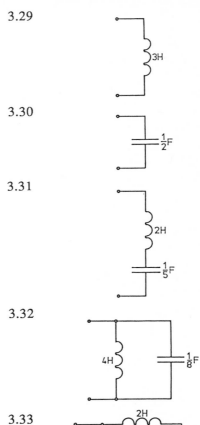

3.34

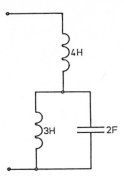

3.35

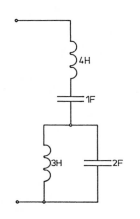

3.36

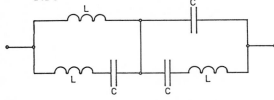

3.37 a. Find the reactance X of this two-
terminal network as a function of

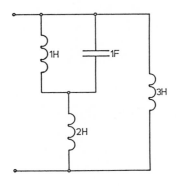

the angular frequency ω.

b. Sketch this function X(ω) and
physically and mathematically
find the equation of the tangent
ω = 0 and ω → ∞.

One now removes the inductor of
3 H.

c. Alter the value of the three other
elements so that the reactance
function remains the same as that
of the original network.

3.38

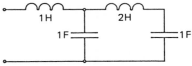

a. Find the reactance X(ω) as a
function of ω.

b. Sketch X(ω) and find the
equation of the tangent in
ω → ∞. Which is the physical
interpretation of this tangent?

c. Give one other circuit with also
four elements which has the same
reactance. Find the value of those
elements.

3.39

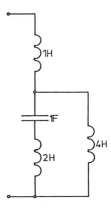

a. Compute the impedance as a
function of the angular frequency
and sketch the reactance X as a

function of the angular
frequency.
b. Find the equation of the tangents
in $\omega = 0$ and $\omega \to \infty$ and explain
your answer physically.
c. Give one network with less
elements with the same reactance
$X(\omega)$ and find the values of those
elements.

3.40

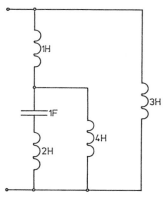

a. Find the reactance $X(\omega)$.
b. Sketch this function. Give the
equation of the tangents in $\omega = 0$
and for $\omega \to \infty$.
Sketch arg Z as a function of ω.
c. Give one network with less
elements which has the same
$X(\omega)$ and find the value of these
elements.

3.41 a. Find the impedance Z of this
network.
b. Find $\lim\limits_{\omega \to 0} \dfrac{Z}{\omega}$ and $\lim\limits_{\omega \to \infty} \dfrac{Z}{\omega}$ and
verify this result by inspecting
the network.
c. Sketch the reactance as a function
of the angular frequency and give
a network with an minimal
number of elements with the

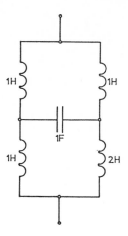

same reactance function. Also
find the value of these elements.

3.42 a. Find the reactance of the network
as a function of the angular
frequency, find the poles and the
zeros and give a sketch of the
function.
b. Give a network with less
elements with the same reactance.
Also find their values.

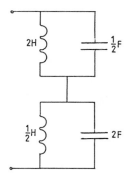

4

Magnetic coupled inductors, transformers

4.1 Introduction

So far we have assumed that there is no magnetic interaction between two inductors in a network. This is the case, however. Part of the magnetic field created by a current in one inductor will be surrounded by the windings of the second inductor. If the field intensity varies, an induction voltage will be created in that inductor. If the circuit is closed an induction current arises which in turn creates a magnetic field of which a part will be surrounded by the first inductor. This interaction, called *mutual induction*, will be dealt with in this chapter.

4.2 Mutual induction

Consider Figure 4.1.

In this figure two coils, with w_1 and w_2 windings respectively, have been drawn. These coils are wrapped on a core. This core can be of iron but this is not necessary. The core may even be absent. Part of the magnetic field created by i_1 will be coupled with coil 2 (see the dotted line of the magnetic flux density B), another part will not go through coil 2 (see the dotted line B_{L_1}). In the same manner the current i_2 will create a field, which is coupled by both coils, a part (B_{L_2}) will not be coupled with coil 1. B_{L_1} and B_{L_2} are called *leakage fields*.

In Figure 4.1 we have *field amplification*: the current i_2 creates a field in coil 2, which has the same direction as the field in coil 2, created by i_1.

The construction in Figure 4.1 is schematically illustrated in Figure 4.2.

L_1 is called the *primary*, L_2 the *secondary* coil. M_{12} is the 'influence' of the current i_2 on

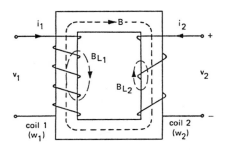

Figure 4.1   Figure 4.2

the flux through the inductor L_1 and M_{21} is the 'influence' of the current i_1 on the flux through L_2.

The dots indicate the sense of winding. In this situation both currents enter near the dot. This means we have field amplification. We obtain the following formulas:

$$\Phi_1 = L_1 i_1 \quad + M_{12} i_2,$$

$$\Phi_2 = M_{21} i_1 + L_2 i_2.$$

We see that M_{12} and M_{21} have the same dimensions as L_1 and L_2, so the unit is the henry (H).

Φ_1 is the coupled flux of inductor L_1.

Φ_2 is the coupled flux of inductor L_2.

If M_{12} and M_{21} are zero we have the formulas for the non-coupled inductors.

With $v = \dfrac{d\Phi}{dt}$:

$$v_1 = L_1 \frac{di_1}{dt} + M_{12} \frac{di_2}{dt}$$

$$v_2 = M_{21} \frac{di_1}{dt} + L_2 \frac{di_2}{dt}$$

when L_1, L_2, M_{12} and M_{21} are constant. It holds that

$$M_{12} = M_{21} = M. \tag{4.1}$$

Proof:

Consider situation 1 where $i_2 = 0$, while we increase i_1 from 0 to I_1 in t_1 second. The energy stored then is

$$W_p = \frac{1}{2} L_1 I_1^2.$$

Next consider situation 2 where we keep i_1 constant at the value of situation 1, i.e. I_1, while we increase i_2 from 0 to the value I_2 in t_2 second. The energy stored in inductor 2 is thus

$$W_s = \frac{1}{2} L_2 I_2^2 .$$

However, this system has been supplied with an *extra* amount of energy W_e, because in situation 2 there is in inductor 1 an induction voltage $M_{12}(di_2/dt)$ V, while the current in that inductor is I_1. So

$$W_e = \int_0^{t_2} M_{12} \frac{di_2}{dt} I_1 \, dt = M_{12} I_1 \int_0^{I_2} di_2 = M_{12} I_1 I_2.$$

So the total of supplied energy is

$$W = \frac{1}{2} L_1 I_1^2 + \frac{1}{2} L_2 I_2^2 + M_{12} I_1 I_2.$$

The sequence of the process described above is arbitrary. If we alter the sequence we get

$$W = \frac{1}{2} L_2 I_2{}^2 + \frac{1}{2} L_1 I_1{}^2 + M_{21} I_2 I_1.$$

In both cases we have the same result.

It follows that $M_{21} = M_{12} = M$.

So Figure 4.2 becomes Figure 4.3.

The corresponding formulas are

$$v_1 = L_1 \frac{di_1}{dt} + M \frac{di_2}{dt} ,$$

$$v_2 = M \frac{di_1}{dt} + L_2 \frac{di_2}{dt} . \tag{4.2}$$

In complex form (if the voltages and the currents are sinusoidal):

$$V_1 = j\omega L_1 I_1 + j\omega M I_2,$$

$$V_2 = j\omega M I_1 + j\omega L_2 I_2. \tag{4.3}$$

So the *impedance matrix* is

$$Z = j\omega \begin{bmatrix} L_1 & M \\ M & L_2 \end{bmatrix} . \tag{4.4}$$

If we alter the sense of winding in either the primary or in the secondary this results in Figure 4.4.

The formulas are:

$$v_1 = L_1 \frac{di_1}{dt} - M \frac{di_2}{dt} ,$$

$$v_2 = -M \frac{di_1}{dt} + L_2 \frac{di_2}{dt} . \tag{4.5}$$

4.3 Current direction, voltage polarity and the mode of winding

Because we are dealing with a two-port here, in which two equations occur simultaneously, it is essential to apply the rules concerning direction, polarity and the mode of winding consistently. The positive current direction can always be chosen at random. Having done that the positive voltage polarities have been fixed: the current flows from plus to minus *through* the element. In other words, *current and voltage are interrelated.*

Figure 4.3   Figure 4.4

The position of the dots near the coils, together with the positive currents result in the sign of M (M itself may be positive or negative).

Example 1 (see Figure 4.5).
For both networks it holds that:

$$V_1 = 6 j\omega I_1 - 3 j\omega I_2,$$

$$V_2 = -3 j\omega I_1 + 8 j\omega I_2.$$

Example 2 (see Figure 4.6).
Find the impedance Z_{ab}.

Solution
In view of the current I drawn the corresponding positive voltages are V_{ac} and V_{cb} (note the sequence of the indices). We find:

$$V_{ac} = j\omega L_1 I - j\omega M I,$$

$$V_{cb} = -j\omega M I + j\omega L_2 I.$$

With $Z_{ab} = \dfrac{V_{ab}}{I}$ and $V_{ab} = V_{ac} + V_{cb}$, $Z_{ab} = j\omega(L_1 + L_2 - 2M)$.

Taking into account the remarks made at the beginning of this section, it is often advisable not to work with mesh currents. The simultaneous occurrence of two (opposite) mesh currents in a particular inductor, together with the arising self-inductance and mutual inductance voltages often create errors in the sign of some terms.

If there is only one mesh current in each inductor, the mesh method can be used without problems.

4.4 The value of $|M|$

The value of the modulus of M is limited. This conclusion can be reached as follows. If both coils are wound on a closed iron core, the greater part of the magnetic field will pass through both coils. The fields H_{L_1} and H_{L_2} are therefore small. The flux that is not coupled with both coils is a *leak*. In an ideal situation this leak is zero and so the induced voltage in one coil caused by the current in another coil will be maximal. Then $|M|$ is also maximal. The coupled fluxes are

$$\left. \begin{array}{l} \Phi_1 = L_1 i_1 + M i_2 = w_1 \Phi_r, \\ \Phi_2 = M i_1 + L_2 i_2 = w_2 \Phi_r, \end{array} \right\} \tag{4.6}$$

where Φ_r is the *real* flux (which is equal in both coils).
So

$$\frac{L_1 i_1 + M i_2}{w_1} = \frac{M i_1 + L_2 i_2}{w_2}$$

or

$$\left(\frac{L_1}{w_1} - \frac{M}{w_2}\right)i_1 = \left(\frac{L_2}{w_2} - \frac{M}{w_1}\right)i_2.$$

This must hold for all currents i_1 and i_2, so

$$\frac{L_1}{w_1} = \frac{M}{w_2} \quad \text{and} \quad \frac{L_2}{w_2} = \frac{M}{w_1} \tag{4.7}$$

or

$$\frac{w_2}{w_1} = \frac{M}{L_1} = \frac{L_2}{M},$$

i.e.

$$\left(\frac{w_2}{w_1}\right)^2 = \frac{L_2}{L_1}. \tag{4.8}$$

Furthermore

$$M = \frac{w_2}{w_1} L_1 = \frac{w_1}{w_2} L_2.$$

So $M^2 = L_1 L_2$

$$|M| = \sqrt{L_1 L_2}. \tag{4.9}$$

In this case we speak of *complete coupling*. For two inductors of 2 H and 18 H, respectively, |M| can be 6 H at the most, so $-6 < M < 6$.

If there is no magnetic coupling, $M = 0$. This can be approximated by winding each coil on a closed iron core or by positioning both coils perpendicular. With the *coefficient of coupling*

$$k = +\sqrt{\frac{M^2}{L_1 L_2}} \tag{4.10}$$

we find

$$0 \le k \le 1. \tag{4.11}$$

The leak is defined as

$$\sigma = 1 - k^2. \tag{4.12}$$

From (4.6) it follows that

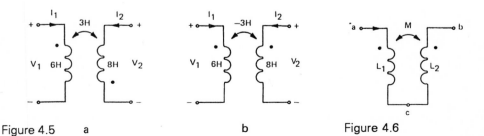

Figure 4.5   a   b   Figure 4.6

$$v_1 = \frac{d\Phi_1}{dt} = w_1 \frac{d\Phi_r}{dt} \text{ and } v_2 = \frac{d\Phi_2}{dt} = w_2 \frac{d\Phi_r}{dt}.$$

Consequently

$$\frac{v_1}{w_1} = \frac{v_2}{w_2}. \qquad (4.13)$$

So: In the case of complete coupling, the voltage is the same in each primary as well as in each secondary winding.

4.5 Hopkinson's formula

Consider Figure 4.7. The core is of iron with high permeability (e.g. $\mu_r = 10^4$). The leak is therefore small and so we shall neglect it.

The numbers of windings are w_1 and w_2 respectively. It is assumed that the breadth b of the iron is small compared with the length of the core l. The magnetic field within the iron core is virtually homogenous.
Apply *Ampère's rule*:

$$\oint \overline{H} \cdot \overline{dl} = I$$

in which the loop is chosen according to the dotted line. It follows that

$$Hl = w_1 i_1 + w_2 i_2,$$

where H is the magnetic field in the iron.
With

$$B = \mu_0 \mu_r H$$

and

$$\Phi_r = BA,$$

where A is the cross section of the iron and B the flux density: we find that

$$\Phi_r = \mu_0 \mu_r AH = \mu_0 \mu_r A \frac{w_1 i_1 + w_2 i_2}{l}$$

or, in another formula

$$\Phi_r = \frac{w_1 i_1 + w_2 i_2}{\dfrac{l}{\mu_0 \mu_r A}}. \qquad (4.14)$$

This is *Hopkinson's formula* or Ohm's law for a magnetic circuit. The numerator $w_1 i_1 + w_2 i_2$ is called the *ampere turns* (AT) and the denominator the *magnetic resistance*

$$R_m = \frac{l}{\mu_0 \mu_r A}, \qquad (4.15)$$

and shows much resemblance with the formula for the resistance of a wire (1.6).
Thus

$$\Phi_r = \frac{(AT)}{R_m} .$$ (4.16)

4.6 The transformer

From (4.6) it follows that

$$\Phi_r = \frac{L_1}{w_1} i_1 + \frac{M}{w_1} i_2 = \frac{w_1 i_1}{\dfrac{w_1^2}{L_1}} + \frac{w_2 i_2}{\dfrac{w_1 w_2}{M}} .$$

From (4.7) it follows that the above denominators are equal and, according to (4.14) and (4.15), also equal to R_m:

$$R_m = \frac{w_1^2}{L_1} = \frac{w_1 w_2}{M} = \frac{w_2^2}{L_2} .$$ (4.17)

Next we idealise the material of the core by assuming that the relative permeability is infinite:

$$\mu_r \to \infty.$$ (4.18)

From (4.15) it follows that

$$R_m \to 0,$$ (4.19)

while the number of ampere turns according to (4.16) will be zero if the flux Φ_r is finite:

$$(AT) = 0.$$ (4.20)

This means that

$$w_1 i_1 + w_2 i_2 = 0.$$ (4.21)

From (4.17) it follows that

$$\left.\begin{array}{l} L_1 \to \infty \\[4pt] L_2 \to \infty \\[4pt] M \to \infty \end{array}\right\}$$ (4.22)

so that we cannot use the formulas with L_1, L_2 and M anymore.

The two-port created in this way is called the ideal *transformer* and the formulas (4.13) and (4.21) hold. The usual symbol is shown in Figure 4.8 where

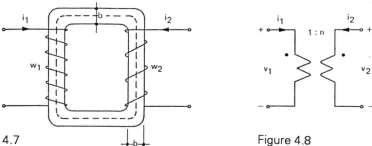

Figure 4.7   Figure 4.8

$$n = \frac{w_2}{w_1}.$$ (4.23)

n is called the *turns ratio*.
The corresponding formulas are

$$v_1 = \frac{v_2}{n}$$ (4.24)

and

$$i_1 + n\, i_2 = 0.$$ (4.25)

Because the voltages and the currents do not appear together in one formula here, as they do in coupled inductors with finite L_1, L_2 and M, it is not strictly necessary to apply the 'voltage and current are inter-related' rule.

If the winding is the other way round Figure 4.9 results, with

$$v_1 = -\frac{v_2}{n}$$

and $\qquad i_1 - n\, i_2 = 0,$

which follow from (4.24) and (4.25) by replacing n by –n.
The transformer is non-energetic, i.e. it does not dissipate energy as a resistor does and it cannot store energy either, as can inductors and capacitors.
The total immediate power supplied to both ports is

$$p = p_1 + p_2 = v_1\, i_1 + v_2\, i_2 = v_1\, i_1 - n\, v_1\, \frac{i_1}{n} = 0, \text{ so}$$

$$p = 0.$$ (4.26)

It is remarkable that the formulas of the transformer do not contain the frequency. In practice, however, one can only construct a transformer if the frequency is at least 10 Hz.

If the voltages and the currents are sinusoidal complex quantities can be introduced. As we have seen above capitals are used then.
We will only use the term *coupled inductors* if L_1, L_2 and M are finite.
The *energy transformer* is used to transform high voltages to lower values, e.g. from 11 kV to 240 V or from 240 V to 6 V (*doorbell transformer*). There are also *transformers for measurement*, while transformers are also often used for the energy supply of electronic devices.
Of great importance is the electric separation between the primary and the secondary. As a safety measure one terminal of the secondary is often connected to earth. Consider Figure 4.10.
If the primary voltage is 240 V, the secondary is 6 V. The non-ideal insulation material between the primary and the secondary can be interpreted by a (large) resistance R. If terminal B is not connected to earth, touching terminal A can be dangerous.
The 'open' voltage of the secondary with respect to earth is 240 V, so that touching A or

B will cause a fatal current through one's body. (That is to say, if R is not too large, which will be the case in a moisty room). If we connect B to earth, the voltage of A with respect to earth is only 6 V. We further note that, from a safety point of view, 'it is better to insulate than to earth'. (The push-button of a door bell is made of insulation material). Moreover, the capacity between the primary and the secondary also plays a part.

Next we shall examine a network in which a transformer is incorporated (see Figure 4.11).

Question: Find $H = \dfrac{V_2}{V_1}$.

Solution

If the secondary voltage is V, the primary voltage is 2V (both plusses near the dot). If the primary current is called I (entering near the dot), the secondary current is 2I (leaving near the dot). Now we have used both transformer equations. Kirchhoff's laws and Ohm's law give:

$$V_1 = 2V - I$$

$$V_2 = V - I$$

$$V_2 = 2I.$$

It follows that $H = \dfrac{2}{5}$.

As a second example we shall take the same network, but with opposite winding mode of the secondary (see Figure 4.12).

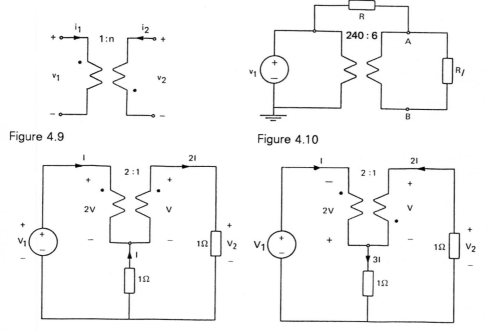

Figure 4.9

Figure 4.10

Figure 4.11

Figure 4.12

The primary voltage and the secondary current are the opposites of the former case. We find

$$V_1 = -2V + 3I$$

$$V_2 = \quad V + 3I$$

$$V_2 = -2I.$$

It follows that $H = -\dfrac{2}{13}$.

4.7 Impedance transformation

Consider the circuit of Figure 4.13 and determine the input impedance Z_i.
We find

$$Z_i = \frac{V_1}{I_1},$$

$$V_1 = \frac{V_2}{n},$$

$$I_1 + n\,I_2 = 0,$$

$$V_2 = -Z\,I_2.$$

Consequently

$$V_1 = -\frac{Z\,I_2}{n} = \frac{Z}{n^2}\,I_1,$$

so

$$Z_i = \frac{Z}{n^2}. \tag{4.27}$$

The impedance Z on the output will be observed as $\dfrac{Z}{n^2}$ on the input.

This is called *impedance transformation*. Note that the winding mode is not relevant. This property enables us to realise maximum power transfer for a given source series resistor R_T and a load R (see Figure 4.14).
We have maximum power transfer if $R_T = \dfrac{R}{n^2}$. We therefore choose

$$n = \sqrt{\frac{R}{R_T}}. \tag{4.28}$$

If we have an impedance Z_T instead of the resistor R_T then the load must also be complex (Z) in order to get maximum power transfer:

$$Z_T = \frac{Z^*}{n^2}.$$

With $Z_T = R_T + jX_T$ and $Z = R - jX$ (note the minus) we get:

$$R_T = \frac{R}{n^2},$$

$$X_T = \frac{X}{n^2}.$$
(4.29)

An extra advantage if n is large is the decreasing *impedance level*. If a connection cable between two networks is long, e.g. in the case of *audio frequencies* (frequencies from 20 up to 16,000 Hz), perturbations will enter the circuit. These perturbations can be interpreted by a *disturbance source* V_d with series impedance Z_d (Figure 4.15).

If $|Z|$ is small, then the perturbance voltage across Z will be small. So it is attractive to choose $|Z|$ small. If $|Z_T|$ is large (and $|Z|$ small) a transformer is placed at the input of the cable, which results in maximum power transfer.

The impedance level decreases and the perturbances become smaller.

In practice perturbations can start at values as low as 100 Ω. If that is the case one must use shielded cables with earthed screen. If one can reduce the impedance level to a few ohms, a screen is not necessary.

Another impedance transformation is shown in Figure 4.16.

Express Z_x in Z and n, so that both two-ports behave in the same way at their ports. For Figure 4.16(a) holds:

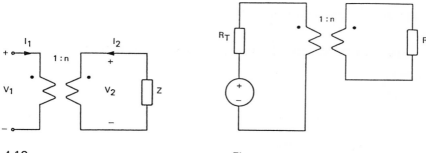

Figure 4.13   Figure 4.14

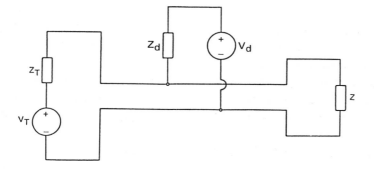

Figure 4.15

$$V_1 = \frac{V_3}{n} , \tag{a}$$

$$V_3 = V_2 - Z I_2, \tag{b}$$

$$I_1 + n I_2 = 0. \tag{c}$$

Consequently

$$V_1 = \frac{V_2}{n} - \frac{Z}{n} I_2. \tag{d}$$

For Figure 4.16(b) holds

$$V_1 = Z_x I_1 + V_4, \tag{e}$$

$$V_4 = \frac{V_2}{n} , \tag{f}$$

$$I_1 + n I_2 = 0. \tag{g}$$

It follows that

$$V_1 = -n Z_x I_2 + \frac{V_2}{n} . \tag{h}$$

The equations with the port quantities, (c) and (g) respectively and (d) and (h) respectively, are equal if $Z/n = n Z_x$. So

$$Z_x = \frac{Z}{n^2} .$$

We find that is the same equation as (4.27).

4.8 Equivalent circuits for magnetic coupled coils

Consider Figure 4.17.
For this figure:

$$V_1 = j\omega L_1 I_1 + j\omega M I_2, \tag{α}$$

$$V_2 = j\omega M I_1 + j\omega L_2 I_2. \tag{β}$$

If we consider these equations to be mesh equations of a two-port in T-configuration Figure 4.18 results.

Both two-ports are identical. However, in the latter there is an electric connection between both ports. One can create a separation of the ports by inserting a transformer with turns ratio 1:1 at the input or at the output.

It is possible that $M > L_1$ or $M > L_2$. In this case one of the self-inductances in Figure 4.18 is negative. The circuit cannot be realised then, but the equivalent circuit can be used for calculation purposes. An equivalent circuit with one less element is shown in Figure 4.19.

We express L_x, L_y and n in L_1, L_2 and M of Figure 4.17.
For Figure 4.19:

$$V_1 = j\omega L_x I_1 + j\omega n L_x I_2, \tag{γ}$$

$$V_2 = n V_1 + j\omega L_y I_2 \tag{δ}$$

Consequently

$$V_2 = j\omega n L_x I_1 + j\omega(n^2 L_x + L_y) I_2. \tag{ε}$$

(γ) and (ε) are identical with (α) and (β) if

$$L_x = L_1 \tag{4.30}$$

and $M = n L_x$, so

$$n = \frac{M}{L_1} \tag{4.31}$$

and $L_2 = n^2 L_x + L_y$, so $L_y = L_2 - n^2 L_1$, or

$$L_y = L_2 - \frac{M^2}{L_1} = \frac{L_1 L_2 - M^2}{L_1}.$$

The leak now equals, see formula (4.12),

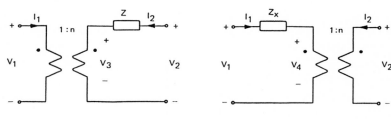

Figure 4.16   a   b

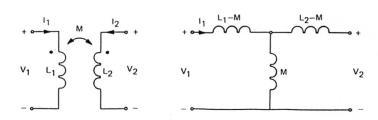

Figure 4.17   Figure 4.18

Figure 4.19

$$\sigma = 1 - k^2 = 1 - \frac{M^2}{L_1 L_2} = \frac{L_1 L_2 - M^2}{L_1 L_2},$$

so

$$L_y = \sigma L_2. \tag{4.32}$$

The equivalent circuit of Figure 4.19 is used in studying the transformer in power technology, which is, of course, not ideal in a practical realisation. For example, in practice $k = 0.99$ so that $\sigma \approx 0.02$.

With impedance transformation two other substitute networks can be derived from Figure 4.19 (see Figures 4.20 and 4.21).

One can also start with Figure 4.22, find L_α, L_β and m and derive two other substitute networks in a similar way as before.

4.9 The voltage and the current transformer

Depending on the turns ratio a transformer simultaneously transforms the primary voltage and the primary current to the secondary voltage and current. So it is not correct to refer to a *voltage transformer* or a *current transformer*. In power technology, however, these names have been given to transformers which have been developed for purposes of measurement.

In Figure 4.23 it is shown how we can measure the voltage of a high voltage system with a low-voltage voltage meter. If the maximum range of the voltage meter is 100 V and the turns ratio is 100:1, voltages of up to 10 kV can be measured in this fashion. The transformer has been made for high voltages and small currents. The secondary current is for instance 10 mA (the meter then consumes 1 W), so that the primary current is only 0.1 mA (thin winding wires).

Note that the voltage meter can be removed without difficulties. A transformer like this is called a *voltage transformer*.

A different matter is the measuring of the current in a conductor with the aid of a transformer (see Figure 4.24).

If the range of the current meter is 5 A a current of up to 500 A can be measured in the primary if the turns ratio is 1:100. If the meter consumes 1 W, the secondary voltage is 200 mV at the most, so that the primary voltage is 2 mV. The transformer has been made for low voltages and large currents (thick wires), and is called a *current transformer*. The meter is not allowed to be switched off. After all, the number of ampere turns will not be zero (there is a primary current only). The flux Φ and thus the flux density B become too large, which will destroy the transformer.

In both circuits the secondaries have been connected to earth in view of safety.

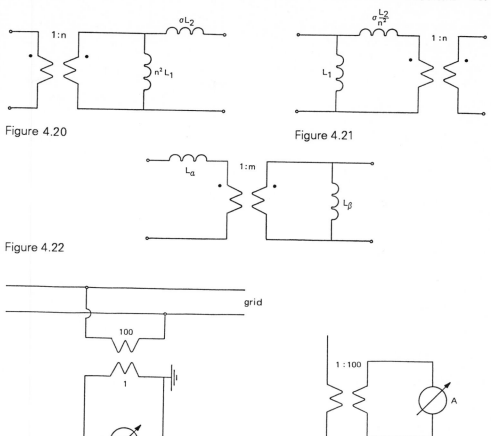

Figure 4.20

Figure 4.21

Figure 4.22

Figure 4.23

Figure 4.24

4.10 Problems

4.1 Find the input impedance.

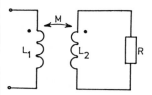

4.2 The winding mode of one of the
coils in 4.1 is changed.
Find the input impedance.

4.3 Find the input impedance.

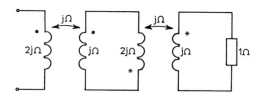

4.4 Find the input impedance.

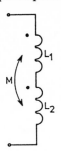

4.5 Find the input impedance.

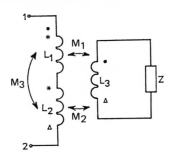

4.6 Find the input impedance.

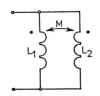

4.7 Find the input impedance.

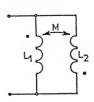

4.8 $\omega = 1$ rad/s.

$M = L$.

$H = \dfrac{V_2}{V_1}$.

Find H(L) and draw the function.

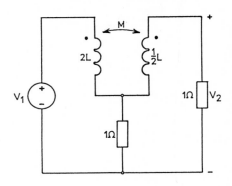

4.9

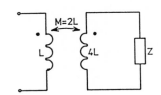

a. Is this a complete coupling?

b. Find the input impedance Z_i and find $\lim\limits_{L\to\infty} Z_i$.

c. Give an equivalent circuit if $L\to\infty$.

4.10 Find the input impedance.

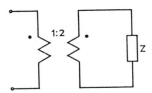

4.11 The mode of winding of the secondary in 4.10 is changed. Find the input impedance.

4.12 The input is measured at 1Ω. Find n.

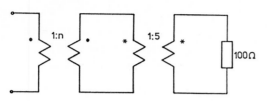

4.13 n is given such a value that maximum power transfer has been realised. Find n. Calculate the power dissipated in the whole network.

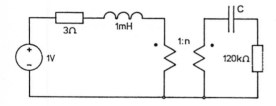

4.14 $\omega = 10^4$ rad/s.
Find n and C so that maximum power transfer is realised.

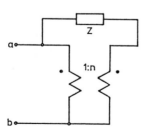

4.15 Given an expression for the impedance at the terminals a and b.

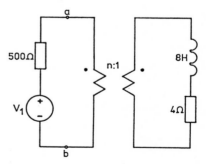

4.16 Find Z_{ab}.

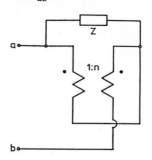

4.17 Find Z_{ab}.

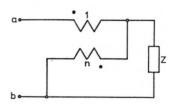

4.18 A voltage source V_1 has series resistance of $500\ \Omega$. We want to get maximum power transfer for a load consisting of an inductor of 8 H and a series resistance of $4\ \Omega$ and a transformer and a capacitor C are to be connected between the nodes a and b. The angular frequency is $\omega = 1$ rad/s.

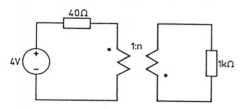

a. Find the Thévenin equivalence seen at the terminals a and b and including C.
b. Find n and C so that you have maximum power transfer.

C is taken away now and $V_1 = 1$ V and $n = 10$ are set.

c. Find the complex power delivered by the source and derive the delivered real power.

4.19 $I_2 = 2$ A.

Draw the vector diagrams of all currents and voltages. Choose 1 A $\triangleq 1$ cm and 1 V $\triangleq 1$ cm.

See Figure at bottom of this page.

4.20   The capacitor C is variable,
$\omega = 1$ rad/s.

One considers $H = \dfrac{V_2}{V_1}$.

Calculate $H(C)$ and draw the function.

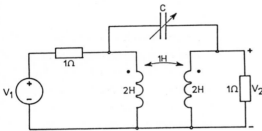

4.21   a. Find the reactance $X_{ab}(\omega)$ at the terminals a and b.

b. Sketch $X_{ab}(\omega)$.

c. Find the tangent equation in the origin through 'physical' observation of the circuit.

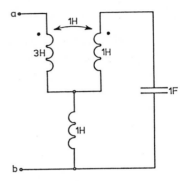

4.22   a. Find the impedance of this one-port as a function of the angular frequency ω.

b. Find the poles and the zeros of the reactance $X(\omega)$.

c. Sketch X as a function of ω and examine how the function behaves in $\omega = 0$ and $\omega \to \infty$.

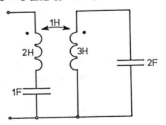

Figure of Problem 4.19

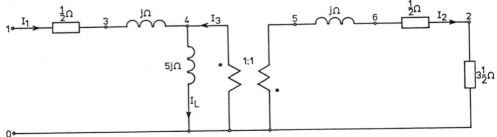

4.23 The source voltage is sinusoidal with amplitude 100 V and has an angular frequency of 1 rad/s.

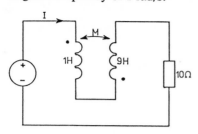

a. Find I as a function of M.
b. Draw the polar plot of I with M as parameter.
c. Which part of the circle can be realised in practice?
d. Which value must M be given, in order for the current to lag 30° behind the voltage?

4.24

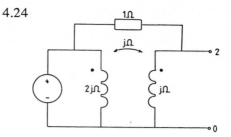

The complex source voltage is 5 V. Find the Thévenin equivalence at the terminals 2 and 0.

4.25 The source voltage is complex. Give the Thévenin equivalence at the nodes a and b. Use two ways to find the Thévenin impedance.

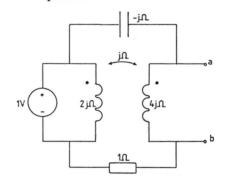

5

Three-phase systems

5.1 The rotating field

In Figure 5.1 the principle of a *rotating* field has been drawn, i.e. a homogenous magnetic field of limited size that rotates with constant velocity.

The vectors drawn represent the magnetic flux density ($\overline{B}$). The field rotates anti-clockwise with angular velocity ω. Such a rotating field could be realised with a rotating horseshoe magnet (Figure 5.2).

Between the north (N) and south (S) poles there is a magnetic field that rotates when the magnet rotates around its spindle. Thus Figure 5.1 is the left-hand view of Figure 5.2.

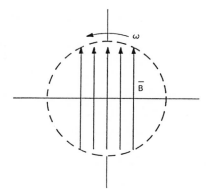

Figure 5.1

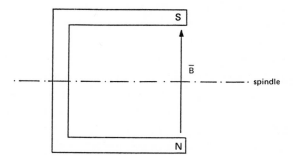

Figure 5.2

5.2 The creation of a rotating field

With the aid of two fixed coils, called *stator*, in which a.c. currents flow, one can obtain a rotating field (Figure 5.3).

The current $I_1 = |I| \sin \omega t$ in the horizontal coils causes a harmonic magnetic field $H_h = |H| \sin \omega t$.

The current $i_2 = |I| \cos \omega t$ in the vertical coils causes a magnetic field $H_v = |H| \cos \omega t$.

Further $\overline{H} = \overline{H}_h + \overline{H}_v$ (vectorial adding).

We find $H_h^2 + H_v^2 = |H|^2$, i.e. $H_v = f(H_h)$ is a circle. So the vectorial sum is a vector of which the head moves along a circle: We have a rotating field.

The circuit of the stator coils is shown in Figure 5.4.

In practice a stator with three coils is used, because we then have symmetry: The mutual phase difference must be 120° (Figure 5.5).

Note

Only the principle has been sketched here. In reality the position of the three coils on the stator is more complicated.

5.3 The principle of the three-phase motor

If the three feeding currents of Figure 5.5 are available (how this is realised in an electric power station will be discussed in Section 5.4), one can quite simply make a *synchronous motor* (Figure 5.6).

In the rotating field, which rotates with angular velocity ω, a *permanent magnet* will rotate with the same velocity. Thus the axis of that magnet can supply mechanical power. A disadvantage of this motor is that the axis cannot suddenly assume the angular velocity ω, starting from the rest position.

Another disadvantage is that the axis will suddenly stop if the torque supplied becomes too large.

A motor that does not have these disadvantages is the *asynchronous* motor.

Its principle is shown in Figure 5.7.

Here we have a homogenous magnetic field (with flux density $\overline{B}$) that rotates anti-clockwise. In the left-hand figure the situation at a certain moment has been drawn, in the right-hand figure we have the situation at a slightly later moment. The wire is a closed copper rectangle and has been drawn in cross-section.

In position 1 the surrounding flux of the loop is zero, in the second position the flux has increased, so that an induction voltage is created. This will result in an induction current that is counter clockwise with respect to the field direction. So the current in the upper wire is forward (indicated by ⊙) and in the lower wire it is backward (indicated by ⊗). The increasing horizontal component of the field is of concern here.

We now have current-carrying wires in a magnetic field. So forces will be exerted on these wires.

On a wire with length *l* in which flows a current $\overline{I}$ and which is in a magnetic field with flux density $\bar{B}$ a force

$$\bar{F} = l\,\bar{I} \times \bar{B}$$

will be exerted.

So the force on the upper wire is to the left and on the lower wire it is to the right. The mechanical torque is therefore anti-clockwise, i.e. the rectangle is going to rotate in the

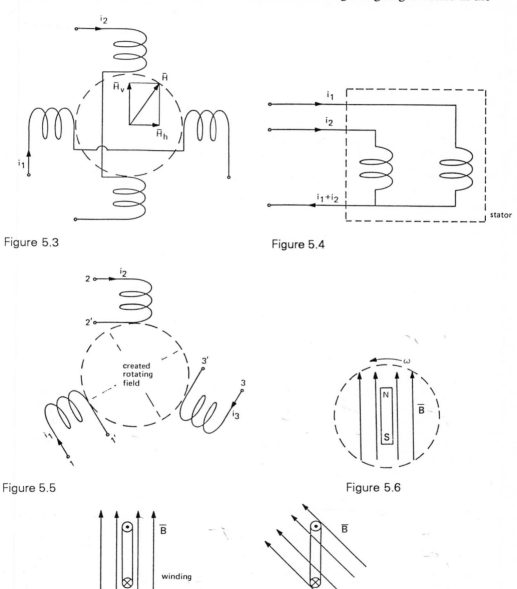

Figure 5.3

Figure 5.4

Figure 5.5

Figure 5.6

Figure 5.7

same direction as the rotating field. The angular velocity of the copper rectangle, however, is always smaller than the velocity of the rotating field, because at equal velocity there will be no induction voltage, the current is zero then and there are no forces.

In this case one speaks of *slip*, i.e. the angular velocity of the axis is smaller than the angular velocity of the rotating field. The motor is self-starting and will not stop suddenly if the torque required is too high. The motor is also very robust (it has no *commutator* with *brushes* as in a *d.c. motor*), and is called an *induction motor*. It is this motor which, together with other advantages, has led to the creation of three-phase systems in the past.

5.4 The principle of the three-phase generator

In Figure 5.8 it is shown how a so-called *three-phase voltage* can, in principle, be created: A permanent bar magnet rotates within a stator with three coils.

The three coils have a mutual geometric angle of 120°, so the induced voltages will also have this mutual phase difference of 120°.

Note that the position of the coils on the stator is more complicated in reality. One must try to approach a sinusoidal voltage as far as possible. The rotor, too, is not a permanent magnet in practice but an electromagnet, fed by a d.c. current. This current is led to the magnet by *slip-rings*.

The resistance of the coils is very small. In an idealised situation one can see the coils as (ideal) voltage sources. In future we shall assume that this is the case. We then speak of a (ideal) *three-phase supply*.

5.5 The three-phase supply

The three voltage sources can be connected in star (Figure 5.9).

Both circuits are electrically equivalent. Figure 5.9(b) *suggests* the phase differences of 120°. Node S is called the *star-point*.

In Figure 5.10 the three voltage sources are connected in delta.

This circuit has the disadvantage that only three voltages are available, whereas there are six in the star circuit. There is still another disadvantage: The sum of the voltages of the three sources must be zero, because otherwise a forbidden network is created. In practice the sources are not ideal (they have a small series impedance), so that a large *circulating current* will arise if the sum of the voltages is not exactly zero. This causes undesired losses.

Finally, if the sum of the ideal source voltage is exactly zero, the network cannot be solved, because both Kirchhoff's laws are met for each value of the circulating current.

The voltages between 1 and S, 2 and S and 3 and S in Figure 5.9 are called *phase voltages*. The voltages between 1 and 2, 2 and 3, 3 and 1 are called *line voltages*.

Often the star-point is earthed. If the amplitudes of the source voltages are not equal and/or the mutual phase difference is not 120° one speaks of an *asymmetric supply* (Figure 5.11).

The three phase voltages as functions of time consecutively reach their maximum value. If this sequence is v_{1S}, v_{2S}, v_{3S}, one says that the *phase sequence* is 1-2-3.
In Figure 5.11(a) the phase sequence is 1-2-3 and in Figures 5.11(b) and (c) it is 3-2-1.

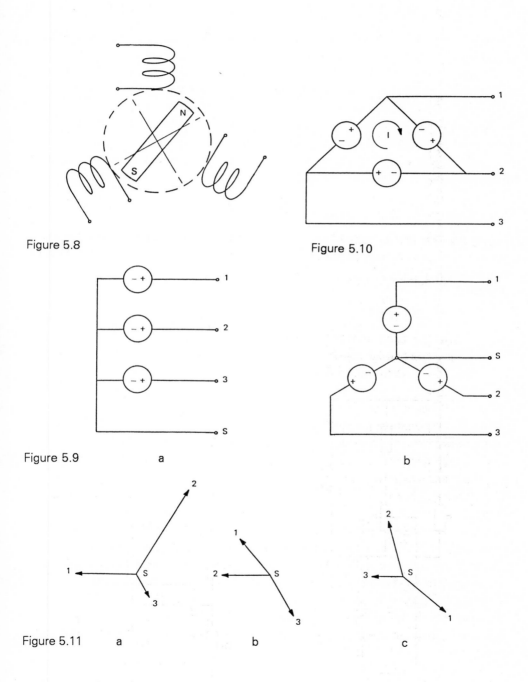

Figure 5.8

Figure 5.10

Figure 5.9   a   b

Figure 5.11   a   b   c

In Figure 5.12 the vector diagram of a symmetrical supply has been drawn.
The phase sequence is 1-2-3.
With the aid of plane geometry we can easily find the relation between the effective value
$|V_l|$ of the line voltage and the effective value $|V_p|$ of the phase voltage:

$$|V_l| = |V_p| \sqrt{3}. \tag{5.1}$$

We have not chosen the amplitude of the voltages on purpose, because in power technology it is usual to choose the effective values.
If the phase voltage is 240 V the line voltage is $240\sqrt{3} = 415$ V. The illumination in factories and in private houses is supplied with the phase voltage of 240 V, while the three-phase motors in the factories are constructed for 415 V line voltage.
In practice we also find the 720 V line voltage and 415 V phase voltage systems. The generator voltage in the central station is much larger. Line voltages of 11 kV, 33 kV, 240 kV, 415 kV and 720 kV are known. With special three-phase transformers this high voltage is transformed to lower values.

5.6 Complex three-phase voltages

The six voltages of a (symmetrical) three-phase supply can be written in a complex form. In the case of a symmetrical supply one voltage can be chosen real, the other five are complex then. In Figure 5.13 V_{1S} is set real.
We find

$$V_{3S} = V_{1S}\, e^{j \cdot 120°} = V_{1S}(-\tfrac{1}{2} + \tfrac{1}{2}j\sqrt{3}),$$

$$V_{2S} = V_{1S}\, e^{-j \cdot 120°} = V_{1S}(-\tfrac{1}{2} - \tfrac{1}{2}j\sqrt{3}),$$

$$V_{13} = V_{1S} - V_{3S} = V_{1S}(\tfrac{3}{2} - \tfrac{1}{2}j\sqrt{3}),$$

$$V_{32} = V_{3S} - V_{2S} = V_{1S}\cdot j\sqrt{3},$$

$$V_{12} = V_{1S} - V_{2S} = V_{1S}(\tfrac{3}{2} + \tfrac{1}{2}j\sqrt{3}).$$

Of course, another position is also possible; then all complex values change.

5.7 The three-phase load

In Figure 5.14 a three-phase generator in star and a load in delta are shown.
The load can be symmetrical, i.e. $Z_1 = Z_2 = Z_3$, or asymmetrical, i.e. at least one impedance has a different value. *Line currents* I_1, I_2 and I_3 and *phase currents* I_4, I_5 and I_6 arise in the load. The vector diagram of all six currents can easily be found with simple calculations. As an example we take a symmetrical supply with phase sequence 1-2-3 and $Z_1 = 1\Omega$, $Z_2 = 2\Omega$, $Z_3 = 3\Omega$.
In Figure 5.15 the vector diagrams are drawn.
In Figure 5.15(a) the given supply voltage vector diagram has been drawn. The vector diagram of Figure 5.15(b) is found as follows:

$$I_4 = \frac{V_{21}}{1} = V_{21}, \quad I_5 = \frac{V_{31}}{2}, \quad I_6 = \frac{V_{23}}{3}.$$

In addition the line currents follow from $I_1 = -I_4 - I_5$, $I_2 = I_4 + I_6$ and $I_3 = I_5 - I_6$. Because the vectors are in a head-to-tail position, $I_1 + I_2 + I_3 = 0$, which just happens to be Kirchhoff's current law for node S.

In Figure 5.16 a load in star has been drawn. The supply is omitted and is equal to that in Figure 5.14.

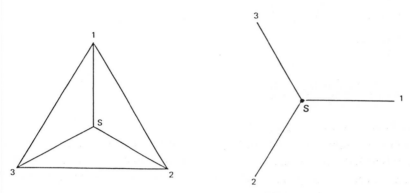

Figure 5.12

Figure 5.13

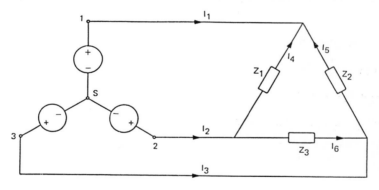

Figure 5.14

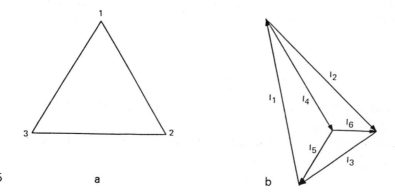

Figure 5.15

a

b

Once more we can distinguish the symmetrical and the asymmetrical loads.
The position of point 0 in the voltage vector diagram must be found through the calculation of the complex voltage V_{10}, V_{20} or V_{30}.
For a symmetrical load ($Z_1 = Z_2 = Z_3 = Z$) we find (using branch currents):

$$V_{12} = ZI_1 - ZI_2,$$

$$V_{23} = ZI_2 - ZI_3,$$

$$I_1 \cdot I_2 + I_3 = 0,$$

$$V_{10} = ZI_1.$$

So,

$$V_{10} = \frac{2V_{12} + V_{23}}{3} = \frac{N}{3}.$$

The construction is shown in Figure 5.17.
The numerator N is found by doubling V_{12} and adding V_{23}. Dividing N by 3 gives the vector V_{10} of which the arrow must be situated in point 1. So point 0 has been found.
We see (as we expected) that point 0 coincides with the star point S of the generator. Both points are in the centre of the triangle 1-2-3.
As an example of an asymmetrical load we choose $Z_1 = -j\Omega$, $Z_2 = 2\Omega$, $Z_3 = 3\Omega$.
We find

$$V_{12} = -jI_1 - 2I_2,$$

$$V_{23} = 2I_2 - 3I_3,$$

$$I_1 + I_2 + I_3 = 0.$$

After some manipulation it follows that

$$I_1 = \frac{2V_{23} + 5V_{12}}{6 - 5j}.$$

Note
This result can, of course, also be found with another method, e.g. with the mesh method or with Thévenin's theorem.

With $V_{10} = -jI_1$,

$$V_{10} = \frac{2V_{23} + 5V_{12}}{5 + 6j} = \frac{N}{D}.$$

(The numerator N of this expression can also be written differently: $N = 2V_{23} + 2V_{12} + 3V_{12} = 2V_{13} + 3V_{12}$).
The denominator D is shown in Figure 5.18, after which the construction of Figure 5.19 follows.

Point 0 is outside the triangle 1-2-3. In some cases of resonance the star point S can be situated infinitely far away.

Except for the delta and star loads the load can be any arbitrary network (provided that is it not forbidden). Besides resistors, inductors and capacitors the load may contain magnetic coupled inductors, transformers and even transactors.

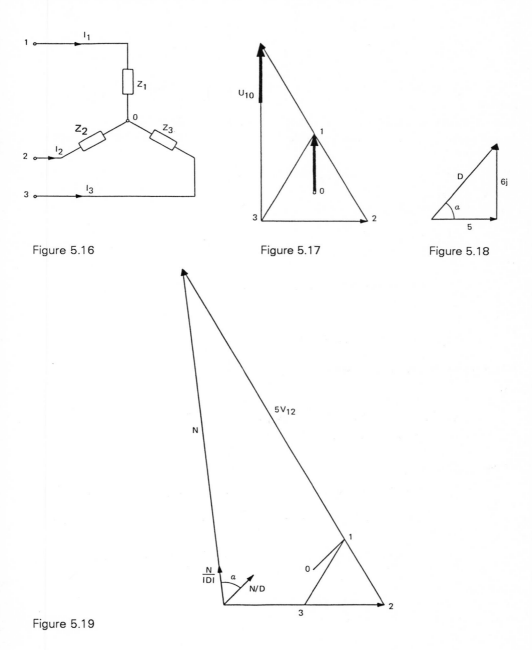

Figure 5.16

Figure 5.17

Figure 5.18

Figure 5.19

5.8 Power in three-phase systems

Consider the *symmetrical* star circuit of Figure 5.20.
The real power consumed by $Z = R + jX$ is $P = |V_{10}|\,|I_1|\cos\varphi$, with $\varphi = \arctan(X/R)$.
(We use effective voltages and currents).
The total power consumed is

$$P_{tot} = 3\,|V_{10}|\,|I_1|\cos\varphi = \frac{3\,|V_{12}|}{\sqrt{3}}\,|I_1|\cos\varphi,$$

$$P_{tot} = \sqrt{3}\,|V_l|\,|I_l|\cos\varphi, \tag{5.2}$$

where $|V_l|$ is the effective value of line voltage and $|I_l|$ the effective value of the line current.

Note: φ is the angle between the *phase* voltage and the *phase* current

In asymmetrical loads one has to calculate the real power in every impedance and add the results.

We now calculate the *immediate* power p(t) consumed by a symmetrical star circuit (Figure 5.20).

The power entering the ports 1-2 and 2-3 is

$$p = v_{12}i_1 + v_{32}i_3.$$

Now

$$v_{12} = \mathrm{Re}(V_{12}\,e^{j\omega t}) = \tfrac{1}{2}(V_{12}\,e^{j\omega t} + V_{12}^{*}\,e^{-j\omega t})$$

and

$$i_1 = \mathrm{Re}(I_1\,e^{j\omega t}) = \tfrac{1}{2}(I_1\,e^{j\omega t} + I_1^{*}\,e^{-j\omega t})$$

and there are analogous expressions for v_{32} and i_3.
So

$$p = \tfrac{1}{4}(V_{12}I_1 + V_{32}I_3)\,e^{2j\omega t} + \tfrac{1}{4}(V_{12}I_1^{*} + V_{12}^{*}I_1 + V_{32}I_3^{*} + V_{32}^{*}I_3)$$
$$+ \tfrac{1}{4}(V_{12}^{*}I_1^{*} + V_{32}^{*}I_3^{*})\,e^{-2j\omega t}.$$

Now

$$V_{12}I_1 + V_{32}I_3 = V_{1s}I_1 + V_{s2}I_1 + V_{3s}I_3 + V_{s2}I_3 = V_{1s}I_1 - V_{2s}(I_1 + I_3) + U_{3s}I_3$$

$$= V_{1s}I_1 + V_{2s}I_2 + V_{3s}I_3 = \frac{1}{Z}(V_{1s}^2 + V_{2s}^2 + V_{3s}^2)$$

$$= \frac{V_{1s}^2}{Z}\{1 + (e^{\frac{2\pi j}{3}})^2 + (e^{\frac{4\pi j}{3}})^2\} = 0 \quad \text{(see Figure 5.21).}$$

Thus also

$$V_{12}^{*}I_1^{*} + V_{32}^{*}I_3^{*} = (V_{12}I_1 + V_{32}I_3)^{*} = 0.$$

So we get

$$p = \frac{1}{4}(V_{12}I_1^* + V_{12}^*I_1 + V_{32}I_3^* + V_{32}^*I_3) = \frac{1}{2}\operatorname{Re}(V_{12}I_1^* + V_{32}I_3^*),$$

which is independent of time!
So we find

$$p = \operatorname{Re} S, \qquad\qquad (8.3)$$

in which S is the complex power consumed. This is an important result. The power consumed by a three-phase motor is *constant* and is for a (small) part consumed as constant losses and for a (large) part as constant mechanical power $T\omega_a$ of the rotating axis, in which T is the torque and ω_a the angular velocity of the axis.
If the desired torque is constant the angular velocity is constant. The combination of motor and tool trembles less than, say, a d.c. motor in which the torque is a function of the place at the circumference and therefore a function of time.

5.9 Phase compensation

A three-phase asynchronous motor is an inductive load, i.e. the phase current is an angle behind the phase voltage.
Figure 5.20 can be regarded as the substitute circuit. In it Z = R + jX, the resistance comprising both the losses and the mechanical power released to the axis of the motor.
The total power consumed by the motor is given by (5.2).
If the angle φ is large, and therefore cos φ small, $|I_l|$ is large for a certain power. In this case we speak of a *bad cos φ*.
One can improve this cos φ with the aid of capacitors. The power remains the same, but the line currents become smaller, which means that one can use thinner cables (saving of copper). In some cases the government demands such a *phase compensation.*

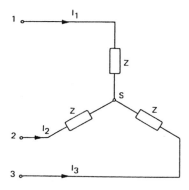

Figure 5.20

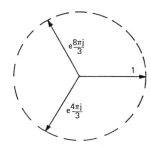

Figure 5.21

In Figure 5.22 the situation for one phase has been drawn, whilst in Figure 5.23 the corresponding vector diagram has been drawn.

The original line current (without capacitor) is I_1 and this current is an angle φ behind the voltage. The power for this phase is

$$P_p = |V_{1s}| \, |I_1| \cos \varphi = |V_{1s}| \, |I_w|.$$

The projection I_w of the current to the voltage vector is sometimes called the *watt component*. If we connect the capacitor now, then the capacitor current I_C arises so that the total current becomes I_1'. In this situation $|I_1'| < |I_1|$ while the watt component, and therefore the power, remains the same. We reach maximum compensation if we choose $|I_C|$ such that $I_1' = I_w$. Phase compensation is, in fact, a resonance problem. Three capacitors are needed for the three phases. These are in star. With star-delta transformation (see Section 1.15) one can derive an equivalent delta circuit. The capacitors then have a smaller capacity, but the capacitor voltages are $\sqrt{3}$ as large as in the case of a star situation.

Example

Suppose $Z = 1 + 2j \, \Omega$, the line voltage is V_l and the frequency is 50 Hz.
Find the value of the three capacitors in delta so that $\cos \varphi = 0.8$.

Solution

The phase voltage is V_{1s}. It is chosen real in accordance with Figure 5.23.

$$I_1 = \frac{V_{1s}}{1 + 2j} \quad \text{and} \quad I_C = \frac{V_{1s}}{-j|X|} \quad \text{with} \quad |X| = \frac{1}{\omega C}.$$

So

$$I_1' = I_1 + I_C = V_{1s} \left(\frac{1}{1 + 2j} + \frac{j}{|X|} \right).$$

It follows that:

$$I_1' = \frac{V_{1s}}{5|X|} \{|X| + j(5 - 2|X|)\}.$$

The argument of I_1' must be $-\arctan \frac{3}{4}$ (Figure 5.24).
So $\dfrac{2|X| - 5}{|X|} = \dfrac{3}{4}$.

From this it follows that $|X| = 4 \, \Omega$. So $\dfrac{1}{\omega C} = 4$, and consequently $C = \dfrac{1}{4\omega} = \dfrac{1}{4 \cdot 2\pi \cdot 50}$ F.

Therefore $C = 796 \, \mu F$.

From (1.36) it follows that:

$$D = \frac{AB}{A + B + C} = \frac{Y^2}{3Y} = \frac{1}{3} Y,$$

in which Y is the admittance of the computed capacitor.

In delta configuration the capacitors become $\frac{796}{3} = 265\ \mu F$. The resulting circuit is shown in Figure 5.25.

Note

If we choose $\frac{5 - 2|X|}{|X|} = \frac{3}{4}$ then I leads the voltage V. If this is the case we speak of *overcompensation*. By calculation the value of 584 μF for the capacitors in delta is found, which is unnecessarily large.

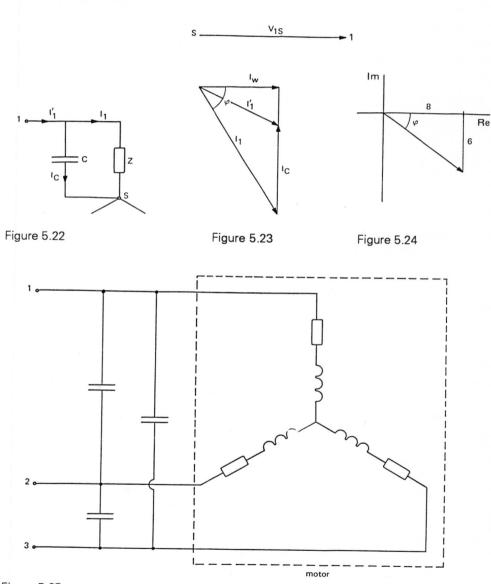

Figure 5.22   Figure 5.23   Figure 5.24

Figure 5.25

5.10 Problems

Note

a. *In this chapter the effective value of the voltages and currents has always been chosen.*

b. *We assume the supply to be symmetrical.*

c. *If not otherwise stated the phase sequence is 1-2-3.*

d. *If not otherwise stated the frequency is 50 Hz.*

5.1   The three voltage sources of a three-phase generator are connected in star.

Find the line voltage if the phase voltage is
a. 138 V, b. 240 V,   c. 415 V.

5.2   The three voltage sources (240 V) of a three-phase generator are connected in delta. Find the current.

5.3   The line voltage is 415 V.
a. Draw the vector diagrams of all voltages and currents.
b. Find the total dissipated power.

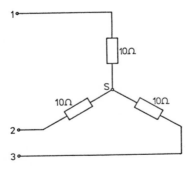

5.4   The line voltage is 415 V. Answer the same questions as in Problem 5.3.

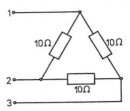

5.5   The line voltage is 240 V. Answer the same questions as in Problem 5.3.

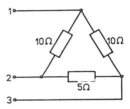

5.6   The line voltage is 415 V. Find the effective value of the line currents.

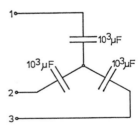

5.7   The line voltage is 415 V.

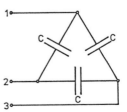

a. Find the effective value of the line currents.
b. Find the value of C so that the same line currents arise as in Problem 5.6.

5.8 The line voltage is 415 V. Give the vector diagram of all voltages.

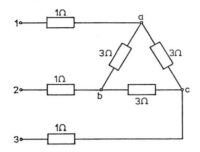

5.9 The line voltage is 415 V. Find |I₀|.
See Figure at bottom of this page.

5.10

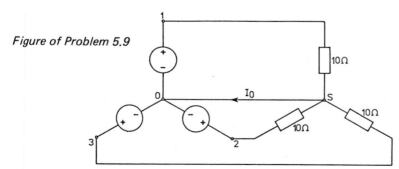

The line voltage is 240 V. Find |I₀|.

5.11 The line voltage is 415 V. Give the vector diagram of the voltages and the current.

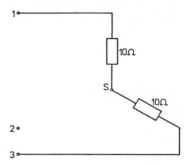

5.12 The line voltage is 415 V. Give the vector diagram of the voltages and the currents.

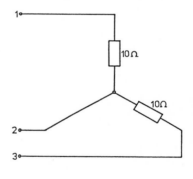

Figure of Problem 5.9

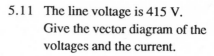

5.13   The line voltage is 415 V.
Find the position of S in the vector
diagram of the voltages.

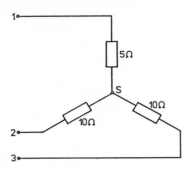

5.14   The line voltage is 415 V.
$Z = 1 + j\,\Omega$.
Give the vector diagram of the
voltages and the currents.

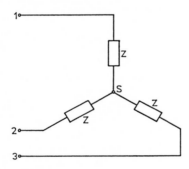

5.15   The line voltage is 415 V.
a. Find the position of S in the
vector diagram.
b. Give a sketch of v_{12} and v_{1S} as
functions of time.

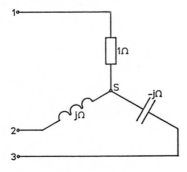

5.16   Prove that the total dissipated power
is

$$P = \mathrm{Re}\,(V_{12}\,I_1{}^* + V_{32}\,I_3{}^*).$$

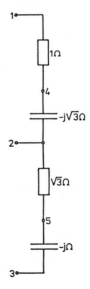

5.17   The line voltage is 240 V.
Prove
a. $|V_{45}| = 360$ V for the phase
sequence 1-2-3.
b. $|V_{45}| = 0$ V for the phase
sequence 3-2-1.

5.18   The line voltage is 240 V.
       Find |I|.

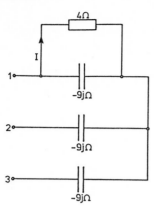

5.19   a. Find V_{ab}.
       Next one short-circuits the nodes a
       and b.
       b. Find the short circuit current *with*
       and *without* Thévenin's theorem.

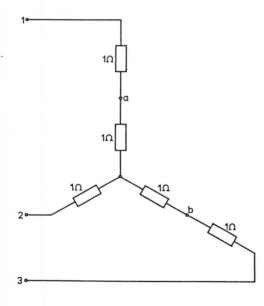

5.20   The line voltage is 200 V. The phase
       sequence is 3-2-1. V_{13} is set to
       positive real. $\omega = 1$ rad/s.

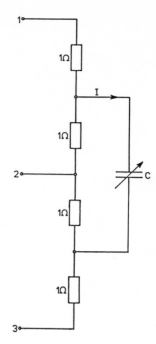

a. Find I as a function of C.
b. Draw the polar plot.
c. For which value of C is there no
   phase difference between v_{23}
   and i?

5.21   In Problem 5.20 an inductor L is
       now chosen instead of a capacitor C.
       Answer the same questions.

5.22   The line voltage is 240 V. Set V_{12} to
       positive real.
       a. Find I_3 with Thévenin's theorem.
       b. Draw the vector diagram.

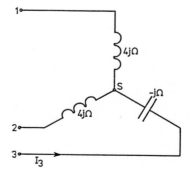

5.23 The line voltage is 240 V. Set V_{30} to positive real.
   a. Find $I(R)$.
   b. For $R = 1\ \Omega$ find the position of S in the vector diagram.
   c. Draw $I(R)$.
   d. Find the dissipated power if $R = 1\ \Omega$.

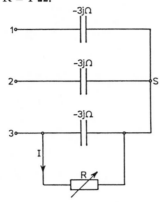

5.24 The line voltage is 100 V. Set V_{23} to positive real. Calculate and draw $V_{S1}(R)$.
   Choose point 1 in the origin of the complex plane.

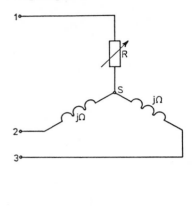

5.25 The line voltage is 415 V. The capacitors are connected in order to meet the authorities' requirements: $\cos \varphi \geq 0.8$.

   a. Determine whether this demand is fulfilled if $C = 400\ \mu F$.
   b. Calculate the real (average) power consumed.

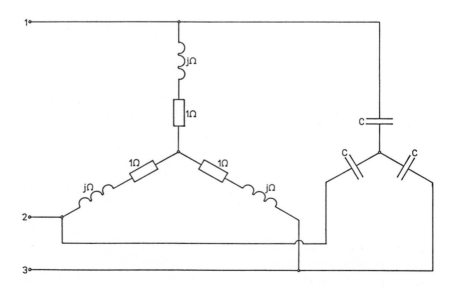

5.26 The line voltage is 415 V. One wants to reduce the angle φ from 45° to 30° by means of three capacitors in delta, which should be as small as possible. Find the value of the capacitors.

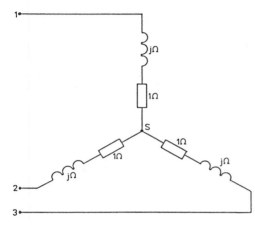

5.28 Of a load connected to a symmetrical three-phase supply with neutral conductor phase 3 is interrupted. One wants to have a terminal with the same potential as node 3. For this purpose one uses two ideal transformers (see the figure).
   a. Determine how to choose n_1 and n_2 so that $V_{a0} = V_{30}$.
   b. Give another circuit with transformers that also suffices.

5.27 The line voltage is 100 V, $\omega = 1$ rad/s. Set V_{23} to positive real. The capacitors interpret the capacity between the cables and earth.
   a. Find I_L.
   Next a short circuit occurs between node 3 and earth.
   b. Again find I_L.
   c. For which value of L is the short circuit current zero?

Note: The inductor is called *Petersen inductor*. In practice it is used to limit damage in the case of short circuits to earth.

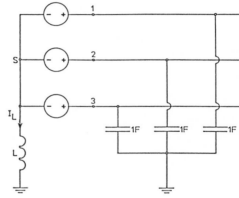

Figure of Problem 5.28

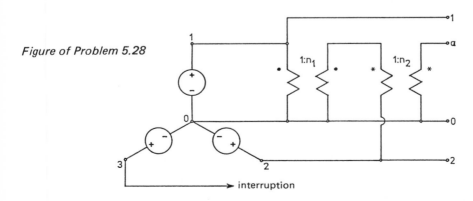

5.29   Node 0 is the star point of the
supply.

   a. Prove

$$V_{1S} = \frac{3V_{10}}{2} \left(1 + \frac{10j}{R}\right).$$

   b. Give a sketch of the set of points
S as a function of R.

   c. Explain the position of S for
$R = 0\ \Omega$ and $R \rightarrow \infty$.

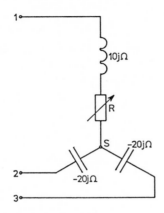

5.30   The angular frequency is $\omega = 1$
rad/s. The star point of the supply is
node 0.

   a. Prove

$$V_{S1} = 3\,\frac{R + j}{2R + 1 + 2j}\,V_{01}.$$

   b. Set $V_{01} = 10$ V (positive real)
and draw the polar plot

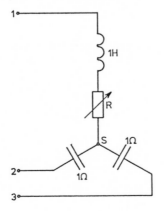

$V_{S1} = f(R)$ in the complex plane.
Set $1\ V \triangleq 1$ cm.
Measure the smallest value of
$|V_{S1}|$ and $|V_{S0}|$.

5.31   The line voltage is 240 V.
Calculate $|V_{45}|$ for both phase
sequences.

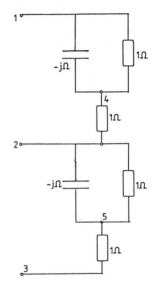

6

Fourier series

6.1 Introduction

So far we have only discussed d.c. voltages and currents and sinusoidal voltages and currents. In this chapter we shall discuss functions that are *periodical*.
In Figure 6.1 some of them are shown.
These functions have the property that in each time interval T the function has the same shape. T is called the *period* and it is always chosen *as small as possible*.
It holds that

$$f(t) = f(t \pm nT) \quad (n = 1,2,...). \tag{6.1}$$

We further note that we have chosen the period at the interval (0,T) here. One can also choose the interval $(-\frac{1}{2}T, \frac{1}{2}T)$ or an arbitrary interval T for the period.

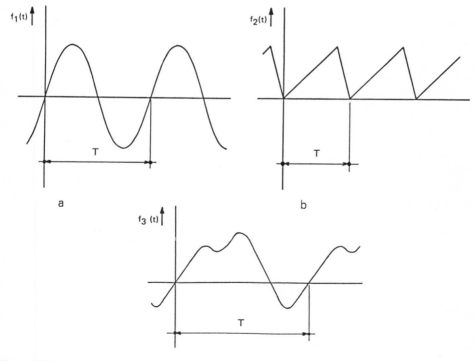

Figure 6.1

6.2 The infinity series of Fourier

If we represent an arbitrary periodic function $f(t)$ as an infinite sum in accordance with

$$f(t) = \sum_{n=0}^{\infty} (A_n \cos n\omega t + B_n \sin n\omega t), \tag{6.2}$$

or in full

$$f(t) = A_0 + A_1 \cos \omega t + A_2 \cos 2\omega t + \ldots + B_1 \sin \omega t + B_2 \sin 2\omega t + \ldots , \tag{6.3}$$

we can try to express the amplitudes A_n and B_n in the properties of the given function. We shall first introduce some names.

ω is called the *fundamental frequency*. The term with A_n or B_n respectively is called the *n-th harmonic*. A_0 is called the *zero harmonic* or (less general) the *d.c. term*. We shall now determine the amplitudes of the harmonics.

In (6.3) both sides are integrated over one period:

$$\int_0^T f(t) \, dt = A_0 \int_0^T dt + A_1 \int_0^T \cos \omega t \, dt + A_2 \int_0^T \cos 2\omega t \, dt + \ldots$$

$$+ \quad B_1 \int_0^T \sin \omega t \, dt + B_2 \int_0^T \sin 2\omega t \, dt + \ldots$$

If we integrate a sinusoidal function over one or more periods the result is zero, so

$$\int_0^T f(t) \, dt = A_0 T,$$

so

$$A_0 = \frac{1}{T} \int_0^T f(t) \, dt. \tag{6.4}$$

The d.c. term equals the *average* value of the given function.

For the determination of A_n we multiply both the left- and right-hand sides of (6.3) by $\cos n\omega t$ and integrate over one period:

$$\int_0^T f(t) \cos n\omega t \, dt = A_0 \int_0^T \cos n\omega t \, dt + A_1 \int_0^T \cos \omega t \cos n\omega t \, dt$$

$$+ A_2 \int_0^T \cos 2\omega t \cos n\omega t \, dt + \ldots$$

$$+ B_1 \int_0^T \sin \omega t \cos n\omega t \, dt + B_2 \int_0^T \sin 2\omega t \cos n\omega t \, dt + \ldots$$

Now examine the following integral I_1:

$$I_1 = \int_0^T \cos n\omega t \cos m\omega t \, dt,$$

with n and m being integers and positive.
Using

$$\cos \alpha \cos \beta = \tfrac{1}{2} \cos (\alpha + \beta) + \tfrac{1}{2} \cos (\alpha - \beta)$$

it follows that

$$I_1 = \tfrac{1}{2} \int_0^T \cos (n + m) \, \omega t \, dt + \tfrac{1}{2} \int_0^T \cos (n - m) \, \omega t \, dt$$

$$= \tfrac{1}{2} \int_0^T \cos (n - m) \, \omega t \, dt.$$

So $I_1 = \tfrac{1}{2} T$ for $n = m$ and $I_1 = 0$ for $n \neq m$.
Examine the integral

$$I_2 = \int_0^T \sin n\omega t \cos m\omega t \, dt.$$

With $\sin \alpha \cos \beta = \tfrac{1}{2} \sin (\alpha + \beta) + \tfrac{1}{2} \sin (\alpha - \beta)$

$$I_2 = \tfrac{1}{2} \int_0^T \sin (n + m) \, \omega t \, dt + \tfrac{1}{2} \int_0^T \sin (n - m) \, \omega t \, dt = 0$$

arises, for all n and m.
So we get

$$\int_0^T f(t) \cos n\omega t \, dt = A_n \cdot \tfrac{1}{2} T,$$

so

$$A_n = \frac{2}{T} \int_0^T f(t) \cos n\omega t \, dt. \qquad n \geq 1 \tag{6.5}$$

Similarly we can derive:

$$B_n = \frac{2}{T} \int_0^T f(t) \sin n\omega t \, dt. \qquad n \geq 1 \tag{6.6}$$

Example (see Figure 6.2).
Given the current i(t), find the Fourier amplitudes. We find

$$A_0 = \frac{1}{T} \int_0^T i \, dt = \frac{1}{T} \int_0^{\frac{1}{2}T} 2 dt = 1.$$

If we subtract this value from the given function a new function $f(t) = i(t) - 1$ arises, of which the average value is zero, see Figure 6.3.
The values A_n and B_n for $n > 0$ are equal for both functions i(t) and f(t) because from (6.3) it follows that

$$f(t) - A_0 = A_1 \cos \omega t + A_2 \cos 2\omega t + \dots + B_1 \sin \omega t + B_2 \sin 2\omega t + \dots$$

Thus the functions f(t) and $f(t) - A_0$ have the same first and higher harmonics. The background of this method becomes clear after the symmetry considerations (Section 6.8) We find that:

$$A_n = \frac{2}{T} \int_0^T i \cos n\omega t \, dt = \frac{4}{T} \int_0^{\frac{1}{2}T} \cos n\omega t \, dt = [\frac{4}{n\omega t} \sin n\omega t \;]_0^{\frac{1}{2}T} = 0$$

for all n.

$$B_n = \frac{2}{T} \int_0^T i \sin n\omega t \, dt = \frac{4}{T} \int_0^{\frac{1}{2}T} \sin n\omega t \, dt = [\frac{4}{n\omega T} \cos n\omega t \;]_{\frac{1}{2}T}^0)$$

$$= \frac{2}{n\pi} (1 - \cos n\pi).$$

So $B_n = 0$ for even n and $B_n = \frac{4}{n\pi}$ for odd n.
The Fourier series becomes

$$i = 1 + \frac{4}{\pi} (\sin \omega t + \tfrac{1}{3} \sin 3\omega t + \tfrac{1}{5} \sin 5\omega t + \dots)$$

$$= 1 + \frac{4}{\pi} \sum_{n=1}^{\infty} \frac{\sin(2n - 1) \; \omega t}{2n - 1}.$$

In this example we see that all cosine terms are missing and that there are only odd harmonics.

6.3 The frequency spectrum

If we plot the amplitudes of the various harmonics as a function of the angular frequency then a *discrete line spectrum* develops. Discrete means that only certain frequencies are present, thus there will be no term with a frequency between the harmonics n and $n + 1$. The n-th harmonic has two components, viz. $A_n \cos n\omega t$ and $B_n \sin n\omega t$.

So the amplitude of the n-th harmonic is $\sqrt{A_n^2 + B_n^2}$ and that is, of course, a positive number.

In Figure 6.4 the line spectrum of the last example has been drawn to the ninth harmonic. If the Fourier series consists merely of cosine or sine terms the negative terms can be drawn downward.

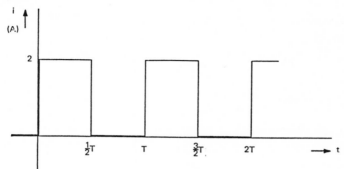

Figure 6.2

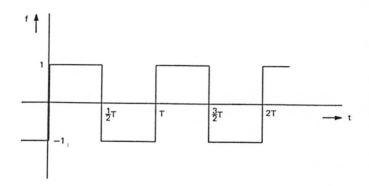

Figure 6.3

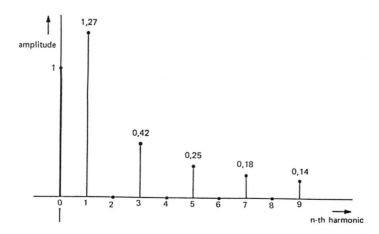

Figure 6.4

6.4 Dirichlet's conditions

In the preceding sections we have assumed that all integrals exist and that the Fourier series converges to the given function. This, however, need not be the case.

Dirichlet formulated conditions which have to be met so that the function can be written as a Fourier series. For instance, the function should have a limited number of finite *discontinuities* in one period. A discontinuity is a leap in the function, i.e. at the moment of time t_1 it holds that

$$\lim_{t \uparrow t_1} f(t) \neq \lim_{t \downarrow t_1} f(t).$$

In Figure 6.5 such a discontinuity is shown.

If the leap is infinite we speak of an infinite discontinuity. Nearly all functions we come across in engineering meet Dirichlet's conditions.

We shall not discuss this any further, but mention that the Fourier series converges to the middle of the leap at the points of discontinuity, see point M in Figure 6.5.

6.5 Gibb's phenomenon

It is interesting to plot the sum of an increasing number of harmonics in a sequence of graphics, beginning with the d.c. term plus the first harmonic. We expect an increasing approach of the original function. We shall do this for the example of Figure 6.3 (see Figures 6.6(a) – (e)).

We see that, as the number of harmonics increases, the function is approached more closely. However, around the discontinuities an oscillation occurs, which does not disappear if the number of harmonics has increased. This is known as *Gibb's phenomenon*. In Figure 6.6(e) only the interval $(0, \frac{1}{4}T)$ has been drawn in order to get the plot of this phenomenon in more detail.

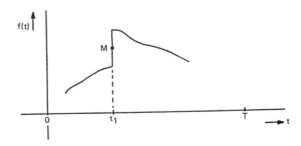

Figure 6.5

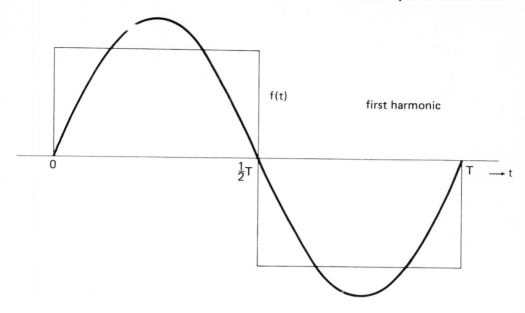

f(t)

first harmonic

Figure 6.6(a)

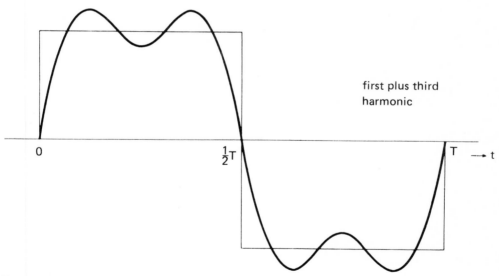

first plus third
harmonic

Figure 6.6(b)

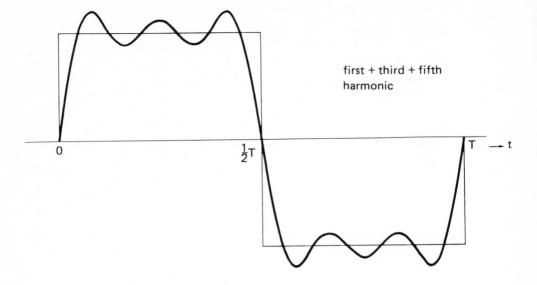

first + third + fifth
harmonic

Figure 6.6(c)

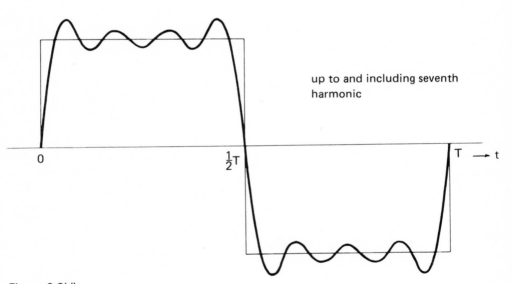

up to and including seventh
harmonic

Figure 6.6(d)

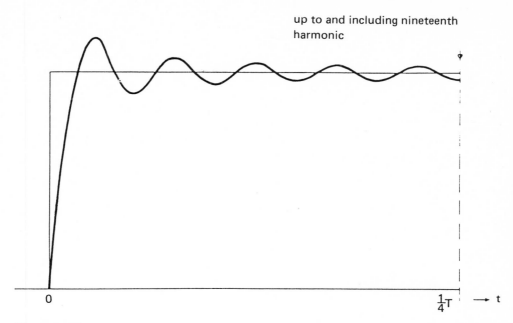

up to and including nineteenth
harmonic

Figure 6.6(e)

6.6 The Fourier series in complex form

We start with the series of (6.2) and write it as follows:

$$f(t) = A_0 + \sum_{n=1}^{\infty} (A_n \cos n\omega t + B_n \sin n\omega t).$$

Now

$$\cos n\omega t = \frac{1}{2} (e^{jn\omega t} + e^{-jn\omega t})$$

and

$$\sin n\omega t = \frac{1}{2j} (e^{jn\omega t} - e^{-jn\omega t}),$$

so

$$f(t) = A_0 + \sum_{n=1}^{\infty} (\frac{A_n}{2} + \frac{B_n}{2j}) e^{jn\omega t} + \sum_{n=1}^{\infty} (\frac{A_n}{2} - \frac{B_n}{2j}) e^{-jn\omega t}$$

$$= A_0 + \sum_{n=1}^{\infty} \frac{A_n - jB_n}{2} e^{jn\omega t} + \sum_{n=1}^{\infty} \frac{A_n + jB_n}{2} e^{-jn\omega t}.$$

Suppose

$$C_0 = A_0,$$

$$C_n = \frac{A_n - jB_n}{2} \qquad (n \neq 0),$$

$$C_{-n} = \frac{A_n + jB_n}{2} = C_n^* \qquad (n \neq 0),$$

(6.8)

then the following arises

$$f(t) = C_0 + \sum_{n=1}^{\infty} C_n e^{jn\omega t} + \sum_{n=1}^{\infty} C_{-n} e^{-jn\omega t}$$

$$= C_0 + \sum_{n=1}^{\infty} C_n e^{jn\omega t} + \sum_{n=-\infty}^{-1} C_n e^{jn\omega t}.$$

So

$$f(t) = \sum_{n=-\infty}^{\infty} C_n e^{jn\omega t}.$$

(6.9)

From (6.4) and (6.8) it follows that:

$$C_0 = \frac{1}{T} \int_0^T f(t) \, dt$$

and from (6.5), (6.6) and (6.8) for $n \neq 0$:

$$C_n = \frac{A_n - jB_n}{2} = \frac{1}{T} \{ \int_0^T f(t) \cos n\omega \, dt - j \int_0^T f(t) \sin n\omega t \, dt \}$$

$$= \frac{1}{T} \int_0^T f(t) e^{-jn\omega t} \, dt,$$

whereas

$$C_{-n} = \frac{A_n + jB_n}{2} = \frac{1}{T} \{ \int_0^T f(t) \cos n\omega t \, dt + j \int_0^T f(t) \sin n\omega t \, dt \}$$

$$= \frac{1}{T} \int_0^T f(t) e^{jn\omega t} \, dt,$$

so that in general it holds for all n (also $n = 0$) that:

$$C_n = \frac{1}{T} \int_0^T f(t) e^{-jn\omega t} \, dt.$$

(6.10)

Starting with the complex form we can find the real form with, see (6.8):

$$A_0 = C_0,$$

$$A_n = C_n + C_n^* = 2 \operatorname{Re} C_n, \qquad (6.11)$$

$$B_n = j(C_n - C_n^*) = -2 \operatorname{Im} C_n.$$

The formulas (6.9) and (6.10) are the basis of the *Fourier transformation*. We shall not discuss it any further.

6.7 The finite Fourier series

In practice one often wants to cut off a Fourier series after a finite number of terms, but then the question arises of how to choose the amplitudes of the harmonics so that the series is a maximal approach of the given function. A proper choice is the method of *least squares*:

$$\frac{1}{T} \int_0^T \{f(t) - g(t)\}^2 \, dt < \epsilon. \qquad (6.12)$$

In this formula f(t) is the given function, g(t) the finite Fourier series and ϵ a number as small as possible. The *square* of the difference is chosen because by doing so both positive and negative differences give a positive contribution.

As an example we shall take the function f(t) of Figure 6.3 and approach it with the first harmonic g(t) only (see Figure 6.7).

We set $g(t) = B_1 \sin \omega t$, of which B_1 has to be found.
So we have to make

$$\eta = \frac{1}{T} \int_0^T \{f(t) - g(t)\}^2 \, dt$$

as small as possible. Consequently

$$\tfrac{1}{2}\eta = \frac{1}{T} \int_0^{\frac{1}{2}T} (1 - B_1 \sin \omega t)^2 \, dt$$

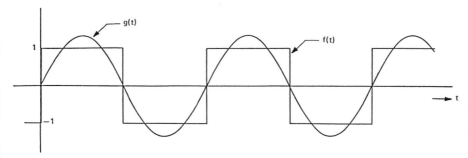

Figure 6.7

$$= \frac{1}{T} \int_0^{\frac{1}{2}T} \{1 - 2B_1 \sin \omega t + \frac{B_1^2}{2} (1 - \cos 2\omega t)\} \, dt$$

$$= \frac{1}{2} + [\frac{2B_1}{\omega T} \cos \omega t \;]_0^{\frac{1}{2}T} + \frac{B_1^2}{4} = \frac{1}{2} - \frac{2B_1}{2\pi} - \frac{2B_1}{2\pi} + \frac{B_1^2}{4}$$

$$= \frac{1}{2} - \frac{4B_1}{2\pi} + \frac{B_1^2}{4} = \frac{2\pi - 8B_1 + \pi B_1^2}{4\pi} = \frac{A}{4\pi}$$

with $A = \pi B_1^2 - 8B_1 + 2\pi$.

This expression is minimal if its derivative is zero, so

$$\frac{dA}{dB_1} = 2\pi B_1 - 8 = 0 \; \text{ so } \; B_1 = \frac{4}{\pi}.$$

We find the same amplitude as in the infinite series.
We are now going to prove this in general and will use the finite series in complex form:

$$g(t) = \sum_{n=-N}^{N} C_n e^{jn\omega t}. \tag{6.13}$$

The given function is f(t), the approach is g(t), so we have to make

$$\eta = \frac{1}{T} \int_0^T \{f(t) - g(t)\}^2 \, dt$$

as small as possible. We find

$$\eta = \frac{1}{T} \int_0^T \{f(t)\}^2 \, dt + \frac{1}{T} \int_0^T \{g(t)\}^2 \, dt - \frac{1}{T} \int_0^T 2f(t) \cdot g(t) \, dt.$$

We shall first examine the second integral.
The integrand is

$$\{g(t)\}^2 = \{ \sum_{n=-N}^{N} C_n e^{jn\omega t}\}^2 = \sum_{n=-N}^{N} \sum_{m=-N}^{N} C_n C_m e^{j(n+m)\omega t}.$$

If $n + m \neq 0$ then the integral over one period is zero.
If $n + m = 0$ for that integral we find

$$\sum_{n=-N}^{N} \sum_{m=-N}^{N} C_n C_m T = T \sum_{n=-N}^{N} C_n C_{-n} = T \sum_{n=-N}^{N} C_n C_n^* = T \sum_{n=-N}^{N} |C_n|^2.$$

We shall now turn our attention to the last integral.
The integrand is

$$2f(t)g(t) = 2f(t) \sum_{n=-N}^{N} C_n e^{jn\omega t} = f(t) \{ \sum_{n=-N}^{N} C_n e^{jn\omega t} + \sum_{n=N}^{-N} C_n e^{jn\omega t} \}.$$

In the second sum the sequence of the terms of that sum has been turned around. We now have

$$2f(t)g(t) = f(t) \{ \sum_{n=-N}^{N} C_n e^{jn\omega t} + \sum_{n=-N}^{N} C_{-n} e^{-jn\omega t} \}$$

$$= f(t) \{ \sum_{n=-N}^{N} C_n e^{jn\omega t} + \sum_{n=-N}^{N} C_n^* e^{-jn\omega t} \}.$$

Thus:

$$\eta = \frac{1}{T} \int_0^T \{f(t)\}^2 \, dt + \sum_{n=-N}^{N} |C_n|^2 - \frac{1}{T} \int_0^T f(t) \sum_{n=-N}^{N} C_n e^{jn\omega t} \, dt - \frac{1}{T} \int_0^T f(t) \sum_{n=-N}^{N} C_n^* e^{-jn\omega t} \, dt,$$

so

$$\eta = \frac{1}{T} \int_0^T \{f(t)\}^2 \, dt + \sum_{n=-N}^{N} \{|C_n|^2 - \frac{C_n}{T} \int_0^T f(t) e^{jn\omega t} \, dt - \frac{C_n^*}{T} \int_0^T f(t) e^{-jn\omega t} \, dt \}.$$

Now consider the next expression:

$$H = \sum_{-N}^{N} \{ [C_n - \frac{1}{T} \int_0^T f(t) e^{-jn\omega t} \, dt][C_{-n} - \frac{1}{T} \int_0^T f(t) e^{jn\omega t} \, dt] \}.$$

To work this out:

$$H = \sum_{-N}^{N} \{ |C_n|^2 - \frac{C_n}{T} \int_0^T f(t) e^{jn\omega t} \, dt - \frac{C_n^*}{T} \int_0^T f(t) e^{-jn\omega t} \, dt$$

$$+ \frac{1}{T^2} \int_0^T f(t) e^{-jn\omega t} \, dt \int_0^T f(t) e^{jn\omega t} \, dt \}.$$

The last two integrands are each other's conjugate complex, so for that product we may write:

$$\frac{1}{T^2} \left| \int_0^T f(t) e^{-jn\omega t} \, dt \right|^2.$$

So

$$H - \frac{1}{T^2} \sum_{n=-N}^{N} \left| \int_0^T f(t) e^{-jn\omega t} \, dt \right|^2$$

$$= \sum_{-N}^{N} \{|C_n|^2 - \frac{C_n}{T} \int_0^T e^{jn\omega t} dt - \frac{C_n^*}{T} \int_0^T f(t) e^{-jn\omega t} dt,$$

with which we find:

$$\eta = \frac{1}{T} \int_0^T \{f(t)\}^2 dt + H - \frac{1}{T^2} \sum_{n=-N}^{N} \left| \int_0^T f(t) e^{-jn\omega t} dt \right|^2.$$

Now H consists of the product of two series of which the terms are each others conjugate complex by twos, so

$$H = \sum_{-N}^{N} \left| C_n - \frac{1}{T} \int_0^T f(t) e^{-jn\omega t} dt \right|^2.$$

Thus η consists of terms that are not negative. Only H is dependent on C_n so that we can minimise η in the choice of C_n by making H zero.
In this way we obtain

$$C_n = \frac{1}{T} \int_0^T f(t) e^{-jn\omega t} dt,$$

which is fully consistent with (6.10).

With this we have proved that the method of least squares for a finite Fourier series results in the same terms as those that are present in the infinite Fourier series.

6.8 Reflections on symmetry

In the calculation of the example of Figure 6.2 we saw that certain terms in the Fourier series are not present. In order to investigate we first determine the d.c. term and subtract it from the given function so that a function f(t) arises, of which the average value is zero. We now assume that f(t) contains one or more of the following symmetries.

a. Even-function symmetry

For a function with even-function symmetry it holds that:

$$f(-t) = f(t). \tag{6.14}$$

An example is shown in Figure 6.8. The uninterrupted line is the function f(t).
Now consider (6.6):

$$B_n = \frac{2}{T} \int_0^T f(t) \sin n\omega t \, dt.$$

The function $\sin n\omega t$ has been drawn with a dotted line in the figure (for a random positive n).
We see that, for the negative values of t, the product $f(t) \sin n\omega t$ is the opposite of the

product for positive values of t, so that the integral over one period is zero.
So $B_n = 0$ for all n. *There will be no sine terms.*

b. Odd-function symmetry

For a function with odd-function symmetry it holds that:

$$f(-t) = -f(t). \tag{6.15}$$

In Figure 6.9 such a function has been drawn. In the plot cos nωt has also been drawn for positive n.

We see that $A_n = \dfrac{2}{T} \int\limits_0^T f(t) \cos n\omega t \, dt$ is zero.

So there will be no cosine terms.

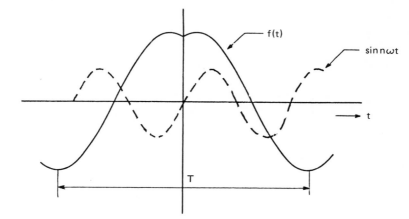

Figure 6.8

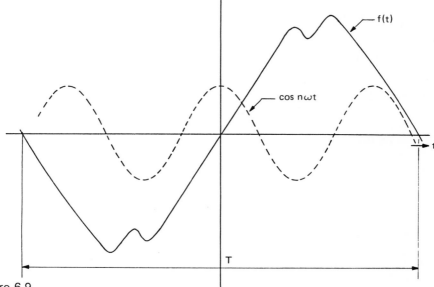

Figure 6.9

c. Half-wave symmetry

If a function has this symmetry it holds that:

$$f(t + \tfrac{1}{2} T) = -f(t). \tag{6.16}$$

(See Figure 6.10).

If we reverse the function over the first half-period and subsequently shift it half a period, the result will cover the second half-period.

The second harmonic of the sine terms has also been drawn in Figure 6.10. We see that the product of that harmonic with the function $f(t)$ and integrated over one period results in zero exactly.

This also happens with the second harmonic of the cosine terms and also with the fourth, sixth, etc. harmonics.

Now consider Figure 6.11.

From Figure 6.11 it appears that the integral over one period of the product $f(t) \cdot B_1 \sin \omega t$ is not zero. All odd harmonics give non-zero results. The conclusion is that, if a function has half-wave symmetry, there *will only be odd harmonics*.

The even and odd symmetry cannot occur at the same time, because all harmonics will be zero. However, the even and half-wave symmetry may occur simultaneously. In this case there will only be odd cosine terms. The odd and half-wave symmetry may also occur simultaneously, in which case there only are odd sine terms. This was the case in the example of Figure 6.3.

6.9 The effective value of a Fourier series

The effective value of a function $i(t)$ with period T is, see (2.8);

$$I_{eff} = \sqrt{\frac{1}{T} \int_0^T i^2 \, dt.}$$

If $i(t)$ can be represented by a finite Fourier series it holds that:

$$i = \sum_{-N}^{N} C_n \, e^{jn\omega t},$$

with which we find

$$I_{eff}^2 = \frac{1}{T} \int_0^T \{\sum_{-N}^{N} C_n \, e^{jn\omega t}\}^2 \, dt = \frac{1}{T} \int_0^T \sum_{-N}^{N} \sum_{-N}^{N} C_r C_s \, e^{j(r+s)\omega t} \, dt,$$

in which r and s range from –N to +N.

We now consider an arbitrary term of the integrand. We then get

$$\int_0^T C_r C_s \, e^{j(r+s)\omega t} \, dt = [\frac{C_r C_s}{j(r+s)\omega} \, e^{j(r+s)\omega t} \,]_0^T = 0$$

with the conditions: $r + s \neq 0$.

If $r + s = 0$ we find:

$$\int_0^T C_r\, C_s\, dt = C_r\, C_s\, T.$$

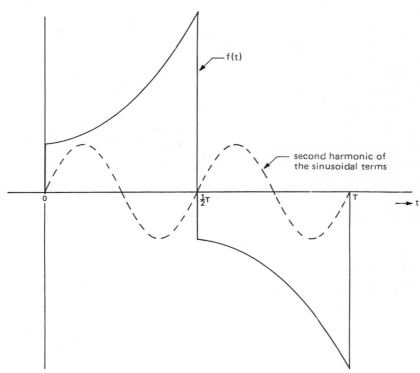

Figure 6.10

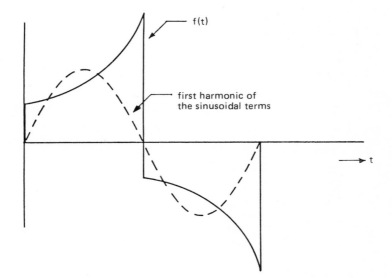

Figure 6.11

So the result, by applying (6.8), is:

$$I_{eff}^2 = \sum_{-N}^{N} C_n C_{-n} = C_0 + 2\sum_{n=1}^{N} C_n C_n^*.$$

So

$$I_{eff}^2 = A_0^2 + \sum_{n=1}^{N} \frac{A_n^2}{2} + \sum_{n=1}^{N} \frac{B_n^2}{2}. \tag{6.17}$$

The conclusion is that the effective value of a function equals the square root of the sum of the squares of the effective values of the separate harmonics.

If there is a current i(t) in a resistor R then the dissipated power is $P = I_{eff}^2 R$, so the powers created by all harmonics separately may be added to find the total power.

However, we have to realise that the law of superposition of powers does not generally hold, but does hold if the frequencies of the separate terms are different, which, of course, is the case here.

6.10 Problems

Find the Fourier series of the 12 following periodical functions.

Also draw the amplitude spectrum.

6.1

6.3

6.2

6.4

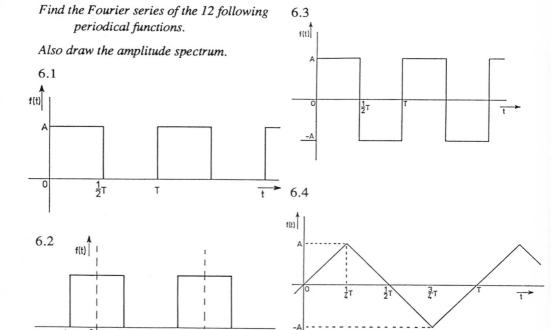

6.5

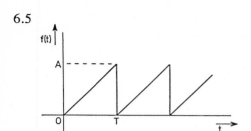

6.6

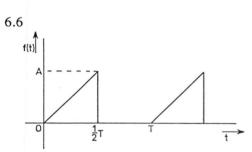

6.7

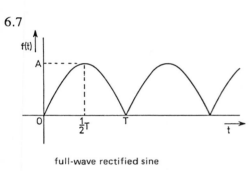

full-wave rectified sine

6.8

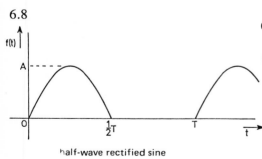

half-wave rectified sine

6.9

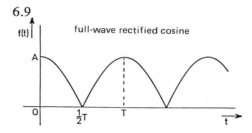

full-wave rectified cosine

6.10

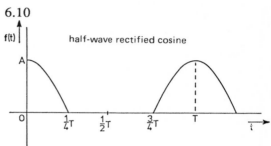

half-wave rectified cosine

6.11

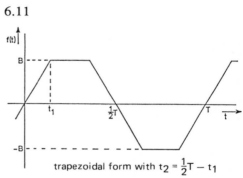

trapezoidal form with $t_2 = \frac{1}{2}T - t_1$

6.12.

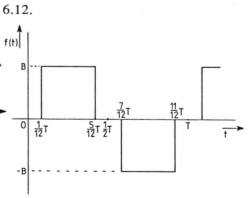

6.13 Find the period of
$$i = 5 \cos 3t + 5 \cos 7t \text{ A}.$$

6.14 Find the first and higher harmonics
of $v = (8 + 3 \cos 2t) \cos 100t$ V.

6.15 Determine, by using the symmetry
reflections, which terms in the
Fourier series will be present in the
following functions:

a.

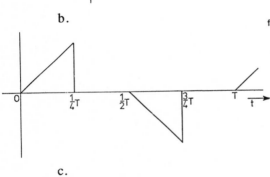

b.

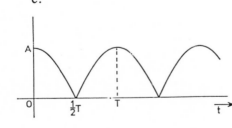

c.

d.

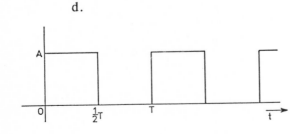

e.

6.16 a. Find the component A_0 of the
Fourier series of this periodical
function without using an
integral.

b. Determine, using the symmetry
reflections, which terms will be
present.

c. Find the Fourier series to the
fourth term.

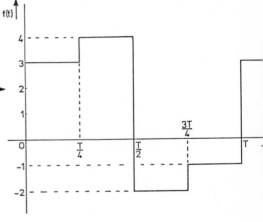

6.17 A periodical function f(t), with
period T, can be written as a Fourier
series:

$$f(t) = A_0 +$$

$$\sum_{n=1}^{\infty} A_n \cos n\omega_1 t + \sum_{n=1}^{\infty} B_n \sin n\omega_1 t.$$

All A_n, B_n and A_0 are constants,
$n = 1,2,3,\dots$ en $\omega_1 = \frac{2\pi}{T}$.

a. Give a sketch of a periodical
function f(t) for which it holds

that $A_0 \neq 0$ and $B_n = 0$ for all n.
Explain your answer.

b. Give a sketch of a function f(t)
for which it holds that $A_0 = 0$,
$A_n = 0$ for all n and $B_n = 0$ for all
even values of n.

6.18 Find the Fourier series.

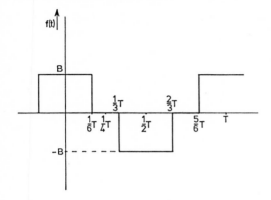

7

The complex frequency

7.1 Introduction

In practice there are voltages and currents as functions of time, of which the shape differs from the shapes we have met so far. For instance, if we discharge a charged capacitor the capacitor voltage will be a particular function of time, also depending on the network connected to the capacitor. In this chapter we shall investigate those special functions.

7.2 A capacitor discharges over a resistor

Consider Figure 7.1.

The switch S is a *make contact*, drawn in an open position and which will be closed at a certain moment. For the moment of closing we choose $t = 0$.

We assume that the capacitor has been charged to the value $v_C = V$ before operating the switch, i.e. $t < 0$.

Next we examine the behaviour of $v_C(t)$ when S is closed, i.e. we determine $v_C(t)$ for $t \geq 0$.

In Figure 7.2 the situation for $t \geq 0$ is shown.

Using Ohm's law we find:

$$v_C = Ri,$$

and with the capacitor formula (note the minus):

$$i = -C \frac{dv_C}{dt}.$$

From these two equations we get

$$RC \frac{dv_C}{dt} + v_C = 0.$$

This is a *linear differential equation of the first order*.

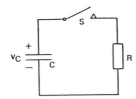

Figure 7.1

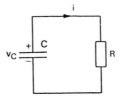

Figure 7.2

It is also called *homogenous* because the right-hand part is zero.
In order to find the solution (Euler) we set:

$$v_C = A\ e^{\lambda t} \qquad \text{(A is constant).}$$

We then get

$$\frac{dv_C}{dt} = \lambda\ A\ e^{\lambda t}.$$

If we substitute this in the differential equation;

$$RC\lambda\ A\ e^{\lambda t} + A\ e^{\lambda t} = 0.$$

So

$$\lambda RC + 1 = 0.$$

This is called the *characteristic equation*, which is an algebraic equation of the *first degree*.
Note that the characteristic equation follows from the differential equation by replacing the first derivative by λ and the zero derivative by 1.
We now solve the characteristic equation and find the *root*:

$$\lambda = -\frac{1}{RC}\ .$$

We find

$$v_C = A\ e^{-\frac{1}{RC}t}.$$

In this deduction we have not used the initial voltage V. It is called the *initial condition* of the differential equation. At $t = 0$, $v_C = V$ while from the above solution $v_{C(t=0)} = A$.
So

$$A = V,$$

with which we finally get

$$v_C = V\ e^{-\frac{t}{RC}} \qquad \text{for } t \geq 0$$

and

$$v_C = V \qquad \text{for } t < 0.$$

This function is shown in Figure 7.3.

Note
In Chapter 9 we shall further discuss the problems about the point $t = 0$.

For $t \to \infty$ we have $v_C \to 0$.
Note that the capacitor *voltage* is *continuous* at $t = 0$.

In the discussion of the Bode diagrams we have introduced the concept 'time constant' (see (3.5)):

$$\tau = RC.$$

At the moment $t = \tau$ we have $v_C = V\,e^{-1}$, i.e. the capacitor voltage has decreased to about 37 % of its original value after τ seconds.

The slope of the tangent is $(\frac{dv_C}{dt})_{t=0} = (-\frac{V}{RC}e^{-\frac{t}{RC}})_{t=0} = -\frac{V}{RC}$ in $t = 0$, which means that the tangent in t=0 passes through the point $t = \tau$.

Now compare the above with the complex expression of a voltage, as introduced in Chapter 2:

$$v = |V|\cos(\omega t + \varphi) = \mathrm{Re}\ |V|\ e^{j(\omega t + \varphi)} = \mathrm{Re}\ |V|\ e^{j\varphi}\ e^{j\omega t}$$

$$= \mathrm{Re}\ V\ e^{j\omega t} = \mathrm{Re}\ V\ e^{\lambda t}$$

with $V = |V|\ e^{j\varphi}$ and $\lambda = j\omega$.

In Chapter 2 a sinusoidal voltage was involved and the exponent of the e-power was imaginary.

Now, in this example of a transient response, a power of e (of which the exponent is real) is involved.

The exponent can also be complex, as will be shown in the following section.

7.3 A capacitor discharges over an inductor with series resistor

As shown in Figure 7.4, the capacitor voltage is V again for $t < 0$. At $t = 0$, S is closed. For $t > 0$ we find:

$$v_C = L\frac{di}{dt} + Ri$$

and

$$i = -C\frac{dv_C}{dt}\ .$$

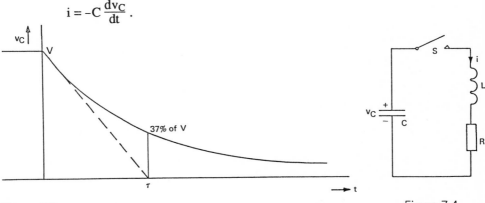

Figure 7.3

Figure 7.4

Subsequently:

$$LC\frac{d^2v_C}{dt^2} + RC\frac{dv_C}{dt} + v_C = 0.$$

This is a *second*-order, linear, homogenous differential equation. In order to solve it we once more set:

$$v_C = A\,e^{\lambda t} \qquad A = \text{constant.}$$

So

$$\frac{dv_C}{dt} = \lambda\,A\,e^{\lambda t}$$

and

$$\frac{d^2v_C}{dt^2} = \lambda^2 A\,e^{\lambda t}.$$

Substitution into the differential equation gives

$$LC\lambda^2 A\,e^{\lambda t} + RC\lambda\,A\,e^{\lambda t} + A\,e^{\lambda t} = 0.$$

So

$$\lambda^2 LC + \lambda RC + 1 = 0.$$

Note that the n-th derivative turns into λ^n with n = 0,1,2.
This is the characteristic equation, which is of the *second* degree here.
It has two roots, λ_1 and λ_2:

$$\lambda_{1,2} = \frac{-RC \pm \sqrt{R^2\,C^2 - 4LC}}{2LC},$$

consequently

$$\lambda_{1,2} = -\frac{R}{2L} \pm \sqrt{\frac{R^2}{4L^2} - \frac{1}{LC}}.$$

Therefore

$$v_C = A_1\,e^{\lambda_1 t}$$

is a solution that can be checked easily: If we substitute the above into the differential equation the left-hand part becomes

$$LC\,\lambda_1^2\,A_1\,e^{\lambda_1 t} + RC\lambda_1\,A_1\,e^{\lambda_1 t} + A_1\,e^{\lambda_1 t} = (LC\,\lambda_1^2 + RC\lambda_1 + 1)\,A_1\,e^{\lambda_1 t}$$

and this is zero because the expression between parentheses just happens to be the characteristic equation for the root $\lambda = \lambda_1$.
So

$$v_C = A_2 \, e^{\lambda_2 t}$$

is also a solution.

Finally, also the sum of both expressions

$$v_C = A_1 \, e^{\lambda_1 t} + A_2 \, e^{\lambda_2 t}$$

is a solution because substitution in the left-hand part of the differential equation gives:

$$LC \, (\lambda_1^2 \, A_1 \, e^{\lambda_1 t} + \lambda_2^2 \, A_2 \, e^{\lambda_2 t}) + RC(\lambda_1 \, A_1 \, e^{\lambda_1 t} + \lambda_2 \, A_2 \, e^{\lambda_2 t}) + A_1 \, e^{\lambda_1 t} + A_2 \, e^{\lambda_2 t}.$$

In another sequence

$$(LC \, \lambda_1^2 + RC\lambda_1 + 1) \, A_1 \, e^{\lambda_1 t} + (LC \, \lambda_2^2 + RC\lambda_2 + 1) \, A_2 \, e^{\lambda_2 t}.$$

Both expressions between parentheses are zero, so that the result is zero.
Now there are three possibilities:

1. $\dfrac{R^2}{4L^2} > \dfrac{1}{LC}$.

2. $\dfrac{R^2}{4L^2} < \dfrac{1}{LC}$.

3. $\dfrac{R^2}{4L^2} = \dfrac{1}{LC}$.

Choose the second one and take the values:

$$C = \frac{5}{26} \ F, L = 5 \ H, R = 2 \ \Omega.$$

We then find the following differential equation:

$$\frac{25}{26} \frac{d^2 v_C}{dt^2} + \frac{5}{13} \frac{d v_C}{dt} + v_C = 0$$

and the characteristic equation:

$$25 \, \lambda^2 + 10 \, \lambda + 26 = 0$$

with the roots

$$\lambda_1 = -0.2 + j,$$

$$\lambda_2 = -0.2 - j.$$

so the solution is

$$v_C = A_1 \, e^{(-0.2 + j)t} + A_2 \, e^{(-0.2 - j)t} \qquad \text{for } t \geq 0$$

and we see that it does indeed result in a complex value of the exponent of the power of e. The roots λ_1 and λ_2 found are called *complex frequencies*.
We shall now determine A_1 and A_2.

For $t = 0$ we have $v_C = V$, so $V = A_1 + A_2$; and for $t = 0$ the current $i = 0$.
So

$$\frac{dv_C}{dt} = 0 \qquad \text{therefore} \qquad \lambda_1 A_1 + \lambda_2 A_2 = 0.$$

From this

$$A_1 = \frac{V}{2} (1 - 0.2j)$$

and

$$A_2 = \frac{V}{2} (1 + 0.2j) = A_1^* .$$

With that result the solution becomes:

$$v_C = \frac{V}{2} e^{-0.2\,t} \{(1 - 0.2j)\, e^{jt} + (1 + 0.2j)\, e^{-jt}\}$$

$$= \frac{V}{2} e^{-0.2\,t} (\cos t + j \sin t - 0.2\, j \cos t + 0.2 \sin t + \cos t - j \sin t$$

$$+ \, 0.2j \cos t + 0.2 \sin t)$$

$$= \frac{V}{2} e^{-0.2\,t} (2 \cos t + 0.4 \sin t)$$

$$= \frac{V}{2} e^{-0.2\,t} \cdot 2.04 \cos (t - 11.3°)$$

$$= 1.02\, V\, e^{-0.2\,t} \cos (t - 11.3°)\ \text{volt} \qquad \text{for} \quad t \geq 0$$

and

$$v_C = V \qquad \text{for} \quad t < 0.$$

In general it is of the form:

$$v_C = V\, e^{\sigma t} \cos (\omega t + \varphi),$$

to which the complex frequency

$$\lambda = \sigma + j\omega$$

and its conjugate belong.
In our example we have $\sigma = -0.2$, which is smaller than zero. This was to be expected because the network in our example is passive and therefore the capacitor voltage will become zero in the long run.
In Figure 7.5 the voltage v_C has been given as a function of time, where $V = 5$ V has been chosen. If $\sigma = 0$:

$$v_C = V \cos (\omega t + \varphi).$$

This can be achieved by choosing the resistance R zero. We then have an undamped sinusoidal voltage of a form that we have discussed thoroughly in Chapters 2 and 3.

If the network contains active components (transactors) we can have $\sigma > 0$. The voltages and currents will then increase unlimited (see Figure 7.6).

Note

As has already been noted we have not thoroughly examined in the above what happens at the moment that $t = 0$.

We shall discuss this delicate question in Chapter 9.

7.4 The complex frequency plane

In Figure 7.7 we have mentioned the various cases in the so-called *complex frequency plane*.

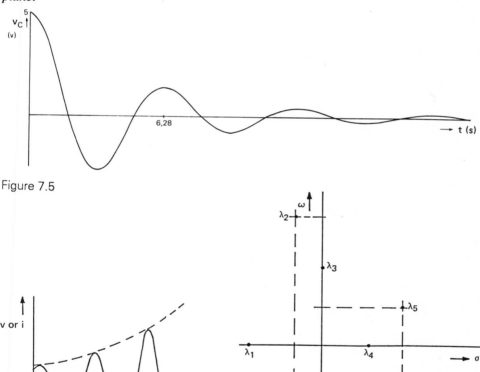

Figure 7.5

Figure 7.6

Figure 7.7

If the roots are situated in the *left-hand half-plane* (LHP) a *damped* oscillation is involved. The point λ_1 lies on the negative real axis and represents a damped function, as was discussed in Section 7.2. The points λ_2 and λ_2^* represent damped oscillations, just as the function of Figure 7.5. The points λ_3 and λ_3^* together represent an undamped vibration. Point λ_4 is an *increasing power of e*, whereas λ_5 and λ_5^* mean a sinusoidal function with *increasing amplitude*.

Finally, the origin is the *d.c.*-case.

Summarising, we may say that the general form of a damped or undamped voltage is

$$v = |V|\, e^{\sigma t} \cos(\omega t + \varphi). \tag{7.1}$$

If we set the complex voltage to

$$V = |V|\, e^{j\varphi}, \tag{7.2}$$

as in Chapter 2, we get

$$v = \operatorname{Re} |V|\, e^{\sigma t}\, e^{j(\omega t + \varphi)} = \operatorname{Re} |V|\, e^{j\varphi}\, e^{(\sigma + j\omega)t}.$$

With

$$\lambda = \sigma + j\omega \tag{7.3}$$

we find

$$v = \operatorname{Re}(V\, e^{\lambda t}). \tag{7.4}$$

So we can regard (2.24) as a special case of (7.4).

7.5 Extension of the meaning of impedance

Now consider an inductor through which a damped sinusoidal current flows (see Figure 7.8). The current is

$$i = |I|\, e^{\sigma t} \cos(\omega t + \varphi), \tag{7.5}$$

so the complex current added is

$$I = |I|\, e^{j\varphi}. \tag{7.6}$$

The inductor voltage is

$$v = L\frac{di}{dt} = \sigma L\, |I|\, e^{\sigma t} \cos(\omega t + \varphi) - \omega L\, |I|\, e^{\sigma t} \sin(\omega t + \varphi)$$

$$= \sigma L|I|\, e^{\sigma t} \cos(\omega t + \varphi) + \omega L\, |I|\, e^{\sigma t} \cos(\omega t + \varphi + \frac{\pi}{2}).$$

Using linearity and the rule of superposition we can add a complex voltage to each of both terms:

$$V = \sigma L|I|\, e^{j\varphi} + \omega L\, |I|\, e^{j(\varphi + \frac{\pi}{2})},$$

so

$$V = \sigma L|I| \, e^{j\varphi} + j\omega L \, |I| \, e^{j\varphi},$$

or

$$V = (\sigma + j\omega) \, L \, |I| \, e^{j\varphi},$$

with which we find

$$V = \lambda LI. \tag{7.7}$$

So the formula $V = j\omega LI$ (2.26) previously derived is also a particular case of (7.7). The relation between voltage and current is again called impedance.

$$Z = \lambda L. \tag{7.8}$$

The time function (7.5) follows from the complex expression (7.6) with

$$i = \operatorname{Re} I \, e^{\lambda t}. \tag{7.9}$$

Similarly

$$v = \operatorname{Re} V \, e^{\lambda t} \tag{7.10}$$

follows from (7.7), because

$$\operatorname{Re} V \, e^{\lambda t} = \operatorname{Re} \lambda LI \, e^{\lambda t} = \operatorname{Re} (\sigma + j\omega) \, L|I| \, e^{j\varphi} \, e^{j\omega t} \, e^{\sigma t}$$

$$= e^{\sigma t} \, L|I| \operatorname{Re} \{(\sigma + j\omega) \, e^{j(\omega t + \varphi)}\}$$

$$= e^{\sigma t} \, L|I| \operatorname{Re} \{(\sigma + j\omega)(\cos(\omega t + \varphi) + j \sin (\omega t + \varphi))\}$$

$$= e^{\sigma t} \, L|I| \, \{\sigma \cos(\omega t + \varphi) - \omega \sin (\omega t + \varphi)\}$$

and this is exactly the expression we found above.

From (7.8) it follows that the impedance of an inductor can now be complex. (If $\omega = 0$ and $\sigma \neq 0$ the impedance of the inductor is even real!)

Similarly for the capacitor we can derive:

$$I = \lambda CV, \tag{7.11}$$

with the admittance

$$Y = \lambda C \tag{7.12}$$

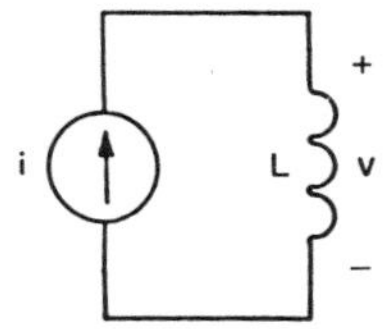

Figure 7.8

and therefore with impedance

$$Z = \frac{1}{\lambda C}.$$ (7.13)

For completeness' sake we also mention the relation for the resistor.
With $i = |I| e^{\sigma t} \cos (\omega t + \varphi)$, $v = R |I| e^{\sigma t} \cos (\omega t + \varphi)$. If we add the complex current I to i the added complex voltage is

$$V = RI.$$ (7.14)

For all three elements we again find linear relations.

7.6 Kirchhoff's laws if the frequency is complex

We shall now prove that Kirchhoff's current law maintains its validity if the amplitudes of the currents are powers of e and the frequency can therefore be introduced as a complex quantity.

The current law is

$$\sum_{n=1}^{b} i_n = 0 \qquad \text{for all time t.}$$

If

$$i_n = |I_n| e^{\sigma t} \cos (\omega t + \varphi_n), \qquad (n = 1,2,...,b).$$

then

$$\sum_{n=1}^{b} |I_n| e^{\sigma t} \cos (\omega t + \varphi_n) = 0 \qquad \text{for all t.}$$

For $t = 0$ we have

$$\sum_{n=1}^{b} |I_n| \cos \varphi_n = 0$$ (a)

and for $\omega t = \frac{1}{2} \pi$, so $t = \frac{\pi}{2\omega}$,

$$\sum_{n=1}^{b} |I_n| e^{\frac{\sigma \pi}{2\omega}} \sin \varphi_n = 0$$

thus also

$$\sum_{n=1}^{b} |I_n| \sin \varphi_n = 0.$$ (b)

Now multiply (b) by j and add that to (a).
We then get:

$$\sum_{n=1}^{b} |I_n| (\cos \varphi_n + j \sin \varphi_n) = 0,$$

so

$$\sum_{n=1}^{b} |I_n| e^{j\varphi_n} = 0,$$

or

$$\sum_{n=1}^{b} I_n = 0.$$

In a similar way one can derive the voltage law:

$$\sum_{m=1}^{l} V_m = 0.$$

The linearity and the validity of Kirchhoff's laws, as was also the case in Section 2.19, have an *important consequence* that all theorems, formulas and rules we used for d.c. remain valid.

Example 1 (Figure 7.9).
The impedance of this one-port is

$$Z = \frac{(\lambda L + R) \cdot \dfrac{1}{\lambda C}}{\lambda L + R + \dfrac{1}{\lambda C}},$$

so

$$Z = \frac{\lambda L + R}{\lambda^2 LC + \lambda RC + 1}.$$

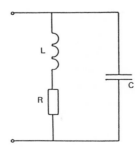

Figure 7.9

Example 2 (Figure 7.10).
The voltage relation of this one-port is

$$H = \frac{V_2}{V_1} = \frac{4 + \dfrac{5}{\lambda}}{\lambda + 4 + \dfrac{5}{\lambda}} = \frac{4\lambda + 5}{\lambda^2 + 4\lambda + 5}.$$

7.7 Poles and zeros

From the above it is evident that we find functions of the form

$$H(\lambda) = \frac{N(\lambda)}{D(\lambda)}.$$ (7.15)

$H(\lambda)$ is the impedance or the admittance of a one-port or a transfer function of a two-port. $N(\lambda)$ is the numerator and $D(\lambda)$ the denominator; both are polynomes in powers of λ with non-negative exponent.

The *zeros* of $H(\lambda)$ are those values of λ for which $H(\lambda) = 0$. These zeros are indicated by $z_1, z_2, \ldots$. We can find these by calculating the roots of the nominator, if it is set to zero. The *poles* of $H(\lambda)$ are those values of λ for which $H(\lambda) \to \infty$. They are indicated by $p_1, p_2, \ldots$ and can be found by calculating the roots of the denominator, if it is set to zero. In Example 2 of the preceding section there is one (finite) zero $z_1 = -5/4$, while there are two poles: $p_1 = -2 + j$ and $p_2 = -2 - j$.

Note

In both examples we also have $H(\lambda) = 0$ for $\lambda \to \infty$, so strictly speaking one zero is $z_2 = \infty$. A zero is then said to be in infinity. We shall not take these poles and zeros in infinity into account. This has certain advantages which shall become clear later.

We can indicate the poles and zeros of a certain function $H(\lambda)$ in the complex plane of $\lambda = \sigma + j\omega$. We then find the so-called *pole-zero plot*.
In Figure 7.11 this is shown for the last example.
We only draw the *finite* poles and zeros and not the poles and zeros in infinity.

The general form of a network function is

$$H(\lambda) = \frac{A_z \lambda^z + A_{z-1} \lambda^{z-1} + \ldots + A_0}{B_p \lambda^p + B_{p-1} \lambda^{p-1} + \ldots + B_0}.$$ (7.16)

Thus there are z zeros and p poles. The factors A_k $(k = 0, \ldots, z)$ and B_l $(l = 0, \ldots, p)$ are real, because the inductances, capacitances and resistances are real.
If we have found the poles and zeros from (7.16) (which may be difficult for a polynome of a degree larger than three) we can write:

$$H(\lambda) = K \frac{(\lambda - z_1)(\lambda - z_2) \ldots (\lambda - z_z)}{(\lambda - p_1)(\lambda - p_2) \ldots (\lambda - p_p)},$$ (7.17)

in which

$$K = \frac{A_z}{B_p}.$$ (7.18)

From this we see that the pole-zero plot does not fully determine the function $H(\lambda)$; if one gives the pole-zero plot, the value of K has to be given to determine $H(\lambda)$.

Because the factors in (7.16) are real, poles and the zeros only occur in conjugated pairs if they are not real.

This can be seen easily for a square form. Suppose the numerator is

$$N(\lambda) = (\lambda - z_1)(\lambda - z_2),$$

while we assume that z_1 and z_2 are complex. If we work out the right-hand part we find:

$$\lambda^2 - (z_1 + z_2)\lambda + z_1z_2,$$

in which $z_1 + z_2$ and z_1z_2 have to be real.
Both demands lead to

$$z_1 = z_2^*.$$

From

$$z_1 + z_2 = A$$

and

$$z_1z_2 = B,$$

in which A and B are real, we find

$$z_1 = \frac{A}{2} \pm \frac{1}{2}\sqrt{A^2 - 4B}$$

$$z_2 = \frac{A}{2} \mp \frac{1}{2}\sqrt{A^2 - 4B}.$$

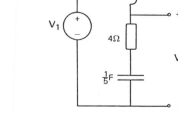

Figure 7.10

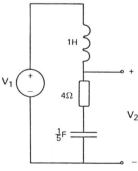

Figure 7.11

For $A^2 \geq 4B$ z_1 and z_2 are real. For $A^2 < 4B$

$$z_1 = \frac{A}{2} \pm \frac{1}{2}j\sqrt{4B - A^2}$$

$$z_2 = \frac{A}{2} \mp \frac{1}{2}j\sqrt{4B - A^2},$$

so that in all cases $z_1 = z_2^*$.

We can factorise a polynomial of a higher degree to square forms (plus a first-degree form if the degree is odd) so that the above supposition is indeed correct.

7.8 Frequency characteristics

We shall now trace how |H| and arg H as functions of the angular frequency ω can be determined from the pole-zero plot. If $\sigma = 0$ we have $\lambda = j\omega$ and so the values of the angular frequencies are on the positive imaginary axis.

From (7.17) it follows that for the modulus of the angular frequency ω_1:

$$|H(j\omega_1)| = K \frac{|j\omega_1 - z_1| \, |j\omega_1 - z_2| \ldots}{|j\omega_1 - p_1| \, |j\omega_1 - p_2| \ldots} . \tag{7.19}$$

Now $j\omega_1 - z$ is a complex number that can be represented by a vector from z to the imaginary number $j\omega_1$, which is the point ω_1 on the imaginary axis, see Figure 7.12.
In the same manner we find the vector, going from p to ω_1, for a pole p.
If we denote the *zero vectors* with A_z and the *pole vectors* with A_p then for (7.19) we find:

$$|H(j\omega_1)| = K \frac{\Pi \, |A_z|}{\Pi \, |A_p|} , \tag{7.20}$$

If there are no zeros, $\Pi \, |A_z| = 1$; if there are no poles, $\Pi \, |A_p| = 1$, in which Π means *product*.
If we call the arguments of A_z and A_p α_z and α_p respectively, then it follows from (7.17) that:

$$\arg \{H(j\omega_1)\} = \Sigma \, \alpha_z - \Sigma \, \alpha_p, \tag{7.21}$$

in which Σ means *sum*.
After all, the argument of a product is the sum of the arguments of the factors, whereas the arguments of the denominator get a negative sign.
K does not occur in (7.21) because arg K = 0.
If we do not take one point $\omega = \omega_1$, as in Figure 7.12, but let ω increase from 0 to infinity (along the imaginary axis) we get *rough* sketches of |H| and arg H both as functions of ω.
We note here that only the construction of vectors for the *finite* poles and zeros has to be done. Besides, the poles and the zeros at infinity would result in infinitely long vectors.

Example 1 (Figure 7.13).

Determine the pole-zero plot of the impedance Z and from that the frequency characteristics.

We find $Z = \dfrac{\lambda + 1}{\lambda + 2}$. The factor K of formula (7.17) is 1. There is one zero $z_1 = -1$. There is one pole $p_1 = -2$. The pole-zero plot is shown in Figure 7.14.

In addition, an arbitrary angular frequency $\omega_1 = 1.7$ rad/s has been indicated in the figure, while the pole and zero vectors have been drawn. For clarity the arrows have not been put at the end of the vectors, but half-way.

By measuring we find $|Z| = \dfrac{2.0}{2.6} = 0.77\ \Omega$ and arg $Z = 60° - 40° = 20°$.

Next we determine $|Z|$ and arg Z both as functions of ω from the pole-zero plot. Let ω move from $\omega = 0$ (the origin) to infinity along the positive imaginary axis.

For $\omega = 0$ we have $|Z| = \dfrac{1}{2}$. For each value $\omega \geq 0$ the length of the zero vector is smaller than that of the pole vector, so that the quotient is smaller than 1. For $\omega \to \infty$ the quotient is 1. Thus we find the plot of Figure 7.15.

For $\omega = 0$ we have arg Z = 0. For each value $\omega > 0$ the angle of the zero vector is larger

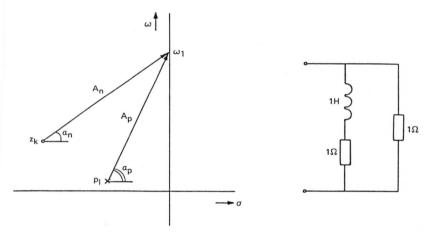

Figure 7.12

Figure 7.13

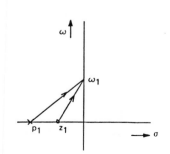

Figure 7.14

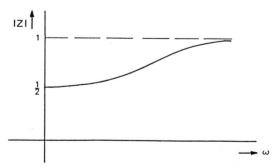

Figure 7.15

than the angle of the pole vector. For $\omega \to \infty$ we have arg $Z = 0$. Thus we find Figure 7.16.

We once more mention here that only rough sketches can be obtained in this way. The correct plots follow with

$$|Z| = \left| \frac{1 + j\omega}{2 + j\omega} \right|$$

and

$$\arg Z = \arg(1 + j\omega) - \arg(2 + j\omega).$$

Example 2 (Figure 7.17).

Examine the transfer function $H = \dfrac{I_2}{I_1}$.

With current division we find $H = \dfrac{\dfrac{1}{\lambda}}{\lambda + 1 + \dfrac{1}{\lambda}}$. So $H = \dfrac{1}{\lambda^2 + \lambda + 1}$. We see $K = 1$.

There are no zeros. Both poles are $p_1 = -\frac{1}{2} + \frac{1}{2} j \sqrt{3}$ and $p_2 = -\frac{1}{2} - \frac{1}{2} j \sqrt{3}$ (see Figure 7.18).

For $\omega = 0$ both vectors have a finite length. Initially the upper pole vector becomes smaller by an increasing frequency, later it becomes larger. In this way we get the rough sketch of $|H|$ (ω) (see Figure 7.19).

From $H = \dfrac{1}{1 - \omega^2 + j\omega}$ we see $|H| = 1$ for $\omega = 0$.

For $\omega = 0$, both angles α and β are equal in magnitude. Since α is negative the sum is zero.

With increasing frequency both angles become larger (α becomes less negative), so that the argument decreases. For $\omega \to \infty$ both angles are $+90°$ so that the total argument is $-180°$ (see Figure 7.20).

Example 3 (Figure 7.21).

Determine $Z(\lambda)$.

We find

$$Z(\lambda) = \frac{(5\lambda + 1) \cdot \dfrac{1}{4\lambda}}{5\lambda + 1 + \dfrac{1}{4\lambda}} = \frac{5\lambda + 1}{20\lambda^2 + 4\lambda + 1} . \text{ So } K = \frac{5}{20} = \frac{1}{4} .$$

There is one zero: $z_1 = -0.2$.

There are two poles: $p_1 = -0.1 + 0.2j$ and $p_2 = -0.1 - 0.2j$.

The pole-zero plot is shown in Figure 7.22.

First determine $|Z|$ (ω).

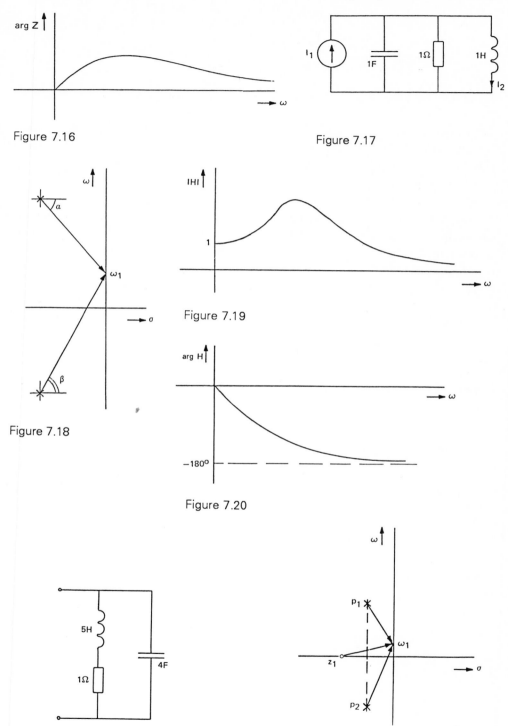

Figure 7.16

Figure 7.17

Figure 7.18

Figure 7.19

Figure 7.20

Figure 7.21

Figure 7.22

For $\omega = 0$, $|Z|$ has a finite value. From the function

$$Z(j\omega) = \frac{1 + 5j\omega}{1 - 20\omega^2 + 4j\omega}$$

we see that this value is one. (Measuring should lead to the value 4, but we must also take into account $K = 0.25$).

Going along the positive imaginary axis, $|Z|$ will have a maximum around p_1, after which the function will go to zero for $\omega \to \infty$. We thus get Figure 7.23. Further calculation shows that the maximum is at $\omega = 0.202$ rad/s.

We now determine arg Z as a function of ω.

For $\omega = 0$ we have arg $Z = 0$.

For a small value of ω (see ω_1 in Figure 7.22) the zero angle has increased, while both pole angles have hardly varied, owing to the greater distance. So arg $Z > 0$ for small ω. However, the total argument is $-90°$ for very large ω (viz. 180° for both poles and 90° for the zero). We thus find Figure 7.24.

The question arises for which value of ω the argument of Z is zero (ω_0).

We can find that from the expression $Z(j\omega)$:

$$Z(j\omega) = \frac{1 + 5j\omega}{1 - 20\omega^2 + 4j\omega},$$

so arg $Z = 0$ if arctan 5ω = arctan $\dfrac{4\omega}{1 - 20\omega^2}$. So, if $\omega = 0$ rad/s and if $5 = \dfrac{4}{1 - 20\omega^2}$; it

follows that $\omega = \omega_0 = 0.1$ rad/s.

7.9 The coincidence of poles and zeros

If two zeros coincide we have to take two coinciding vectors into account. This means that we have to square the length for the modulus and to double the angle for the argument. The same holds for two coinciding poles.

Example (Figure 7.25).

Determine $Z(\lambda)$. We find $Z(\lambda) = \dfrac{(\lambda + 2) \cdot \frac{1}{\lambda}}{\lambda + 2 + \frac{1}{\lambda}} = \dfrac{\lambda + 2}{\lambda^2 + 2\lambda + 1}$.

Zero: $z_1 = -2$.
Poles: $p_1 = -1$ and $p_2 = -1$ (see Figure 7.26).
Determine $|Z|$ and arg Z for $\omega = \omega_1 = 2$ rad/s.
By measuring we find

$$|Z| = \frac{2.8}{(2.2)^2} = 0.57 \ \Omega.$$

and

$$\arg Z = 45° - 2\cdot 63\tfrac{1}{2}° = -82°.$$

If a zero and a pole coincide they cancel each other. After all, we have equal factors in the numerator and in the denominator, which can both be neglected. However, this has consequences for the *order* of a network.

7.10 The order of a network

We can store energy in the capacitors and the inductors of a network. The order concerns the *capacitor voltages* and the *inductor currents* at a certain moment (t = 0). Together with the sources they cause branch voltages and branch currents in the network. If we make

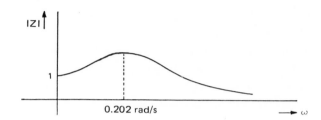

Figure 7.23

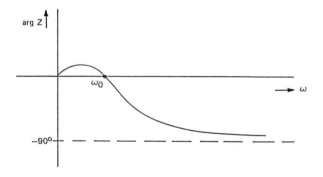

Figure 7.24

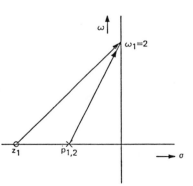

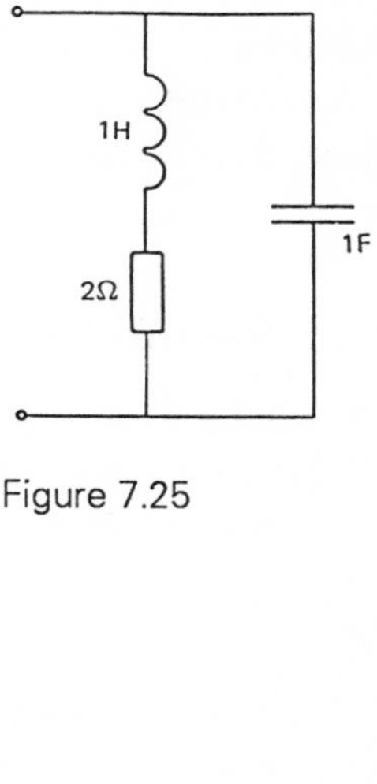

Figure 7.25

Figure 7.26

the source intensities zero, the resulting voltages and currents are called *natural oscillations*. The complexity of these natural oscillations depends on the *order* of the network, i.e. the number of the inductor currents plus the number of the capacitor voltages which can be chosen freely while the source intensities have been set to zero. Thus the network of Section 7.2 is of the first order, whereas the network of Section 7.3 is of the second order (both the capacitor voltage and the inductor current can be freely chosen). It must be emphasised that it concerns the inductor currents and the capacitor voltages at a certain *moment* while the *function* of those quantities as a function of time is not relevant. Thus if the inductor current has been chosen the inductor voltage is indefinite and (dual) if the capacitor voltage has been chosen the capacitor current is indefinite. The order of a network is not necessarily equal to the number of *reactive elements* (inductors and capacitors), which Figure 7.27 clarifies.

Both capacitor voltages are equal according to Kirchhoff's voltage law, therefore the order of this network is one. In Figure 7.28 the dual network is shown.

According to Kirchhoff's current law one inductor current can be chosen at random, due to which the other has been determined.

If there are more inductors and capacitors in a network, determination of the order is more difficult. If the network does not contain transformers one has to search for loops of capacitors and cut-sets of inductors.

After all, in a loop of capacitors all capacitor voltages except one can be chosen freely while the last capacitor voltage is fixed by the voltage law. Dual reasoning holds for inductors.

If the network contains transformers, determination of the order requires a more advanced theory that will not be treated here.

7.11 Natural oscillations of a one-port

The natural frequencies of a one-port (with resistors, inductors and capacitors) are also determined by the type of the input source. If the source is a voltage source we have to short-circuit the terminals for the examination of the natural oscillations. If the source is a current source we must use open terminals for the natural oscillations.

In Figure 7.29 the general form of a one-port is shown.

The impedance is given by $Z = \frac{V}{I}$. On the other hand, Z consists of a numerator and a denominator, that are both functions of λ.

So

$$Z = \frac{N}{D} = \frac{V}{I},$$ (7.22)

or

$$NI = DV.$$ (7.23)

If a voltage source has been connected to the one-port the voltage must be set to zero for the natural frequencies. If the polynome $D(\lambda)$ is finite, and therefore λ limited, we have:

$$NI = 0. \tag{7.24}$$

This means that I may differ from zero if $N = 0$ and therefore there may be a (complex) gate current.

This means that there are branch voltages and branch currents as functions of time in the one-port which all meet Kirchhoff's laws at the same moment and which are called natural oscillations. Thus

$$N(\lambda) = 0 \tag{7.25}$$

is the *characteristic equation*.
In a similar way one can derive that

$$D(\lambda) = 0 \tag{7.26}$$

is the characteristic equation if a current source is connected to the one-port. If we start with the admittance Y the equations (7.25) and (7.26) will change places because $Y = Z^{-1}$.

The above method is known as *immittance method* (immittance is the collective name of impedance and admittance).

Example 1

Consider the network of Figure 7.13. If there is a voltage source connected, we find as the characteristic equation $\lambda + 1 = 0$. This equation is of the first degree; the order of the network is one. Only the series resistor plays a role, because the parallel resistor is short-circuited. If a current source is connected we find $\lambda + 2 = 0$. Both resistances are in series now. The network is of the first order again.

Example 2

Consider the network of Figure 7.21. If the input is short-circuited, the order is one (the capacitor voltage cannot be chosen at random). The characteristic equation is $5\lambda + 1 = 0$. For open terminals the order is two, the characteristic equation is $20\lambda^2 + 4\lambda + 1 = 0$.

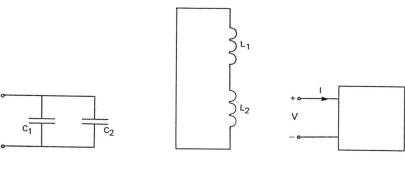

Figure 7.27   Figure 7.28   Figure 7.29

Example 3

Consider a Zobel network. We choose the network of Figure 3.54 which is also shown in Figure 7.30.

This network is of the second order, both for open and for short-circuited terminals.
The impedance is

$$Z = \frac{(4\lambda + 2)(\frac{1}{\lambda} + 2)}{4\lambda + \frac{1}{\lambda} + 4} = \frac{(4\lambda + 2)(2\lambda + 1)}{4\lambda^2 + 4\lambda + 1} = 2\,\Omega.$$

We find a numerator and a denominator both of degree zero. The cause is clear: there are poles and zeros that cancel each other. The characteristic equation for open terminals can be found as follows with a trick (see Figure 7.31).
We find

$$Z_1 = 4\lambda + 4 + \frac{1}{\lambda} = \frac{4\lambda^2 + 4\lambda + 1}{\lambda}.$$

• The numerator gives the characteristic equation if both upper nodes are short-circuited, which just happens to be the case for open terminals in Figure 7.30. So the result is that Figure 7.30 is a network of which the characteristic equation equals

$$4\lambda^2 + 4\lambda + 1 = 0.$$

for both open and short-circuited nodes.

Example 4 (see Figure 7.32).
We find

$$Z = \lambda + 2 + \frac{1}{4\lambda} = \frac{4\lambda^2 + 8\lambda + 1}{4\lambda}.$$

For open terminals the characteristic equation is the denominator set to zero, thus $4\lambda = 0$. This is a first degree equation. The inductor has open terminals, so the inductor current cannot be chosen at random. The capacitor determines the order. The natural frequency is not a real oscillation, but is the d.c. voltage of the capacitor, which originates at the terminals of the one-port.

For short-circuited terminals the characteristic equation is $4\lambda^2 + 8\lambda + 1 = 0$, which is a second degree equation, in accordance with the order of the network.

At this point we will return for a moment to the loss-less one-port of Figure 3.49(a). The numerator of $X(\omega)$, set to zero, is the characteristic equation of the network with *short-circuited* port, so that the zero of $X(\omega)$ can be determined in this manner.

With the so-called *graph theory* one can derive *matrix relations* for more complicated networks, with which it is possible to find exactly the characteristic equation. This is also possible with the *state equations* theory. These concepts are outside the scope of this book.

7.12 The location of the poles and zeros for an immitance

We assume that the one-port only contains resistors, inductors and capacitors. This means that the natural oscillations are damped or, at least, will not increase. This holds for both the one-port with open terminals and for one with short-circuited terminals.
Consider the impedance

$$Z = K \frac{(\lambda - z_1)(\lambda - z_2) \ldots}{(\lambda - p_1)(\lambda - p_2) \ldots}.$$

For *short-circuited* terminals the characteristic equation is

$$(\lambda - z_1)(\lambda - z_2) \ldots = 0$$

with the roots (i.e. the zeros of Z): $z_1, z_2, \ldots$
According to Euler's formula the solution is

$$v = A_1 e^{z_1 t} + A_2 e^{z_2 t} + \ldots$$

in which v is an arbitrary voltage in the network (an arbitrary current is, of course, also possible). The constants $A_1, A_2, \ldots$ are given by the initial conditions.
Now, for the exponent in the k-th term:

$$z_k = \sigma_k + j\omega_k,$$

so that the k-th term becomes:

$$A_k e^{\sigma_k t} e^{j\omega_k t} = A_k e^{\sigma_k t} (\cos \omega_k t + j \sin \omega_k t).$$

We note here in passing that A_k is complex if $\omega_k \neq 0$ because the result must be real. Because the vibration does not increase we have

$$\sigma_k \leq 0.$$

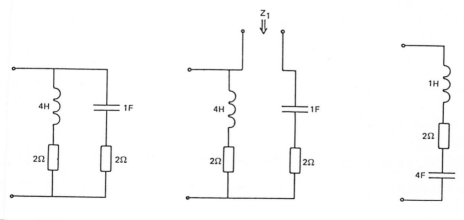

Figure 7.30   Figure 7.31   Figure 7.32

So the zeros are situated in the *left half-plane* (LHP) of the complex frequency plane.
A similar discussion can also be held for an *open* one-port, so *poles and zeros of an immittance of a passive one-port are situated in the left half-plane*:

$$\sigma \leq 0. \tag{7.27}$$

Poles and zeros of two-ports will be discussed in the next chapter. See Section 8.6.

7.13 The number of poles for an immittance

If the one-port only contains resistors, inductors and capacitors, then the real part of the impedance or the admittance is not smaller than zero.
To understand this consider the impedance:

$$Z = R + jX, \tag{7.28}$$

in which R and X are functions of λ.
The power supplied is not negative, so

$$P \geq 0.$$

With $V = ZI$ the complex power supplied is $S = \frac{1}{2} VI^* = \frac{1}{2} Z |I|^2$. $P = \frac{1}{2} R |I|^2$, so we must have

$$R \geq 0. \tag{7.29}$$

this means that

$$|\arg Z| \leq \frac{\pi}{2}. \tag{7.30}$$

See Figure 7.33, in which two complex values of an impedance have been drawn (Z_1 and Z_2).
Starting in the pole-zero plot with $\omega = 0$ and for increasing ω each zero contributes to the total argument of Z, which contribution increases from zero to $\frac{\pi}{2}$ rad maximum and each pole supplies a contribution decreasing from zero to $-\frac{\pi}{2}$ rad. The difference between the number of poles and zeros can at most be one:

$$|P - Z| \leq 1 \tag{7.31}$$

in which P is the number of finite (!) poles and Z the number of finite zeros. If there are, for example, two more zeros than there are poles there is a finite frequency for which $\arg Z > \frac{\pi}{2}$, which is in contradiction with (7.30).

7.14 Poles and zeros on the imaginary axis

First consider a zero z_k which is near the imaginary axis in the left half-plane (see Figure 7.34).

The influence of z_k on the function concerned is largest if the frequency is at the level of z_k. This results in the modulus having a small value for ω_1 (see Figure 7.35).

The argument changes from α (which is negative) to β (which is positive) around ω_1. This results in an increase of nearly 180° for the argument. (see Figure 7.36).

If the zero is at the imaginary axis, the modulus is zero for that frequency and the argument makes a leap if +180°.

(Note that the leap is −180° if we consider the zero in the right half-plane; however, we choose +180° in accordance with Figure 7.36).

For a pole at the imaginary axis the modulus is infinite for that frequency (asymptote, see Section 3.5, loss-less one-ports), while the leap of the argument will be −180°.

7.15 The amplitude surface

We consider a (bent) surface V above the complex plane of λ, of which the height equals $|H(\lambda)|$ in each point $\lambda = \sigma + j\omega$.

In the zeros V touches the λ-surface, in the poles V has an infinite height. In the points $\lambda \to \infty$ (the edge that is infinitely far away) the height of V is determined by $\lim\limits_{\lambda \to \infty} |H(\lambda)|$.

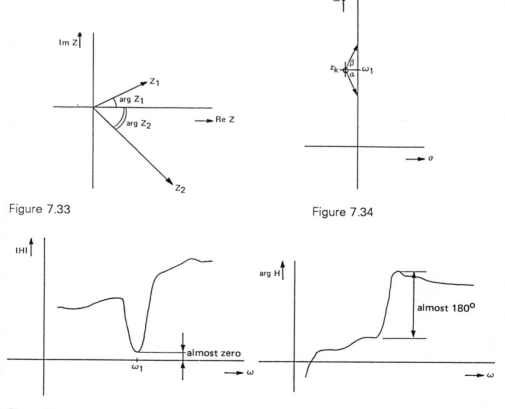

Figure 7.33

Figure 7.34

Figure 7.35

Figure 7.36

The amplitude characteristic $|H(j\omega)|$ is the intersection of V with the plane through the positive ω-axis perpendicular to the λ-plane.

We thus have a clear impression of the amplitude characteristic. For simple cases one can (depending on one's degree of geometric insight) determine a sketch of $|H(j\omega)|$.

Example 1

For a single inductor we have $Z = \lambda L$.

Zero $z_1 = 0$ (see Figure 7.37).

In addition we have $\lim_{\lambda \to \infty} |Z(\lambda)| = \infty$. So the plane V must be attached at an infinite height and it touches the λ-plane in the origin.

A cone results, of which the intersection with the vertical plane through the ω-axis is a straight line (see Figure 7.38).

Example 2

For the capacitor $Z = \dfrac{1}{\lambda C}$.

Pole: $p_1 = 0$ (see Figure 7.39).

We have $\lim_{\lambda \to \infty} |Z(\lambda)| = 0$.

So the plane V must be attached in infinity at height zero and it has an infinite height in the origin. The intersection is $|Z(\omega)|$ and is shown in Figure 7.40.

Example 3

The admittance of an inductor with series resistance is $Y = \dfrac{1}{R + \lambda L}$.

Pole: $p_1 = -\dfrac{R}{L}$ (see Figure 7.41).

$\lim_{\lambda \to \infty} |Y| = 0$.

The intersection is shown in Figure 7.42.

Note that a similar construction for the argument is not possible.

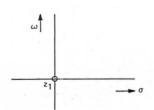

Figure 7.37

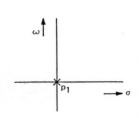

Figure 7.39

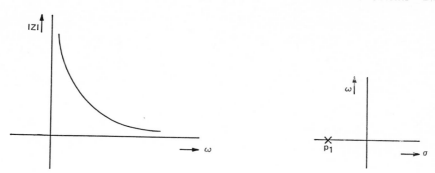

Figure 7.40

F,gure 7.41

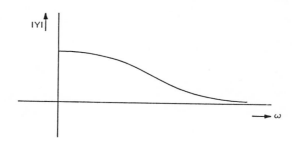

Figure 7.42

7.16 Problems

7.1 Find the pole-zero plot of the impedance of the following one-ports.

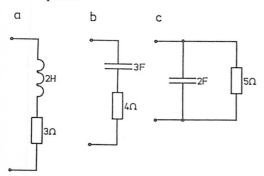

7.2 Find the pole-zero plot for the admittance for these one-ports.

7.3 a. Give the pole-zero plot of I.
   b. Graphically find |I| for $\omega = 2$ rad/s.

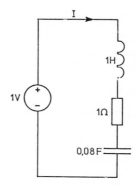

7.4   a. Give the pole-zero plot of Z.
      b. Graphically find IZI and arg Z for
         ω = 2 rad/s.

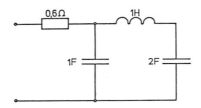

7.5   Sketch both IZI and arg Z as
      functions of ω for the one-ports of
      Problem 7.1.

7.6   Give the pole-zero plot of Z for
      a. R = 3 Ω   C = 1 F   L = 2 H.
      b. R = 1 Ω   C = 2 F   L = 5 H.
      c. R = 2 Ω   C = 1 F   L = 1 H.

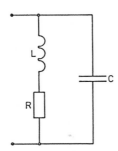

7.7   Sketch IZI and arg Z as functions of
      ω for Problem 7.6.b and determine
      for which angular frequency IZI is
      maximal and for which angular
      frequency arg Z is zero.

7.8   a. Give the pole-zero plot of Z.

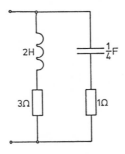

b. Find the order of the network for
   open and for short-circuited
   terminals.

7.9   Answer the same questions as in
      Problem 7.8.

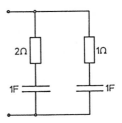

7.10 Find the characteristic equation of
      the network of Problem 7.3.

7.11 Find the characteristic equation of
      the network of Problem 7.8 for open
      and for short-circuited terminals.

7.12 a. Find the characteristic equation
         for open terminals.
      b. Find the characteristic equation
         for short-circuited terminals.
      c. Give the pole-zero of the
         impedance.

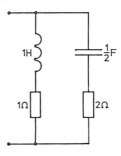

7.13 Give the pole-zero of the impedance.

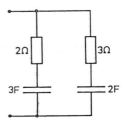

7.14 Give the pole-zero plot of the impedance.

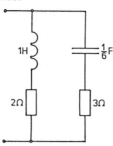

7.15 Give the pole-zero plot of the admittance.

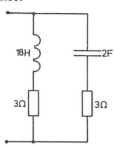

7.16 Graphically determine |H| and arg H of the following functions:

a. $H = \dfrac{1}{\lambda + 2}$

b. $H = \dfrac{\lambda + 3}{\lambda + 2}$

c. $H = \dfrac{(\lambda + 3)(\lambda + 2)}{\lambda + 1}$

d. $H = \dfrac{1}{\lambda - 5}$

e. $H = \dfrac{\lambda + 2}{\lambda - 3}$

f. $H = \dfrac{1}{(\lambda + 2)(\lambda + 3)}$

g. $H = \dfrac{\lambda - 2}{\lambda + 2}$

h. $H = \dfrac{1}{2\lambda^2 + 6\lambda + 9}$

i. $H = \dfrac{9}{2\lambda^2 + 6\lambda + 9}$

j. $H = \dfrac{\lambda^2 - \lambda + 2}{\lambda^2 + \lambda + 2}$

7.17 Which of the functions of Problem 7.16 can be an immittance function of a passive one-port?

7.18 Give the pole-zero plot of the impedance of the one-port of Problem 3.34.

7.19 a. Give the pole-zero plot of $H = \dfrac{V_2}{V_1}$.

   b. Sketch both |H| and arg H as functions of ω.

   c. Find the value of arg H for $\omega = 0$ rad/s, $\omega = \sqrt{2}$ rad/s and $\omega \to \infty$.

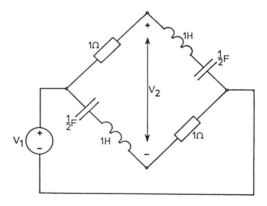

7.20 a. Find the transfer function $H(\lambda) = \dfrac{V_2}{V_1}$.

   In addition $R = 1250\ \Omega$, $L = 1$ H and $C = 400\ \mu F$ are given.

   b. Determine, using the pole-zero plot, at about which angular

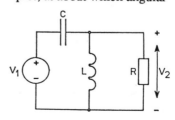

frequency ω the modulus $|H(j\omega)|$
has its largest value. How large
is that maximal value
approximately?
Sketch (roughly to scale) $|H(j\omega)|$
as a function of ω; indicate the
value of $\omega \to \infty$ in this sketch.

7.21 a. Find the voltage transfer function
$H(\lambda) = \dfrac{V_2}{V_1}$.

b. Using the pole-zero plot, sketch
the function $|H(j\omega)|$ as a function
of ω.

c. Sketch arg $\{H(j\omega)\}$ as a function
of ω.

d. Find the band width.

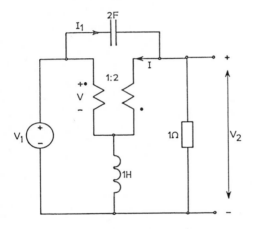

7.22 a. Give the pole-zero plot of
$H = \dfrac{V_2}{V_1}$.

b. Sketch both $|H|$ and arg H as
functions of ω.

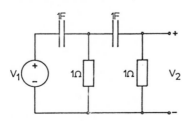

7.23 a. Find the impedance $Z(\lambda)$ of this
one-port as a function of the
complex frequency λ.

b. Find $\lim\limits_{\lambda \to 0} Z$ and $\lim\limits_{\lambda \to \infty} Z$ and give a
physical interpretation of these
results.

c. Give the pole-zero plot of $Z(\lambda)$.

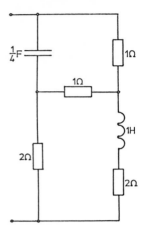

7.24 a. Find the impedance Z of this one-
port as a function of the complex
frequency λ, determine the poles
and the zeros and draw them in
the complex frequency plane.
Give a rough sketch of $|Z|$ as
function of ω.

b. Examine for which frequency
arg Z is zero, find the sign of
arg Z for small and also for large
values of ω and finally sketch
arg Z as a function of ω.

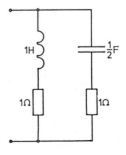

7.25 a. Find I as a function of λ.
      b. Give the pole-zero plot of I and
         sketch |I| and arg I as functions of
         ω, using this plot.
      c. Sketch the polar diagram of I as a
         function of ω.
      d. The voltage source intensity as a
         function of time is $v_1 = \cos 2t$ v
         and the current source intensity is
         $i_2 = \cos 2t$ A, find the current in
         the resistor as a function of time.

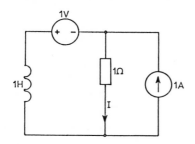

7.26 a. Find the transfer function $H = \dfrac{V_2}{I_1}$
         as a function of the complex
         frequency λ.
      b. Calculate the poles and the zeros
         of H and draw the pole-zero plot.
      c. Sketch arg H as a function of ω
         and calculate the value of ω for
         which arg H = 0.

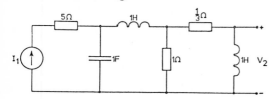

7.27 a. Find $H = \dfrac{V_2}{V_1}$ as a function of the
         complex frequency λ.
      b. Give the pole-zero plot of H and
         find |H| and arg H for $\omega = \frac{1}{2}$ rad/s
         by measuring.

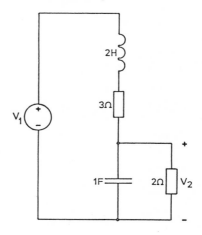

7.28 a. Find $H = \dfrac{V}{I}$ as a function of the
         complex frequency λ.
         Give the pole-zero plot of H.
      b. Sketch both |H| and arg H as
         functions of ω.
      c. Calculate the extreme values
         (maxima and minima) of |H| (ω).

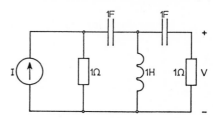

7.29 a. Find V_2 as a function of the
         complex frequency λ.
         Give the pole-zero plot.
      b. Sketch both |V_2| and arg V_2 as
         functions of ω. Find arg V_2 for
         ω = 0 , for ω = 1 and for
         ω = 2 rad/s.

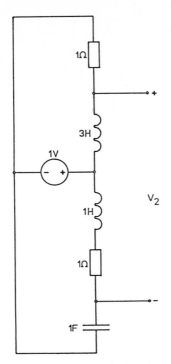

c. Why do the poles, but not the zeros, of this function lie in the left half-plane?

7.30 Consider the voltage ratio $\frac{V_2}{V_1} = H$.

a. Find $H(\lambda)$.
b. Find the poles and the zeros of $H(\lambda)$.
   Draw the pole-zero plot to scale.

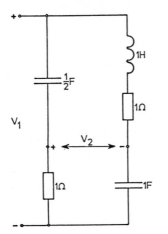

c. Graphically find $|H|$ for $\omega = 1$ rad/s.
d. Does the position of one of the zeros cause instability in the circuit?

7.31 For the gyrator it holds that

$$V_1 = -RI_2$$

$$V_2 = RI_1 \quad \text{with } R = 1\ \Omega.$$

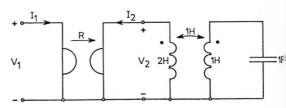

a. Prove that the input impedance of this network as a function of the complex frequency λ is given by

$$Z = \frac{\lambda^2 + 1}{\lambda(\lambda^2 + 2)}.$$

One next sets: $\lambda = j\omega$.
b. Find the reactance X as a function of ω.
c. Sketch this function $X = f(\omega)$.
d. Find a network without a gyrator and without coupled inductors with the same $X = f(\omega)$. Calculate the value of the elements used.

7.32

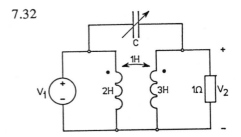

a. Find $H = \frac{V_2}{V_1}$ as a function of the complex frequency λ and the value of C.

b. Find the poles and the zeros of H for C = 1 F, draw them in the complex plane and sketch arg H = f(ω).

Next one sets λ = j rad/s and chooses C as a variable.

c. Find H(C) and sketch this function. For which value of C is H imaginary?

Give the pole-zero plot of H.

b. Graphically find IHI for ω = 0.8 rad/s and check it with a calculation.

7.35 Consider the voltage ratio H = $\dfrac{V_0}{V_1}$.

a. Find H as a function of the complex frequency λ.

b. Give the pole-zero plot of H.

c. Sketch IHI and arg H as a function of ω.

d. Find the angular frequency for which v_0 has an opposite phase compared to v_1.

7.33

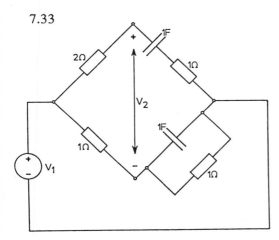

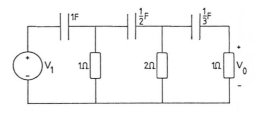

a. Find H = $\dfrac{V_2}{V_1}$ as a function of λ.

b. Give the pole-zero plot of H.

c. Sketch IHI and arg H as functions of ω.

d. Find $\displaystyle\lim_{\omega\uparrow 1}$ (arg H) and $\displaystyle\lim_{\omega\downarrow 1}$ (arg H).

7.34

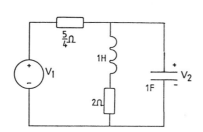

a. Find H = $\dfrac{V_2}{V_1}$ as a function of λ.

8
Two-ports, filters

8.1 Introduction

In Chapter 1, Section 12, we discussed two-ports. There we met concepts such as *port*, *port condition*, *resistance matrix* (which can be extended to *impedance matrix* with the theory of Chapter 2) and *reciprocity*. Further, the *conductance matrix* (which can be extended to *admittance matrix*) followed.

In Sections 1.17 and 1.18 we came across the *active* two-ports, namely the *transactors* and the *NIC*, *NII* and *gyrator* derived from it. Finally in Chapter 4 still other two-ports followed, viz. the *magnetic coupled inductors* and the *transformer*. It is useful to study two-ports theory in more detail here with the newly acquired knowledge.

In Figure 8.1 the general diagram of a two-port is shown.

8.2 Two-port matrices

There are relations between the four variables of a two-port. We distinguish:

a. The Z-matrix (*impedance matrix*).

$$V_1 = Z_{11}I_1 + Z_{12}I_2,$$

$$V_2 = Z_{21}I_1 + Z_{22}I_2.$$

In matrix notation: $\mathcal{V} = \mathcal{Z} I$ (8.1)
with

$$\mathcal{V} = \begin{bmatrix} V_1 \\ V_2 \end{bmatrix}, \quad I = \begin{bmatrix} I_1 \\ I_2 \end{bmatrix}, \quad \mathcal{Z} = \begin{bmatrix} Z_{11} & Z_{12} \\ Z_{21} & Z_{22} \end{bmatrix}.$$

Example (see Figure 8.2).

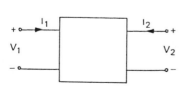

Figure 8.1

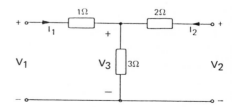

Figure 8.2

$$V_1 = 4I_1 + 3I_2$$

$$V_2 = 3I_1 + 5I_2 \qquad \text{so} \quad Z = \begin{bmatrix} 4 & 3 \\ 3 & 5 \end{bmatrix}.$$

Note: $Z_{21} = Z_{12}$ (reciprocity).

Z_{11} is called the *input impedance*, Z_{22} the *output impedance* and Z_{12} and Z_{21} the *transfer impedance*.

b. The $\mathcal{Y}$-matrix (*admittance matrix*).

$$I_1 = Y_{11}V_1 + Y_{12}V_2,$$

$$I_2 = Y_{21}V_1 + Y_{22}V_2.$$

In matrix notation:

$$I = \mathcal{Y} V. \tag{8.2}$$

Example

We use the node method in the preceding example:

$$I_1 = V_1 \qquad\qquad -V_3,$$

$$I_2 = \tfrac{1}{2}V_2 \qquad -\tfrac{1}{2}V_3,$$

$$0 = -V_1 - \tfrac{1}{2}V_2 + (1 + \tfrac{1}{2} + \tfrac{1}{3})\,V_3.$$

After elimination of V_3 follows:

$$I_1 = \frac{1}{11}\,(\,5V_1 - 3V_2),$$

$$I_2 = \frac{1}{11}\,(-3V_1 + 4V_2).$$

So

$$\mathcal{Y} = \frac{1}{11}\begin{bmatrix} 5 & -3 \\ -3 & 4 \end{bmatrix}.$$

Note: $Y_{21} = Y_{12}$.

From $\mathcal{V} = Z\,I$, $I = Z^{-1}\,\mathcal{V}$ so

$$\mathcal{Y} = Z^{-1}, \tag{8.3}$$

provided that Z is non-singular, i.e. provided that $|Z|$ (= determinant of Z) $\neq 0$. With this we can find $\mathcal{Y}$ directly from the Z already calculated in the above example.

An example of a two-port for which $|Z| = 0$ is sketched in Figure 8.3. For this it holds that

$$V_1 = V_2 = R(I_1 + I_2).$$

So

$$\begin{bmatrix} V_1 \\ V_2 \end{bmatrix} = \begin{bmatrix} R & R \\ R & R \end{bmatrix} \begin{bmatrix} I_1 \\ I_2 \end{bmatrix}.$$

Consequently

$$Z = \begin{bmatrix} R & R \\ R & R \end{bmatrix}.$$

So the admittance matrix does not exist.
Further:

$$Z = \mathcal{Y}^{-1}, \tag{8.4}$$

provided that $\mathcal{Y}$ is not singular.
c. The $\mathcal{H}$-matrix (*hybrid matrix*)

$$\begin{bmatrix} V_1 \\ I_2 \end{bmatrix} = \mathcal{H} \begin{bmatrix} I_1 \\ V_2 \end{bmatrix}. \tag{8.5}$$

d. The G-matrix (*reverse hybrid matrix*)

$$\begin{bmatrix} I_1 \\ V_2 \end{bmatrix} = G \begin{bmatrix} V_1 \\ I_2 \end{bmatrix}. \tag{8.6}$$

So

$$G = \mathcal{H}^{-1} \quad \text{provided } |\mathcal{H}| \neq 0$$

and

$$\mathcal{H} = G^{-1} \quad \text{provided } |G| \neq 0. \tag{8.7}$$

e. The $\mathcal{K}$-matrix (*cascade matrix*)

$$\begin{bmatrix} V_1 \\ I_1 \end{bmatrix} = \mathcal{K} \begin{bmatrix} V_2 \\ -I_2 \end{bmatrix}. \qquad \text{(Note the minus sign!)} \tag{8.8}$$

f. The $\mathcal{J}$-matrix (*reverse cascade matrix*).

$$\begin{bmatrix} V_2 \\ I_2 \end{bmatrix} = \mathcal{J} \begin{bmatrix} V_1 \\ -I_1 \end{bmatrix}. \qquad \text{Note: } \mathcal{J} \neq \mathcal{K}^{-1}. \tag{8.9}$$

The elements of a certain two-port matrix can be expressed in the elements of another matrix, provided the demanded matrix exists.

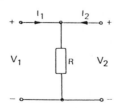

Figure 8.3

For example, given Z, what is K?

Solution

$$V_1 = Z_{11}I_1 + Z_{12}I_2,$$

$$V_2 = Z_{21}I_1 + Z_{22}I_2.$$

From the last equation:

$$I_1 = \frac{V_2}{Z_{21}} - \frac{Z_{22}}{Z_{21}} I_2.$$

This substituted into the first results in:

$$V_1 = \frac{Z_{11}}{Z_{21}} V_2 - \frac{Z_{11}Z_{22}}{Z_{21}} I_2 + Z_{12}I_2.$$

So we have:

$$V_1 = \frac{Z_{11}}{Z_{21}} V_2 - \frac{Z_{11}Z_{22} - Z_{12}Z_{21}}{Z_{21}} I_2,$$

$$I_1 = \frac{1}{Z_{21}} V_2 - \frac{Z_{22}}{Z_{21}} I_2.$$

If we compare this with

$$V_1 = K_{11}V_2 - K_{12}I_2,$$

$$I_1 = K_{21}V_2 - K_{22}I_2,$$

then

$$K_{11} = \frac{Z_{11}}{Z_{21}} \qquad K_{12} = \frac{\det Z}{Z_{21}} = \frac{|Z|}{Z_{21}}$$

$$K_{21} = \frac{1}{Z_{21}} \qquad K_{22} = \frac{Z_{22}}{Z_{21}}.$$

Det Z is the determinant of Z and it is $Z_{11}Z_{22} - Z_{12}Z_{21} = |Z|$.
Thus we get the conversion table of two-port parameters shown on page 255.

The elements of a matrix can be found in still another way.
Consider, for instance:

$$V_1 = K_{11}V_2 - K_{12}I_2$$

$$I_1 = K_{21}V_2 - K_{22}I_2.$$

We see that

$$K_{11} = \left(\frac{V_1}{V_2}\right)_{I_2 = 0}. \qquad\qquad (8.10.a)$$

That is the voltage ratio for an open output.

In the example of Figure 8.2 we find $K_{11} = \frac{4}{3}$.
We further have

$$K_{12} = (-\frac{V_1}{I_2})v_2 = 0.$$ (8.10.b)

Conversion table for two-port parameters

FROM → TO ↓	$\mathcal{Z}$		$\mathcal{Y}$		$\mathcal{H}$		$\mathcal{G}$		$\mathcal{K}$		$\mathcal{J}$	
$\mathcal{Z}$	z_{11}	z_{12}	$\frac{Y_{22}}{|\mathcal{Y}|}$	$\frac{-Y_{12}}{|\mathcal{Y}|}$	$\frac{|\mathcal{H}|}{H_{22}}$	$\frac{H_{12}}{H_{22}}$	$\frac{1}{G_{11}}$	$\frac{-G_{12}}{G_{11}}$	$\frac{K_{11}}{K_{21}}$	$\frac{|\mathcal{K}|}{K_{21}}$	$\frac{J_{22}}{J_{21}}$	$\frac{1}{J_{21}}$
	z_{21}	z_{22}	$\frac{-Y_{21}}{|\mathcal{Y}|}$	$\frac{Y_{11}}{|\mathcal{Y}|}$	$\frac{-H_{21}}{H_{22}}$	$\frac{1}{H_{22}}$	$\frac{G_{21}}{G_{11}}$	$\frac{|\mathcal{G}|}{G_{11}}$	$\frac{1}{K_{21}}$	$\frac{K_{22}}{K_{21}}$	$\frac{|\mathcal{J}|}{J_{21}}$	$\frac{J_{11}}{J_{21}}$
$\mathcal{Y}$	$\frac{z_{22}}{|\mathcal{Z}|}$	$\frac{-z_{12}}{|\mathcal{Z}|}$	Y_{11}	Y_{12}	$\frac{1}{H_{11}}$	$\frac{-H_{12}}{H_{11}}$	$\frac{|\mathcal{G}|}{G_{22}}$	$\frac{G_{12}}{G_{22}}$	$\frac{K_{22}}{K_{12}}$	$\frac{-|\mathcal{K}|}{K_{12}}$	$\frac{J_{11}}{J_{12}}$	$\frac{-1}{J_{12}}$
	$\frac{-z_{21}}{|\mathcal{Z}|}$	$\frac{z_{11}}{|\mathcal{Z}|}$	Y_{21}	Y_{22}	$\frac{H_{21}}{H_{11}}$	$\frac{|\mathcal{H}|}{H_{11}}$	$\frac{-G_{21}}{G_{22}}$	$\frac{1}{G_{22}}$	$\frac{-1}{K_{12}}$	$\frac{K_{11}}{K_{12}}$	$\frac{-|\mathcal{J}|}{J_{12}}$	$\frac{J_{22}}{J_{12}}$
$\mathcal{H}$	$\frac{|\mathcal{Z}|}{z_{22}}$	$\frac{z_{12}}{z_{22}}$	$\frac{1}{Y_{11}}$	$\frac{-Y_{12}}{Y_{11}}$	H_{11}	H_{12}	$\frac{G_{22}}{|\mathcal{G}|}$	$\frac{-G_{12}}{|\mathcal{G}|}$	$\frac{K_{12}}{K_{22}}$	$\frac{|\mathcal{K}|}{K_{22}}$	$\frac{J_{12}}{J_{11}}$	$\frac{1}{J_{11}}$
	$\frac{-z_{21}}{z_{22}}$	$\frac{1}{z_{22}}$	$\frac{Y_{21}}{Y_{11}}$	$\frac{|\mathcal{Y}|}{Y_{11}}$	H_{21}	H_{22}	$\frac{-G_{21}}{|\mathcal{G}|}$	$\frac{G_{11}}{|\mathcal{G}|}$	$\frac{-1}{K_{22}}$	$\frac{K_{21}}{K_{22}}$	$\frac{-|\mathcal{J}|}{J_{11}}$	$\frac{J_{21}}{J_{11}}$
$\mathcal{G}$	$\frac{1}{z_{11}}$	$\frac{-z_{12}}{z_{11}}$	$\frac{|\mathcal{Y}|}{Y_{22}}$	$\frac{Y_{12}}{Y_{22}}$	$\frac{H_{22}}{|\mathcal{H}|}$	$\frac{-H_{12}}{|\mathcal{H}|}$	G_{11}	G_{12}	$\frac{K_{21}}{K_{11}}$	$\frac{+|\mathcal{K}|}{K_{11}}$	$\frac{J_{21}}{J_{22}}$	$\frac{-1}{J_{22}}$
	$\frac{z_{21}}{z_{11}}$	$\frac{|\mathcal{Z}|}{z_{11}}$	$\frac{-Y_{21}}{Y_{22}}$	$\frac{1}{Y_{22}}$	$\frac{-H_{21}}{|\mathcal{H}|}$	$\frac{H_{11}}{|\mathcal{H}|}$	G_{21}	G_{22}	$\frac{1}{K_{11}}$	$\frac{K_{12}}{K_{11}}$	$\frac{|\mathcal{J}|}{J_{22}}$	$\frac{J_{12}}{J_{22}}$
$\mathcal{K}$	$\frac{z_{11}}{z_{21}}$	$\frac{|\mathcal{Z}|}{z_{21}}$	$\frac{-Y_{22}}{Y_{21}}$	$\frac{-1}{Y_{21}}$	$\frac{-|\mathcal{H}|}{H_{21}}$	$\frac{-H_{11}}{H_{21}}$	$\frac{1}{G_{21}}$	$\frac{G_{22}}{G_{21}}$	K_{11}	K_{12}	$\frac{J_{22}}{|\mathcal{J}|}$	$\frac{J_{12}}{|\mathcal{J}|}$
	$\frac{1}{z_{21}}$	$\frac{z_{22}}{z_{21}}$	$\frac{-|\mathcal{Y}|}{Y_{21}}$	$\frac{-Y_{11}}{Y_{21}}$	$\frac{-H_{22}}{H_{21}}$	$\frac{-1}{H_{21}}$	$\frac{G_{11}}{G_{21}}$	$\frac{|\mathcal{G}|}{G_{21}}$	K_{21}	K_{22}	$\frac{J_{21}}{|\mathcal{J}|}$	$\frac{J_{11}}{|\mathcal{J}|}$
$\mathcal{J}$	$\frac{z_{22}}{z_{12}}$	$\frac{|\mathcal{Z}|}{z_{12}}$	$\frac{-Y_{11}}{Y_{12}}$	$\frac{-1}{Y_{12}}$	$\frac{1}{H_{12}}$	$\frac{H_{11}}{H_{12}}$	$\frac{-|\mathcal{G}|}{G_{12}}$	$\frac{-G_{22}}{G_{12}}$	$\frac{K_{22}}{|\mathcal{K}|}$	$\frac{K_{12}}{|\mathcal{K}|}$	J_{11}	J_{12}
	$\frac{1}{z_{12}}$	$\frac{z_{11}}{z_{12}}$	$\frac{-|\mathcal{Y}|}{Y_{12}}$	$\frac{-Y_{22}}{Y_{12}}$	$\frac{H_{22}}{H_{12}}$	$\frac{|\mathcal{H}|}{H_{12}}$	$\frac{-G_{11}}{G_{12}}$	$\frac{-1}{G_{12}}$	$\frac{K_{21}}{|\mathcal{K}|}$	$\frac{K_{11}}{|\mathcal{K}|}$	J_{21}	J_{22}

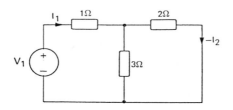

Figure 8.4

The corresponding circuit for our example is shown in Figure 8.4.
We find

$$I_1 = \frac{V_1}{1 + \frac{6}{5}} = \frac{5}{11} V_1,$$

so

$$-I_2 = \frac{3}{5} \cdot \frac{5}{11} V_1 = \frac{3}{11} V_1. \qquad \text{So } K_{12} = \frac{11}{3}.$$

In a similar manner we find K_{21} and K_{22}.
The other matrix elements can also be found in this way. However, this method is only handy if one or at most two elements of a matrix have to be determined. If the total matrix has to be calculated, it is better to write down the mesh, branch or node equations and from that determine the matrix.

8.3 Interconnection of two-ports

– *Series interconnection* of two-ports.

Both inputs as well as both outputs are connected in series (see Figure 8.5).
It is assumed that all four ports meet the port condition.
We then have:

$$\mathcal{V}_a = \left[\begin{array}{c} V_{1a} \\ V_{2a} \end{array} \right] = Z_a \left[\begin{array}{c} I_1 \\ I_2 \end{array} \right] \text{ and } \mathcal{V}_b = \left[\begin{array}{c} V_{1b} \\ I_{2b} \end{array} \right] Z_b \left[\begin{array}{c} I_1 \\ I_2 \end{array} \right],$$

So:

$$\mathcal{V} = \left[\begin{array}{c} V_1 \\ V_2 \end{array} \right] = \left[\begin{array}{c} V_{1a} \\ V_{2a} \end{array} \right] + \left[\begin{array}{c} V_{1b} \\ V_{2b} \end{array} \right] = (Z_a + Z_b) \left[\begin{array}{c} I_1 \\ I_2 \end{array} \right] = (Z_a + Z_b) \, I = ZI.$$

Therefore

$$Z = Z_a + Z_b. \tag{8.11}$$

So we have:
The impedance matrix of the series connection equals the sum of the individual impedance matrices.
In practice it is often inconvenient to verify whether the port condition for both ports has been met. If one is not certain of this it is better to calculate the impedance matrix of the whole network directly.
For *three-terminal* networks in series it is allowed to add the impedance matrices regardless of the port condition (Figure 8.6).
If we interconnect two of these networks in series the port voltages and the port currents do not change (Figure 8.7).
The relations between the voltages and the currents remain the same for each two-port, so the above deduction remains valid, therefore

$$Z = Z_a + Z_b.$$

– *Parallel interconnection* (Figure 8.8)

If the port condition for each port has been met it holds that:

$$\mathcal{Y} = \mathcal{Y}_a + \mathcal{Y}_b. \tag{8.12}$$

For two three-terminal networks in parallel the port condition is automatically met.

Example (see Figure 8.9).

These are two T-networks in parallel . The admittance matrix of a T-network like this has already been found (Section 8.2).

So for the total $\mathcal{Y}$-matrix we find:

$$\mathcal{Y} = 2 \, \mathcal{Y}_a = \frac{2}{11} \begin{bmatrix} 5 & -3 \\ -3 & 4 \end{bmatrix}.$$

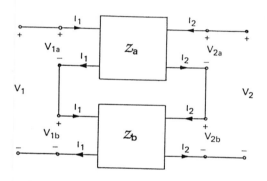

Figure 8.5

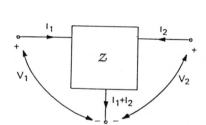

Figure 8.6

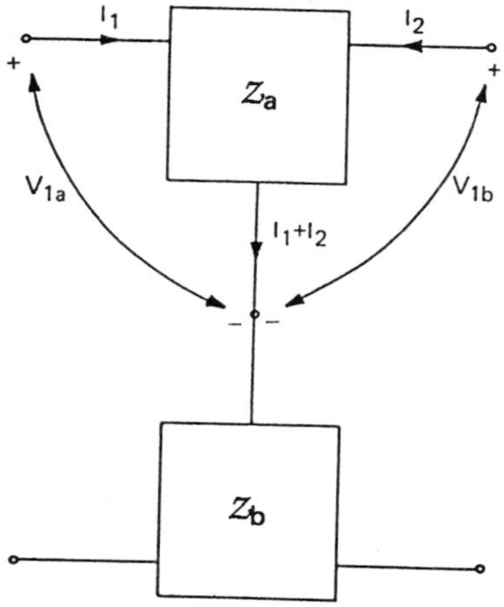

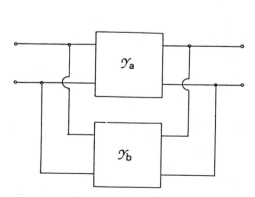

Figure 8.7

Figure 8.8

– The *series-parallel interconnection* (Figure 8.10).

For this figure:

$$\mathcal{H} = \mathcal{H}_a + \mathcal{H}_b. \tag{8.13}$$

– The *parallel-series interconnection* (Figure 8.11)

For this figure:

$$\mathcal{G} = \mathcal{G}_a + \mathcal{G}_b. \tag{8.14}$$

In all the above situations one must thoroughly check whether the port condition for each port has been met.

– The *cascade interconnection* (Figure 8.12)

$$\begin{bmatrix} V_1 \\ I_1 \end{bmatrix} = \mathcal{K}_a \begin{bmatrix} V_2 \\ -I_2 \end{bmatrix} = \mathcal{K}_a \begin{bmatrix} V_3 \\ I_3 \end{bmatrix},$$

$$\begin{bmatrix} V_3 \\ I_3 \end{bmatrix} = \mathcal{K}_b \begin{bmatrix} V_4 \\ -I_4 \end{bmatrix}, \text{ so } \begin{bmatrix} V_1 \\ I_1 \end{bmatrix} = \mathcal{K}_a \mathcal{K}_b \begin{bmatrix} V_4 \\ -I_4 \end{bmatrix}.$$

So $\mathcal{K} = \mathcal{K}_a \cdot \mathcal{K}_b.$ (8.15)

Example (see Figure 8.13).

For the two-port left of the dotted line we find for the cascade matrix.

$$\mathcal{K}_1 = \begin{bmatrix} \frac{4}{3} & 1 \\ \frac{1}{3} & 1 \end{bmatrix},$$

and for the right-hand part

$$\mathcal{K}_2 = \begin{bmatrix} 1 & 2 \\ 0 & 1 \end{bmatrix},$$

so that the cascade matrix for the whole two-port becomes

$$\mathcal{K} = \mathcal{K}_1 \mathcal{K}_2 = \begin{bmatrix} \frac{4}{3} & 1 \\ \frac{1}{3} & 1 \end{bmatrix} \begin{bmatrix} 1 & 2 \\ 0 & 1 \end{bmatrix} = \frac{1}{3} \begin{bmatrix} 4 & 11 \\ 1 & 5 \end{bmatrix}.$$

This result can, of course, also be determined by considering the whole two-port. The meaning of the minus sign in (8.8) is clarified by this.

– The *reverse cascade interconnection.*

For this one we can derive

$$\mathcal{I} = \mathcal{I}_a \cdot \mathcal{I}_b.\tag{8.16}$$

Note that the port condition is always met in the cascade interconnection.

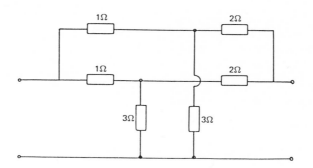

Figure 8.9

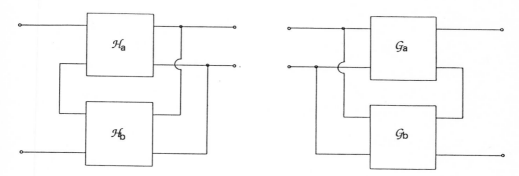

Figure 8.10

Figure 8.11

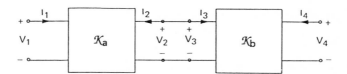

Figure 8.12

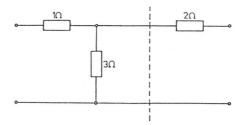

Figure 8.13

8.4 Reciprocity

In (1.26) we found that $R_{21} = R_{12}$ for a reciprocal two-port. This means that the resistance matrix is symmetrical:

$$\mathcal{R} = \mathcal{R}^T.$$

$\mathcal{R}^T$ is the transpose of the matrix $\mathcal{R}$.
In general, for the impedance matrix of a reciprocal two-port holds:

$$Z = Z^T \tag{8.17}$$

and for the admittance matrix it holds that:

$$\mathcal{Y} = \mathcal{Y}^T. \tag{8.18}$$

From the conversion table in Section 8.2 we also see that

$$H_{21} = -H_{12} \tag{8.19}$$

and

$$G_{21} = -G_{12} \tag{8.20}$$

(note the minus sign!).

There is no such simple relation for the cascade matrix.
Determine the determinant of $\mathcal{K}$. From the table we find:

$$|\mathcal{K}| = \frac{Z_{11}Z_{22}}{Z_{21}{}^2} - \frac{|Z|}{Z_{21}{}^2} = \frac{Z_{11}Z_{22} - Z_{11}Z_{22} + Z_{12}Z_{21}}{Z_{21}{}^2} = 1,$$

if $Z_{21} = Z_{12}$, i.e. for a reciprocal two-port. So for a reciprocal two-port

$$|\mathcal{K}| = 1. \tag{8.21}$$

We can also determine that for a reciprocal two-port:

$$|\mathcal{J}| = 1. \tag{8.22}$$

From the two-port matrix of the transactors (see Section 1.17) we can immediately see that they are not reciprocal. For instance, for the current-current transactor:

$$\mathcal{K} = \begin{bmatrix} 0 & 0 \\ 0 & -\dfrac{1}{\alpha} \end{bmatrix}$$

so that $|\mathcal{K}| = 0$.
In the following outline the principal matrices of the various two-ports are shown.

Outline

	$\mathcal{K}$	$\mathcal{Z}$	$\mathcal{Y}$	symbol
vvt	$\begin{matrix} \frac{1}{\mu} & 0 \\ 0 & 0 \end{matrix}$	X	X	
iit	$\begin{matrix} 0 & 0 \\ 0 & -\frac{1}{a} \end{matrix}$	X	X	
vit	$\begin{matrix} 0 & -\frac{1}{G} \\ 0 & 0 \end{matrix}$	X	$\begin{matrix} 0 & 0 \\ G & 0 \end{matrix}$	
ivt	$\begin{matrix} 0 & 0 \\ \frac{1}{R} & 0 \end{matrix}$	$\begin{matrix} 0 & 0 \\ R & 0 \end{matrix}$	X	
gyrator	$\begin{matrix} 0 & R \\ \frac{1}{R} & 0 \end{matrix}$	$\begin{matrix} 0 & -R \\ R & 0 \end{matrix}$	$\begin{matrix} 0 & \frac{1}{R} \\ -\frac{1}{R} & 0 \end{matrix}$	
nic	$\begin{matrix} \mu & 0 \\ 0 & -\frac{1}{\mu} \end{matrix}$	X	X	nic
nii	$\begin{matrix} 0 & -R \\ \frac{1}{R} & 0 \end{matrix}$	$\begin{matrix} 0 & R \\ R & 0 \end{matrix}$	$\begin{matrix} 0 & \frac{1}{R} \\ \frac{1}{R} & 0 \end{matrix}$	nii
transformator	$\begin{matrix} \frac{1}{n} & 0 \\ 0 & n \end{matrix}$	X	X	

For the transactors the voltage is indicated by v and the current by i. Thus a current-voltage transactor is *ivt*, etc.

We note here that all two-ports are active in this outline, except the *gyrator* and the *transformer*. A *gyrator* is *passive* because the sum of the powers supplied to the input and to the output is:

$p = v_1 i_1 + v_2 i_2 = v_1 i_1 + (R i_1)\cdot(-v_1/R) = 0$ (see (1.41)).

So the gyrator does not supply power via its ports, which a NIC (Negative Impedance Converter, see Section 1.18), for example, can: $p = v_1 i_1 + v_2 i_2 = v_1 i_1 + (v_1/\mu)\,\alpha v_2$ $= \frac{\alpha}{\mu}\,v_1 v_2$, which can be positive, zero or negative (see (1.44)).

8.5 Restoring the port condition

Consider the series interconnection (see Figure 8.14).

If the port condition for both two-ports has not been met, then in general

$$Z = Z_a + Z_b$$

does *not* hold for the total impedance matrix.

The port condition can be restored by connecting a transformer to all four ports (see Figure 8.15), so that the above formula becomes valid.

Because the turns ratio is 1:1 the primary and the secondary voltages, as well as the primary and secondary currents are equal. The electric separation caused by the transformers leads to equality of the input and output currents. The same result will be achieved by omitting both lower transformers (see Figure 8.16).

Subsequently it is possible to remove still another transformer (see Figure 8.17).

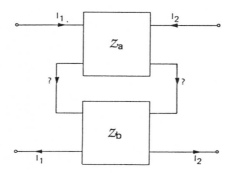

Figure 8.14

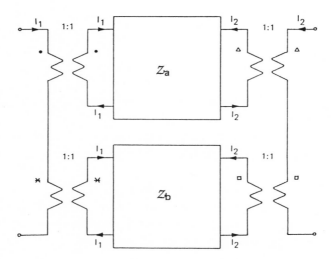

Figure 8.15

With the cut-set indicated by the dotted line one can see that $I_a = I_2$. So here, too, the port condition has been met.

Note that the internal currents in both two-ports will change when the transformer(s) are inserted.

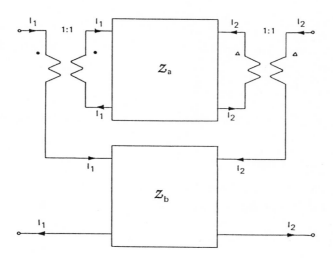

Figure 8.16

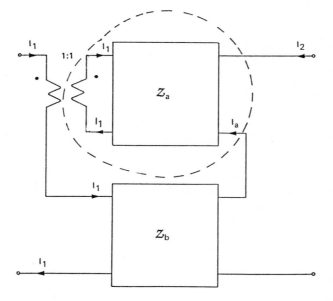

Figure 8.17

8.6 Poles and zeros of a transfer function

Consider a two-port with resistors, inductors and capacitors only (see Figure 8.18). The voltage ratio is

$$H = \frac{V_2}{V_1} = \frac{N}{D},$$

so

$$DV_2 = NV_1.$$

Because a voltage source has been connected to the input the following holds for the natural oscillations:

$$D(\lambda) = 0,$$

i.e. the *poles* p_1, p_2,... of $H(\lambda)$ represent the natural oscillations. For the output voltage they are:

$$v_2 = A_1 e^{p_1 t} + A_2 e^{p_2 t} + \dots$$

This voltage will not increase unlimited, i.e. the *poles are situated in the left half-plane*. The zeros may be situated in the right half-plane as demonstrated in the following example (Figure 8.19).
Determine $H = \dfrac{V_{ab}}{V_1}$.

We find $H = \dfrac{1}{3\lambda + 1} - \dfrac{\frac{1}{\lambda}}{\lambda + 1 + \frac{1}{\lambda}}$.

Consequently

$$H = \frac{\lambda(\lambda - 2)}{(3\lambda + 1)(\lambda^2 + \lambda + 1)}.$$

The zeros are $z_1 = 0$ and $z_2 = 2$.
The poles are

$$p_1 = -\frac{1}{3}, \ p_2 = -\frac{1}{2} + \frac{1}{2} j \sqrt{3} \text{ and } p_3 = -\frac{1}{2} - \frac{1}{2} j\sqrt{3}.$$

The zero z_2 is in the right half-plane. All three poles are in the left half-plane.
A pole of a network with a *transactor* can be situated in the right half-plane, as illustrated by the network of Figure 8.20.
Determine $H = \dfrac{V_2}{V_1}$.
We find

$$V_2 = V,$$

$$V_1 = I_1 + V,$$

$$3V = I_2 + V,$$

$$V = \frac{1}{\lambda} (I_1 + I_2).$$

From this it follows that:

$$H = \frac{1}{\lambda - 1},$$

thus H is a function with a pole $p_1 = 1$, lying in the right half-plane.
A pole in the right half-plane causes natural oscillations with an increasing amplitude. We then speak of *instability*.

8.7 The operational amplifier (opamp) once more

With the aid of an opamp we shall now construct an integrator, i.e. the output voltage is (a constant factor times) the integrated input voltage (see Figure 8.21).
The node equation for node a is

$$0 = (G + \lambda C)V_a - GV_1 - \lambda CV_2.$$

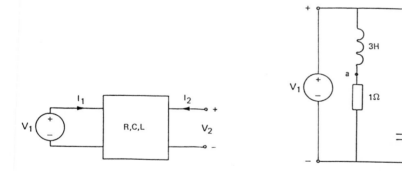

Figure 8.18

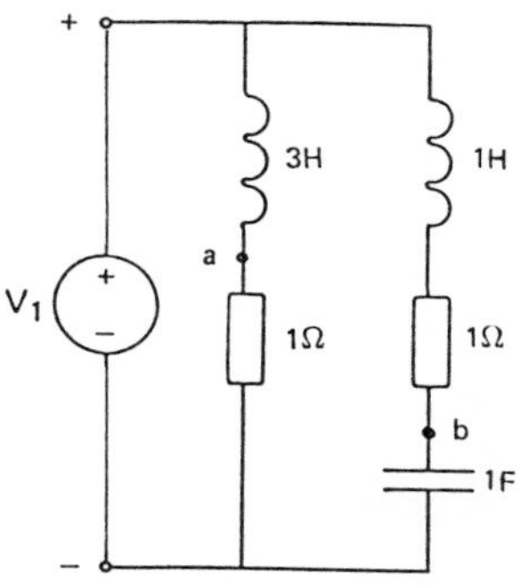

Figure 8.19

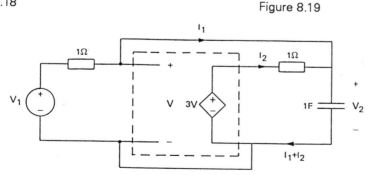

Figure 8.20

We further have

$$V_2 = \mu V_a.$$

Subsequently

$$V_2 = \frac{\mu G}{G + \lambda C(1 - \mu)} V_1. \tag{8.23}$$

The transfer function $H = \frac{V_2}{V_1}$ has a pole

$$p_1 = \frac{-G}{C(1 - \mu)} = \frac{G}{C(\mu - 1)}. \tag{8.24}$$

For $\mu > 1$ and thus for large μ we have $p_1 > 0$ so that this circuit is unstable.
Feeding back from the output to the negative input terminal results in a stable circuit, see Figure 8.22.
For node b we find

$$0 = (G + \lambda C)V_b - GV_1 - \lambda CV_2.$$

Also

$$V_2 = -\mu V_b.$$

It follows that

$$V_2 = -\frac{\mu G}{G + \lambda C(1 + \mu)} V_1. \tag{8.25}$$

The pole of the transfer function $H = \frac{V_2}{V_1}$ is

$$p_1 = -\frac{G}{(\mu + 1)G} \tag{8.26}$$

and is thus negative for all positive μ. In other words, the network is *stable*.

This introduction makes it clear why there has been a feedback to the negative input terminal in Section 1.20. The connection wires of a resistor inevitably create a capacitor. It is called the *parasitic* capacity of the resistor. The feedback to the positive input terminal causes instability as a result of this parasitic capacity.
For $\mu \to \infty$ we find

$$V_2 = -\frac{1}{\lambda RC} V_1. \tag{8.27}$$

Formulated differently

$$V_1 = -RC\lambda V_2. \tag{8.28}$$

The output voltage evidently resulted from the differential equation:

$$v_1 = -RC \frac{dv_2}{dt} .$$

(8.29)

So the output voltage is the integrated input voltage:

$$v_2 = -\frac{1}{RC} \int v_1 \, dt.$$

(8.30)

The integration constant is set to zero.

The symbolic representation is shown in Figure 8.23.

Adding and integrating can be realised with the following network (see Figure 8.24).

For node b we find:

$$0 = (G_1 + G_2 + \lambda C)V_b - G_1 V_1 - G_2 V_2 - \lambda C V_3.$$

Further

$$V_b = 0.$$

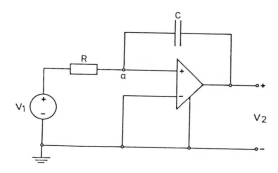

Figure 8.21

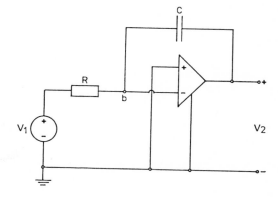

Figure 8.22

Figure 8.23

From this

$$V_3 = -\frac{G_1V_1 + G_2V_2}{\lambda C}.$$

In the time region:

$$v_3 = \int (-\frac{1}{R_1C} \ v_1 - \frac{1}{R_2C} \ v_2) \ dt.$$

See Figure 8.25 for the block diagram.

(In practice p is often used instead of λ.)

The *operator* p is defined with

$$px = \frac{d}{dt} x \tag{8.31}$$

p^n is the n-th derivative.

In this way the differential equation

$$\frac{d^2v_2}{dt^2} + A_1 \frac{dv_2}{dt} + A_0v_2 = B_1 \frac{dv_1}{dt} + B_0v_1 \tag{8.32}$$

turns into

$$(p^2 + A_1p + A_0)v_2 = (B_1p + B_0)v_1.$$

The ratio of the voltages can be written as

$$\frac{v_2}{v_1} = \frac{B_1p + B_0}{p^2 + A_1p + A_0}.$$

This expression only has meaning if one considers the denominator on the right-hand side as an operator on v_2 and the numerator on the right-hand side as an operator on v_1. Similarly one can introduce the operator p^{-1}:

$$p^{-1}x = \frac{1}{p} x = \int x \ dt. \tag{8.33}$$

The integration constant has been set to zero. p^{-n} is the n-fold integral.

Thus the voltage ratio of the above differential equation can be written as follows:

$$\frac{v_2}{v_1} = \frac{B_1p^{-1} + B_0p^{-2}}{1 + A_1p^{-1} + A_0p^{-2}}. \tag{8.34}$$

8.8 The realisation of differential equations

With adders and integrators differential equations can be realised. In (8.34) we wrote the differential equation (8.32) as a quotient with integrations.

First consider

$$\frac{w}{v_1} = \frac{1}{1 + A_1p^{-1} + A_0p^{-2}}, \tag{8.35}$$

thus

$$(1 + A_1p^{-1} + A_0p^{-2})w = v_1 \tag{8.36}$$

or

$$w = v_1 - A_1p^{-1}w - A_0p^{-2}w. \tag{8.37}$$

This results in the block diagram of Figure 8.26.
It further holds that

$$v_2 = (B_1p^{-1} + B_0p^{-2})w. \tag{8.38}$$

With this we find the total block diagram of the given differential equation (8.32) (see Figure 8.27).

With this we have got a so-called *canonic* realisation, i.e. the number of integrators equals the order of the differential equation. We further note that more canonic realisations of one differential equation are possible.

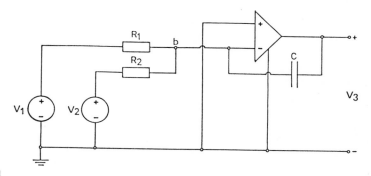

Figure 8.24

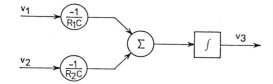

Figure 8.25

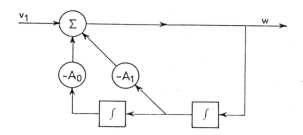

Figure 8.26

8.9 Filters

In general a *filter* is a two-port of which the relation between the input and output quantity depends on the frequency.

A filter of which the relation between input and output quantity can be written as a differential equation is called *analogous* or *time-continuous* (A *discrete filter* is characterised by a *difference equation*).

The *low-pass filter* is one through which the lower frequencies pass, the *high-pass filter* is one through which the higher frequencies pass and the *band-pass* filter through which a certain interval of the frequency spectrum passes or is attenuated.

A simple low-pass filter is shown in Figure 8.28.
We find

$$H(p) = \frac{1/p}{1 + 1/p} = \frac{1}{1 + p} = \frac{v_2}{v_1}.$$ (8.39)

So

$$H(j\omega) = \frac{1}{1 + j\omega}$$

thus

$$|H(j\omega)|^2 = \frac{1}{1 + \omega^2}.$$ (8.40)

This function is shown in Figure 8.29 (uninterrupted line).

ω_c is called the *cut-off frequency*.

That is the (angular) frequency for which $|H|^2 = \frac{1}{2}$. Here we find $\omega_c = 1$ rad/s.

The function $H(p)$ does not have a zero and has one pole $p_1 = -1$.

We now define

$$\tilde{H}(p) = H(p) \cdot H(-p).$$ (8.41)

Here we find

$$\tilde{H}(p) = \frac{1}{1 - p^2}.$$ (8.42)

Note that $\tilde{H}(p)$ is found from (8.40) by setting $\omega = p/j$. The function $\tilde{H}(p)$ has two poles lying symmetrically with regard to the imaginary axis, as in Figure 8.30.

We start with (8.39) for the realisation with adders and integrators.
From this formula it follows that

$$\frac{v_2}{v_1} = \frac{p^{-1}}{p^{-1} + 1},$$

consequently

$$v_2 = p^{-1}(v_1 - v_2).$$

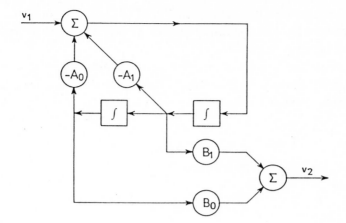

Figure 8.27

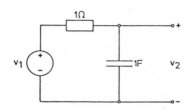

Figure 8.28

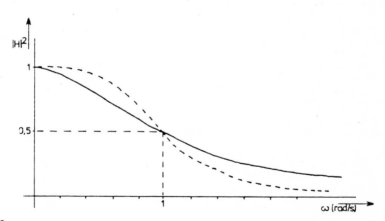

Figure 8.29

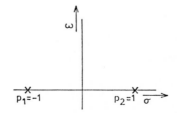

Figure 8.30

This results in the block diagram of Figure 8.31.

Note that the transfer function equals the 'straight forward amplification' divided by one minus the 'loop amplification'.

With respect to the network of Figure 8.28 the network of Figure 8.31 has the great advantage that the transfer function does not depend on the load connected to the output. One can hardly speak of a filter in the above. A better filter arises if we choose

$$|H|^2 = \frac{1}{1 + \omega^4} \cdot \tag{8.43}$$

The dotted line in Figure 8.29 is its plot.

We consider

$$\widetilde{H}(p) = \frac{1}{1 + p^4} \cdot \tag{8.44}$$

Note the sign in the denominator compared to the earlier network (formula (8.42))

Again $\omega = -jp$ holds.

This function has four poles which have been drawn in Figure 8.32.

The poles are situated on a circle with radius 1. The poles on the imaginary axis have been avoided due to the plus sign in the denominator.

Just as in the former network, the poles in the left half-plane are used in order to realise the filter (which will then be stable).

The transfer function of the filter is thus

$$H(p) = \frac{1}{(p - p_2)(p - p_3)} \cdot \tag{8.45}$$

After calculation we find

$$H(p) = \frac{1}{p^2 + p\sqrt{2} + 1} \cdot$$

Here we have a second-order filter. Note that

$$H(p) \cdot H(-p) = \frac{1}{p^2 + p\sqrt{2} + 1} \cdot \frac{1}{p^2 - p\sqrt{2} + 1} = \frac{1}{1 + p^4}$$

so it holds again that

$$\widetilde{H}(p) = H(p) \cdot H(-p).$$

Again it turns out that

$$|H(j\omega)|^2 = \widetilde{H}(j\omega). \tag{8.46}$$

For a third-order filter we start with

$$\widetilde{H}(p) = \frac{1}{1 - p^6} \cdot \tag{8.47}$$

The poles of this function are shown in Figure 8.33.

The poles are $p_{k+1} = e^{jk\frac{\pi}{6}}$ with k = 0, 1, 2, 3, 4, 5.

We again use the poles in the left half-plane to realise the filter:

$$H(p) = \frac{1}{(p - p_3)(p - p_4)(p - p_5)} .$$ (8.48)

Calculation results in

$$H(p) = \frac{1}{p^3 + 2p^2 + 2p + 1} .$$ (8.49)

It further holds that

$$|H|^2 = \frac{1}{1 + \omega^6} .$$ (8.50)

Without much trouble we now can increase the filter's order and so get an increasingly better amplitude characteristic.

The filters discussed here are called *Butterworth filters*.

These filters meet the condition

$$\tilde{H}(p) = \frac{1}{1 + (-1)^n p^{2n}}$$ (8.51)

in which the order of the filter is n and for which holds

$$|H(j\omega)|^2 = \frac{1}{1 + \omega^{2n}} .$$ (8.52)

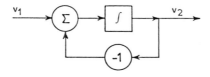

Figure 8.31

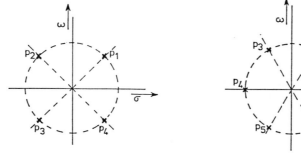

Figure 8.32   Figure 8.33

Finally we shall construct the third order Butterworth filter. We find

$$H(p) = \frac{p^{-3}}{1 + 2p^{-1} + 2p^{-2} + p^{-3}} = \frac{v_2}{v_1}.$$

We first construct part of the filter in accordance with

$$\frac{1}{1 + 2p^{-1} + 2p^{-2} + p^{-3}} = \frac{w}{v_1}$$

and after that $v_2 = p^{-3}w$ (see figure 8.34).

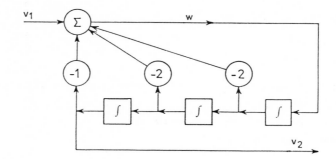

Figure 8.34

8.10 Problems

8.1   Find the matrices $\mathcal{Z}$, $\mathcal{Y}$ and $\mathcal{K}$.

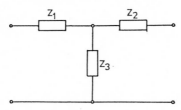

8.2   Find $\mathcal{Z}$ and $\mathcal{Y}$.

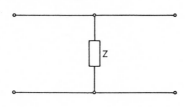

8.3   Find $\mathcal{Z}$ and $\mathcal{Y}$.

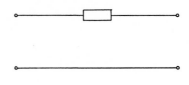

8.4   Find $\mathcal{K}$.

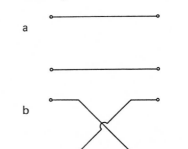

a

b

8.5 Find $\mathcal{Z}$, $\mathcal{Y}$ and $\mathcal{K}$.

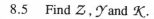

8.6 Find $\mathcal{Z}$. Under which condition does this two-port not have a $\mathcal{Y}$-matrix? What is the physical importance of this?

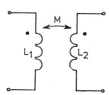

8.7 Prove, by using the result of Problem 8.6, that the ideal transformer does not have an impedance and an admittance matrix.

8.8 a. Find I_2'.

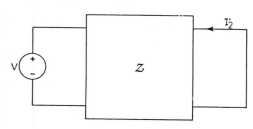

b. Find I_1''.

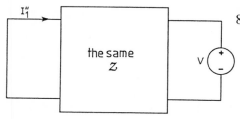

8.9 Find the Thévenin equivalence at the terminals a and b.

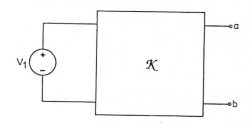

8.10 Find the admittance matrix.

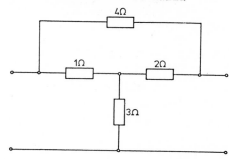

8.11 Find $\mathcal{Z}$.

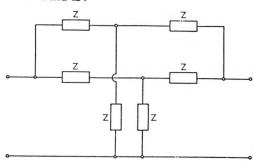

8.12 a. Find $\mathcal{Z}_a$.

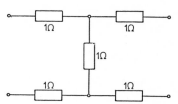

b. Find Z_b.

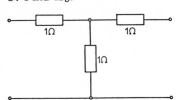

c. The above two-ports are connected in series. Is
$Z = Z_a + Z_b$?
Explain your answer.

8.13

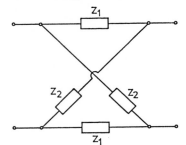

Find K_1.   Find K_2.

Find K of the cascade interconnection. Also find K if one alters the sequence of both two-ports.

8.14   Find the cascade matrix of a transformer.

8.15   For a symmetrical two-port holds

$$K = \frac{1}{H}\begin{bmatrix} A & D \\ C & B \end{bmatrix}$$

with B = A, and C ≠ 0.

a. Find the input impedance Z_i.
b. Under which condition is

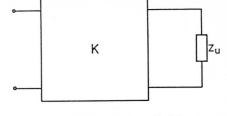

$Z_i = Z_0$? ($Z_i = Z_0$ is called *image impedance*.)

8.16   Find the impedance matrix.

8.17   Express the elements of J in the elements of K.

8.18   Which of the functions of Problem 7.16 can be transfer functions of a passive two-port?

8.19   There is no load on the terminals 2 and 0.
a. Find V_{10} so that $V_{20} = 1$ V. Subsequently a resistor $R_x = 21\ \Omega$ is connected between the nodes 1 and 2.
b. Find V_{10} so that $V_{20} = 1$ V.

Figure of Problem 8.19

8.20 A two-port has the cascade matrix

$$\mathcal{K} = \begin{bmatrix} 3 & 2 \\ 4 & 3 \end{bmatrix}.$$

Subsequently both lower terminals of both ports are connected, while a resistor of 1 Ω is connected between the upper terminals:

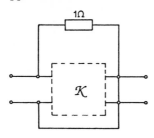

Find the admittance matrix of the total network.

8.21 a. Find the cascade matrix of this two-port.

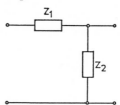

Next one chooses an inductance L for Z_1 and a capacitance C for Z_2. Subsequently three of those two-ports are interconnected in cascade. The angular frequency is $\omega_0 = 1/\sqrt{LC}$.

b. Find the cascade matrix of the total two-port and find a simpler equivalence.

8.22 a. Find the cascade matrix of this two-port.
Next one connects a voltage source

$V_{10} = 8$ V to the the input.

b. Find the Thévenin equivalence on the terminals 2 and 0.

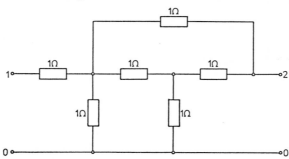

8.23 All resistances are 1 Ω.
Find the voltage V_{20}.

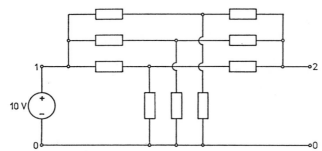

8.24 The admittance matrix of a two-port is given with $Y_{21} = Y_{12} = 0$.
A voltage source V_1 is connected to the input, the output is open.
a. Find the output voltage V_2.
Subsequently the following two-ports are interconnected in parallel:

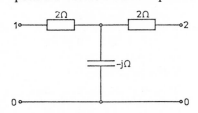

two-port A

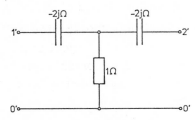

two-port B

So 1 is connected to 1',
2 is connected to 2',
0 is connected to 0'.

b. Find the admittance matrix of the two-port thus created. Give the elements of the matrix in the form $A + jB$.

c. Finally, give the cascade matrix of two-port A.

8.25 Find the input resistance if an infinite number of these two-ports are interconnected in cascade.

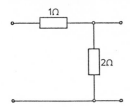

8.26 The cascade matrix of a reciprocal passive two-port is given:

$$\mathcal{K} = \begin{bmatrix} K_{11} & K_{12} \\ K_{21} & K_{22} \end{bmatrix}.$$

a. Give an expression for the impedance Z_i measured at the input while the output is open.
Subsequently the following two-port T consisting of two resistors is given:

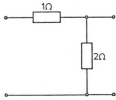

b. Find the cascade matrix (call it $\mathcal{K}_A$) of the two-port which is created when one interconnects four two-ports T in cascade. Also find the input-resistance (call it R_4) when the output is open.
Next n two-ports T are interconnected in cascade. The input resistance for an open output is R_n.
Now the network T is interconnected in front of this two-port, so that in total $n + 1$ two-ports T are interconnected in cascade. The input resistance is R_{n+1}.

c. Express R_{n+1} in R_n and prove that

$$R_n = \frac{2^{2n+1} + 1}{2^{2n} - 1}.$$

8.27 The cascade matrix of a two-port is given:

$$\mathcal{K} = \begin{bmatrix} K_{11} & K_{12} \\ K_{21} & K_{22} \end{bmatrix}.$$

a. Express the transfer function

$H = (\frac{V_2}{V_1})_{I_2 = 0}$ into the elements

of the matrix $\mathcal{K}$.

Next the following two-port T is given.

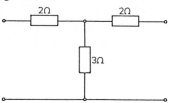

b. Find the cascade matrix (call it $\mathcal{K}(1)$) and the transfer function (call it $H(1)$) of this two-port.

The cascade matrix of n two-ports in cascade is called $\mathcal{K}(n)$ and the ratio of the open output voltage and the input voltage $H(n)$. In order to find $H(n)$ we replace each network T by three two-ports interconnected in cascade, with the cascade matrices being $\mathcal{K}_A$, $\mathcal{K}_B$ and $\mathcal{K}_A{}^{-1}$ respectively:

$$\mathcal{K}_A = \begin{bmatrix} 4 & -4 \\ 1 & 1 \end{bmatrix}$$

and $\mathcal{K}_B = \begin{bmatrix} 3 & 0 \\ 0 & \frac{1}{3} \end{bmatrix}$:

c. Are $\mathcal{K}_A$ and $\mathcal{K}_B$ cascade matrices of reciprocal two-ports? If so, give a realisation.

d. Prove the equivalence between $K(1)$ and $\mathcal{K}_A \mathcal{K}_B \mathcal{K}_A{}^{-1}$ and find $H(n)$.

8.28

a. Find the cascade matrix of two of these two-ports interconnected in cascade. Examine whether the total two-port is reciprocal.

Next a voltage-voltage transactor with voltage amplification +1 is interconnected in cascade between both two-ports.

b. Find the cascade matrix of the two-port created and examine the reciprocity.

c. For both cases find the ratio of the open output voltage and the input voltage.

8.29 a. The admittance matrix of a two-port is given. At the input a voltage source with series resistance has been connected, at the output a resistor. See circuit A.

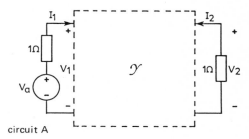

circuit A

Find $H = \frac{V_2}{V_a}$ as a function of the elements of $\mathcal{Y}$.

Subsequently the following network is given, see circuit B. The complex frequency is λ.

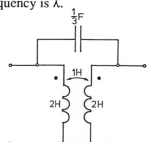

circuit B

b. Find the admittance matrix $\mathcal{Y}_1$ of the magnetic coupled inductors, then find the admittance matrix $\mathcal{Y}_2$ of the two-port consisting of only the capacitor (and which has been interconnected in parallel) and finally find the $\mathcal{Y}$-matrix of the total two-port as a function of λ.

Finally for the two-port $\mathcal{Y}$ of circuit A the two-port of circuit B is taken.

c. Find $H = \dfrac{V_2}{V_a}$ as a function of λ.

d. Find the poles and zeros of H and give the pole-zero plot.

e. Sketch $|H|$ as a function of ω.

8.30 Find the Thévenin equivalence at the nodes 4 and 0.

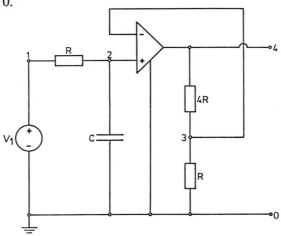

8.31 One first considers this network with open output.

a. Find $H = \dfrac{V_4}{V_1}$ as a function of the complex frequency λ, give the pole-zero plot and sketch $|H|^2$ as a function of ω.

Calculate the cut-off frequency.

b. Find the input impedance Z_i as a function of λ, measured to the right of the terminals 1 and 0. Next one connects a 1 Ω resistor to the output.

c. Find $H = \dfrac{V_4}{V_1}$ again as a function of λ.

d. Find $v_4(t)$ in the steady state if $v_1 = 4 \sin 3t$ V.

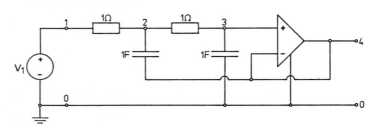

8.32 Find the Thévenin equivalence at the terminals 5 and 0 and find the input impedance Z_i. The complex frequency is λ. Use the node method.

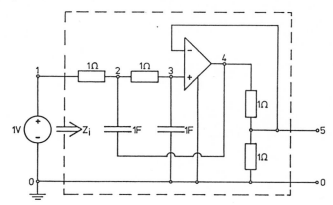

8.33 Find $\dfrac{V_5}{V_1}$.

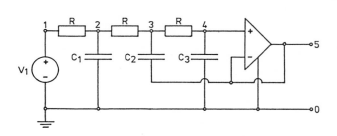

8.34 The complex frequency is λ. Find the Thévenin equivalence at the terminals 2 and 0. Use two methods to find the Thévenin impedance.

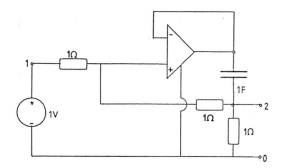

9

Networks containing switches

9.1 Introduction

In Chapter 7 we came across switches several times. Such a switch can be a *make contact* or a *break contact* (see Figure 9.1).

We speak of *transient response* if one or more switches alter their state in a network. We assume that all switches in a network change their positions simultaneously. It is not necessary that the switches are mechanical (manually operated or as contacts on a relay), they also may consist of electronic components (valve, transistor, thyristor). Also sources of which the intensity has a discontinuity cause transient response (see figure 9.2).

The whole theory of transient response is too extensive to be discussed thoroughly here. However, in this chapter we shall discuss some basic concepts.

9.2 Discontinuity

In Section 7.2 we considered a capacitor which discharged over a resistor. Here we give the results in summary (see Figure 9.3).

We found

$$v_C = Ve^{-t/RC} \qquad \text{for } t \geq 0$$

and

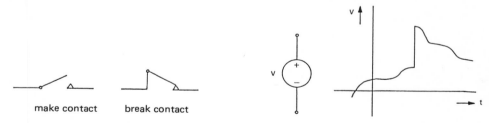

make contact   break contact

Figure 9.1

Figure 9.2

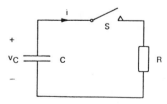

Figure 9.3

$$v_C = V \qquad\qquad \text{for } t < 0,$$

if S was closed at the moment $t = 0$ and the initial voltage of the capacitor was V.
This function has been sketched in Figure 9.4.

With $i = -C \dfrac{dv_C}{dt}$ for $t > 0$:

$$i = \frac{V}{R} e^{-t/RC},$$

while the current is zero for $t < 0$ (see Figure 9.5).

We see that v_C is continuous at $t = 0$, whereas i is discontinuous at $t = 0$. If we call the moment immediately before switching $t = 0^-$ and the moment immediately after switching $t = 0^+$ (in which 0^- and 0^+ approach 0) we have

$$v_C(0^+) = v_C(0^-)$$

and

$$i(0^+) \neq i(0^-).$$

We have calculated the solution of the network problem in the interval $t \geq 0$ so that this solution does not enable us to find the values at $t = 0^-$. On the other hand, the initial condition (here $v_C(0^-)$) is a value in the interval $t < 0$ so that here we come across the *first main problem* of transient response. In simple cases (such as in the above network) the solution can be found easily, in more intricate cases one has to use the *transformation of Laplace* or the *continuity theorem*.

9.3 The continuity theorem

The energy stored in a capacitor is $W_C = \frac{1}{2} CV^2$, see Section 2.13. Variation of the capacitor voltage thus means variation of the energy stored. In order to change the capacitor voltage in the infinitely short interval from $t = 0^-$ to $t = 0^+$ an infinitely large power is required. If we assume that the infinitely large power is not available the following theorem holds:

Each capacitor voltage is continuous.

The dual rule says:

Each inductor current is continuous.

In practice, where every voltage source has a series resistance and every current source has a parallel conductance, this rule applies.

However, in theory a source can supply an infinite power. Consider Figure 9.6, where V is a d.c. voltage source.

If $v_C(0^-) = 0$ V and V is a d.c. voltage unequal to zero, then v_C will change from 0 to V at the moment of switching.

The capacitor voltage is discontinuous. The source supplies an infinite power at the moment of switching because the charge changes in an infinitely short time and so the load current is infinite.

Further examination reveals that the following continuity theorem holds:

In a network consisting of resistors, inductors, capacitors, switches and sources with continuous intensity the following holds if the source intensities have been made zero:

a. the *capacitor voltages* are continuous in those capacitors which do not form a *loop* with *make* contacts only,

b. the *inductor currents* are continuous in those inductors which do not form a *cut-set* with *break* contacts only.

It must be emphasised that it does not follow that a capacitor voltage is discontinuous if such a loop does exist.

Consider Figure 9.6. If we make the source intensity zero (it becomes a short circuit then), we have a loop consisting of a capacitor and a make contact. However, the capacitor voltage is continuous if $v_C(0^-) = V$ and discontinuous if $v_C(0^-) \neq V$. A definite answer about continuity is only possible if one knows the initial condition!

Now consider Figure 9.3. There is no loop of make contacts and capacitors only. So the capacitor voltage is continuous *despite the initial condition.*

We can easily see that the capacitor voltage and the inductor current are continuous in the network of Figure 7.4.

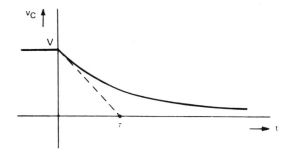

Figure 9.4

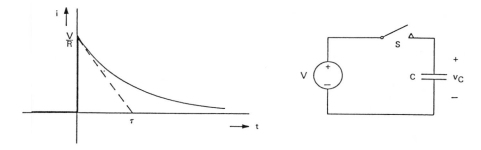

Figure 9.5

Figure 9.6

The proof of the theorem will not be discussed in this book.*

We further note that a continuous capacitor voltage does not necessarily mean that the capacitor *current* is continuous (see Figure 9.7).

The voltage v_C is continuous, but has a 'twist', so that the current $i_C = C\dfrac{dv_C}{dt}$ is discontinuous. Also compare Figures 9.4 and 9.5.

9.4 Initial conditions

In Section 7.3 we have examined what happens if a charged capacitor discharges over an inductor with series resistance. This network is shown in Figure 9.8 with an extra circuit added.

The voltage source V_h together with R give an initial condition for the inductor current (which can, of course, also be zero).

If we operate switches S_1 and S_2 simultaneously we once more have the problem of Section 7.3, with any initial condition desired for the inductor current.

Of course, the solution will not be the same. In order to solve the differential equation ,

$$\frac{25}{26}\frac{d^2v_C}{dt^2} + \frac{5}{13}\frac{dv_C}{dt} + v_C = 0 \qquad \text{for } t \geq 0,$$

one has to know the value of $v_C(0^+)$, which is immediately known (viz. $v_C(0^-)$) and $(\dfrac{dv_C}{dt})_{t=0^+}$, which, however, is not immediately known.

In the case of the series connection there is a simple relation between the initial condition of the inductor current and the derivative of v_C:

$$i = -C\frac{dv_C}{dt} ,$$

so

$$i(0^-) = i(0^+) = -C\left(\frac{dv_C}{dt}\right)_{t=0^+}.$$

For more intricate networks (certainly if they are of a higher order) there is no such simple relation, so that here we come across the *second main problem*:

How to derive the desired initial conditions in order to solve the differential equation from the given initial values of the inductor currents and capacitor voltages.

The names *initial condition* and *initial value* require some explanation. In general an initial condition is a zero, first or higher derivative at $t = 0^+$ whereas an initial value is an inductor current or a capacitor voltage at $t = 0^-$ (or at $t = 0^+$).

For a first-order problem both concepts are the same if we choose the inductor current i_L or the capacitor voltage v_C as variables:

$$\frac{di_L}{dt} + Ai_L = f(t)$$

---

* See A. Henderson, *Continuity of inductor currents and capacitor voltages in linear networks containing switches*, International Journal of Electronics, 1971.

or

$$\frac{dv_C}{dt} + Bv_C = f(t).$$

For higher-order networks there is a solution (mentioned briefly at the end of Section 7.11) called *state equations*, in which one does not start with one differential equation of the n-th order but with n *simultaneous* first-order differential equations. The variables in those first-order equations are the n inductor currents and capacitor voltages, so that in that case, too, the names initial condition and initial value are the same.

9.5 A second-order transient problem

Consider the network of Figure 9.9.
Determine v(t) for t ≥ 0 if the switch S is closed at t = 0.

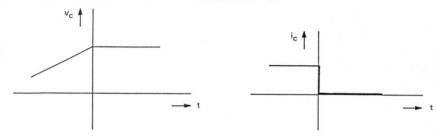

Figure 9.7   a   b

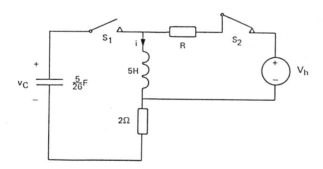

Figure 9.8

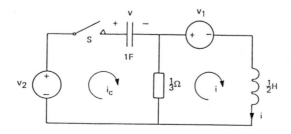

Figure 9.9

For the two meshes:

$$v_2 = v + \frac{1}{3}(i_C - i),$$ (a)

$$-v_1 = \frac{1}{2}\frac{di}{dt} + \frac{1}{3}(i - i_C).$$ (b)

Further

$$i_C = \frac{dv}{dt}.$$ (c)

We shall now solve these equations.
(c) in (a) and (b) results in:

$$v_2 = v + \frac{1}{3}\frac{dv}{dt} - \frac{1}{3}i,$$ (d)

$$-v_1 = \frac{1}{2}\frac{di}{dt} + \frac{1}{3}i - \frac{1}{3}\frac{dv}{dt}.$$ (e)

From (d) it follows that:

$$i = -3v_2 + 3v + \frac{dv}{dt}.$$ (f)

Differentiated:

$$\frac{di}{dt} = -3\frac{dv_2}{dt} + 3\frac{dv}{dt} + \frac{d^2v}{dt^2}.$$ (g)

Substitute (f) and (g) in (e):

$$-v_1 = -\frac{3}{2}\frac{dv_2}{dt} + \frac{3}{2}\frac{dv}{dt} + \frac{1}{2}\frac{d^2v}{dt^2} - v_2 + v + \frac{1}{3}\frac{dv}{dt} - \frac{1}{3}\frac{dv}{dt}.$$

Subsequently

$$\frac{d^2v}{dt^2} + 3\frac{dv}{dt} + 2v = -2v_1 + 2v_2 + 3\frac{dv_2}{dt}.$$ (h)

We have avoided integration in this deduction so that difficulties with possible integration constants will not occur.
We have found a second-order differential equation of the form:

$$a_2\frac{d^2v}{dt^2} + a_1\frac{dv}{dt} + a_0v = f(t),$$ (9.1)

in which f(t) is a linear combination of the source intensities and their derivatives.
If the source intensities are set to zero, the right-hand side becomes zero and we have a homogenous differential equation.
Such a deduction can also be given for higher order networks: Setting the source intensities to zero results in a homogenous differential equation.

We further note that the equations are all linear: Higher derivatives do occur, but there are no squares, roots etc.

The solution of (9.1) and, in principle, also that of higher order equations is as follows: Determine the solution v_h (h= homogenous) of the homogenised equation. The determination of v_h is performed by means of the characteristic equation

$$a_2\lambda^2 + a_1\lambda + a_0 = 0. \tag{9.2}$$

After determination of the roots λ_1 and λ_2 we find:

$$v_h = A_1 e^{\lambda_1 \tau} + A_2 e^{\lambda_2 \tau}. \tag{9.3}$$

Suppose v_p is a particular solution of (9.1).
Then it is evident that:

$$a_2 \frac{d^2v_p}{dt^2} + a_1 \frac{dv_p}{dt} + a_0 v_p = f(t). \tag{9.4}$$

If we subtract (9.4) from (9.1) we get:

$$a_2 \frac{d^2}{dt^2} (v - v_p) + a_1 \frac{d}{dt} (v - v_p) + a_0 (v - v_p) = 0. \tag{9.5}$$

The homogenous differential equation is:

$$a_2 \frac{d^2v_h}{dt^2} + a_1 \frac{dv_h}{dt} + a_0 v_h = 0 \tag{9.6}$$

so that $v_h = v - v_p$ or

$$v = v_h + v_p. \tag{9.7}$$

So: The total solution is the sum of the homogenous solution and a particular solution.
The third main problem is: How can we find a particular solution?
In practice it is often useful to determine the particular solution for the limit $t \to \infty$ (steady state).
If we use d.c. sources in our example, e.g. $v_1 = 1$ V and $v_2 = 2$ V, then from (h) for $t > 0$:

$$\frac{d^2v}{dt^2} + 3\frac{dv}{dt} + 2v = 2. \tag{i}$$

For $t \to \infty$, v will become a constant, so that $\frac{dv}{dt}$ and $\frac{d^2v}{dt^2}$ are zero:

$$v_p = 1 \text{ V}. \tag{j}$$

We can also find this value from the network for $t \to \infty$: The inductor voltage is zero then, so that v_C is 1 V.
The characteristic equation is

$$\lambda^2 + 3\lambda + 2 = 0 \tag{k}$$

with the solution

$$\lambda_1 = -1 \quad \text{and} \quad \lambda_2 = -2, \tag{l}$$

so that the total solution becomes:

$$v = A_1 e^{-t} + A_2 e^{-2t} + 1. \tag{m}$$

We can now find the constants A_1 and A_2 if the initial conditions are known. If we set

$$v(0^-) = 0 \text{ V}, \tag{n}$$

then, according to the continuity theorem,

$$v(0^+) = 0 \text{ V} \tag{o}$$

and if we further assume that the network exists infinitely long in the situation drawn, so that switch operations in the past no longer have any effect, we find

$$i(0^-) = -3 \text{ A} \tag{p}$$

and again with the continuity theorem

$$i(0^+) = -3 \text{ A}. \tag{q}$$

(o) substituted in (m) results in:

$$0 = A_1 + A_2 + 1. \tag{r}$$

In order to be able to use (q) we consider the network at the moment $t = 0^+$, so immediately after the switch operation (see Figure 9.10).

The voltage across the resistor is 2 V (because of the source voltage of 2 V and the capacitor voltage of 0 V). The current through the resistor is thus 6 A, and going downward. From this it follows that the capacitor current is 3 A:

$$i_C(0^+) = \left(\frac{dv}{dt}\right)_{t=0^+} = 3 \text{ A}. \tag{s}$$

Differentiating (m) and setting $t = 0^+$ results in:

$$\left(\frac{dv}{dt}\right)_{t=0^+} = -A_1 - 2A_2, \tag{t}$$

so that together with (s) it results in:

$$3 = -A_1 - 2A_2. \tag{u}$$

(r) and (u) give

$$A_1 = 1 \qquad A_2 = -2 \tag{v}$$

with which the total solution has been found:

$$v = e^{-t} - 2e^{-2t} + 1 \text{ V} \qquad t \geq 0. \tag{w}$$

The interval in which this formula is valid is $t > 0$ on account of the calculation and $t = 0$ because of the continuity.

9.6 Natural oscillations

We have seen that setting the source intensities to zero results in a homogenous differential equation. Setting the source intensities to zero also means the creation of natural oscillations. So the characteristic equation can be found directly with, for instance, the immittance method (see Section 7.11).

In order to determine the characteristic equation of the network of Figure 9.9 we consider the one-port of Figure 9.11.

Short-circuiting the input results in the same network as setting the source intensities of the network of Figure 9.9 to zero.

For the impedance we find

$$Z = \frac{1}{\lambda} + \frac{\frac{1}{2}\lambda \cdot \frac{1}{3}}{\frac{1}{2}\lambda + \frac{1}{3}} = \frac{1}{\lambda} + \frac{\lambda}{3\lambda + 2} = \frac{\lambda^2 + 3\lambda + 2}{\lambda(3\lambda + 2)}.$$

The numerator set to zero is the characteristic equation:

$$\lambda^2 + 3\lambda + 2 = 0.$$

Now the homogenous solution is known. We can find a particular solution by inspection of the network for $t \to \infty$, so that the total solution can be found without first determining the differential equation of the network.

As was said at the end of Section 7.11, the graph theory or the state equations are necessary to find the characteristic equation *exactly*.

9.7 The impulse function

The following is a classic problem (see Figure 9.12).

The left-hand capacitor has been charged, for instance up to 10 V, the right-hand capacitor has not been charged.

At $t = 0$ we close S. Both capacitor voltages now become 5 V.

The energy stored in a capacitor is

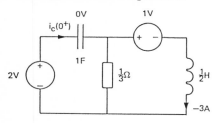

Figure 9.10

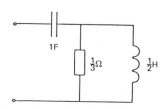

Figure 9.11

$$W_C = \frac{1}{2} CV^2,$$

so that the total energy *before* switching is

$$W_b = \frac{1}{2} \cdot 1 \cdot 10^2 = 50 \text{ J}$$

while *after* switching it is

$$W_a = \frac{1}{2} \cdot 1 \cdot 5^2 + \frac{1}{2} \cdot 1 \cdot 5^2 = 25 \text{ J}.$$

The question arises of what happens to the energy difference?

To solve that problem we introduce a resistor R and let R approach zero after the calculation (see Figure 9.13).

Given: $v_1(0^-) = 10$ V, $v_2(0^-) = 0$ V. S is closed at $t = 0$, calculate $i(t)$ for $t > 0$.

First note that both capacitor voltages are now continuous.

We find

$$i = -\frac{dv_1}{dt}, \qquad (a)$$

$$i = \frac{dv_2}{dt}, \qquad (b)$$

$$v_1 - v_2 = Ri. \qquad (c)$$

It follows that

$$v_1 - v_2 = -R \frac{dv_1}{dt}$$

and

$$v_1 - v_2 = R \frac{dv_2}{dt}.$$

From the former it follows that

$$v_2 = v_1 + R \frac{dv_1}{dt}.$$

If we substitute this and its derivative in the latter, it follows that

$$R^2 \frac{d^2v_1}{dt^2} + 2R \frac{dv_1}{dt} = 0$$

so

$$R \frac{d^2v_1}{dt^2} + 2 \frac{dv_1}{dt} = 0 \qquad (R \neq 0).$$

The characteristic equation is

$$R\lambda^2 + 2\lambda = 0, \qquad (d)$$

consequently

$$\lambda_1 = 0 \quad \text{and} \quad \lambda_2 = -\frac{2}{R}, \tag{e}$$

so that

$$v_1 = A_1 + A_2 e^{-2t/R}$$

with which we find

$$i = \frac{2A_2}{R} e^{-2t/R}.$$

At $t = 0^+$ the current is the difference of the initial voltage of the capacitors divided by R:

$$i(0^+) = \frac{10}{R},$$

so

$$\frac{2A_2}{R} = \frac{10}{R} \tag{f}$$

so that

$$i = \frac{10}{R} e^{-2t/R} \text{ A} \qquad t > 0. \tag{g}$$

$$\text{For } t < 0, i = 0. \tag{h}$$

The dissipated power is

$$p = i^2 R \tag{i}$$

or

$$p = \frac{100}{R} e^{-4t/R} \text{ W}. \tag{j}$$

The dissipated energy is

$$W = \int_0^\infty p \, dt \tag{k}$$

so that

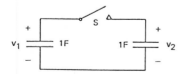

Figure 9.12

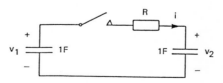

Figure 9.13

$$W = \int_0^\infty \frac{100}{R} e^{-4t/R} dt = [\frac{100}{R} \cdot \frac{R}{4} \cdot e^{-4t/R})]_\infty^0 = 25 \text{ J} \qquad (1)$$

and is independent of R(!).

Also if we let R approach zero the dissipated energy is 25 J, which happens to be the difference between the energy stored before and after switching exactly. In Figure 9.14 $i(t)$ is shown for two different values R.

Call the small resistance R_s and the large resistance R_l. Then their maximums are $10/R_s$ and $10/R_l$ respectively and the time constants are $\tau_s = R_sC$ and $\tau_l = R_lC$.

If we make the resistance even smaller the maximum will become higher and the time constant will become smaller. For the limit $R \rightarrow 0$ the maximum will be infinite while the time constant becomes zero. Such a function is called the *impulse function*. Note that the capacitor voltage is not continuous for $R \rightarrow 0$. We can now answer the question asked in the beginning of this section: *The energy difference will be dissipated.*

9.8 Sine and cosine functions with decreasing amplitude

In the example of Section 9.5 we found terms in the solution which are all damped powers of e. However, the natural oscillations of a network that contains an inductor and a capacitor can have sine and cosine terms, as described in Section 7.3. If we choose $\frac{1}{2}\Omega$ for the resistance in the example of Section 9.5 the characteristic equation becomes:

$$\lambda^2 + 2\lambda + 2 = 0.$$

The roots are complex: $\lambda_{1,2} = -1 \pm j$ and the solution becomes

$$v = A_1 e^{(-1+j)t} + A_2 e^{(-1-j)t} + 1 \text{ volt.}$$

The imaginary part of the exponents results (via Euler) in harmonic functions, while the real part of the exponents means a decreasing amplitude. We shall illustrate this with the network of Figure 9.15.

The source V_b is a d.c. source. We choose the initial values as follows:

$$v_C(0^-) = -11 \text{ V}, \; i(0^-) = 2 \text{ A}.$$

At $t = 0$, S is closed.

Determine $i(t)$ and $v_C(t)$ for $t > 0$.

Solution: On account of continuity we have

$$v_C(0^+) = v_C(0^-)$$

and

$$i(0^+) = i (0^-).$$

The characteristic equation is found with the immittance method:

$$Z = 2 + \lambda + \frac{50}{\lambda} = \frac{\lambda^2 + 2\lambda + 50}{\lambda} = \frac{N}{D} = \frac{V}{I} .$$

The numerator set to zero is the characteristic equation:

$$\lambda^2 + 2\lambda + 50 = 0.$$

The roots are $\lambda_{1,2} = -1 \pm 7j$.
Determine $v_C(t)$. A particular solution is $v_{Cp} = 12$ V (viz. for $t \to \infty$).
Thus we get.

$$v_C = A_1 e^{(-1+7j)t} + A_2 e^{(-1-7j)t} + 12 \text{ V}.$$

We have $v_C(0^+) = -11$ V, so

$$-11 = A_1 + A_2 + 12,$$

further

$$\frac{dv_C}{dt} = (-1 + 7j) A_1 e^{(-1+7j)t} + (-1 - 7j) A_2 e^{(-1-7j)t},$$

therefore

$$i = C \frac{dv_C}{dt} = \frac{-1 + 7j}{50} A_1 e^{(-1+7j)t} + \frac{-1 - 7j}{50} A_2 e^{(-1-7j)t}.$$

The initial condition substituted results in:

$$2 = \frac{-1 + 7j}{50} A_1 + \frac{-1 - 7j}{50} A_2.$$

From this follow $A_1 = \dfrac{-23 - 11j}{2}$ and $A_2 = \dfrac{-23 + 11j}{2}$.

So

$$v_C = \frac{-23 - 11j}{2} e^{(-1+7j)t} + \frac{-23 + 11j}{2} e^{(-1-7j)t} + 12 \text{ V}.$$

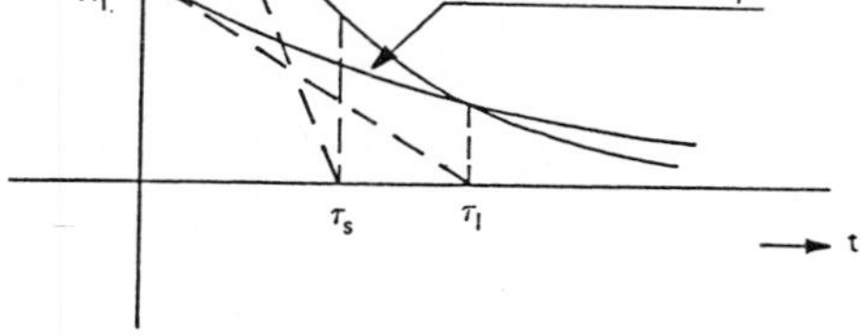

Figure 9.14

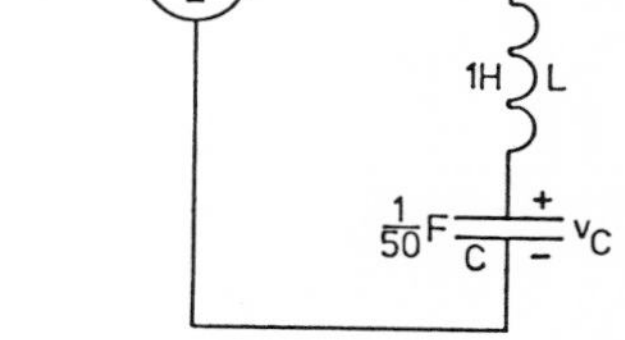

Figure 9.15

Both first terms are each other's conjugated complex, so

$$v_C = 2Re \ \{\frac{-23 - 11j}{2} \ e^{(-1+7j)t}\} + 12 \ V,$$

$$v_C = e^{-t} \cdot Re\{(-23 - 11j) \ (cos \ 7t + j \ sin \ 7t)\} + 12 \ V,$$

$$v_C = e^{-t} \ (-23 \ cos \ 7t + 11 \ sin \ 7t) + 12 \ V \qquad t \geq 0.$$

We further find

$$i = e^{-t} \ (2 \ cos \ 7t + 3 \ sin \ 7t) \ A \qquad t \geq 0.$$

Note that instead of

$$v = A_1 e^{(-1+7j)t} + A_2 e^{(-1-7j)t} + 12 \ V$$

we may also write

$$v = e^{-t} \ (F_1 \ cos \ 7t + F_2 \ sin \ 7t) + 12 \ V,$$

because further calculation of the first form results in the second form, according to Euler. The constants F_1 and F_2 can be found with the usual method. In Figure 9.16 $v_C(t)$ and in Figure 9.17 $i(t)$ are shown.

9.9 Summarising example

Consider a network in which a great part of the problems in this chapter have been concentrated (see Figure 9.18).
Given that $v_1 = 7 \ cos \ 2t \ V$. The network exists infinitely long. At $t = 0$, S is opened. One wants to compute $v(t)$ for $t \geq 0$.

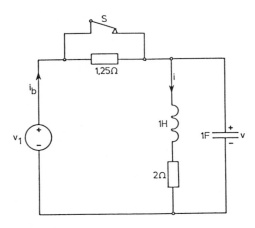

Figure 9.18

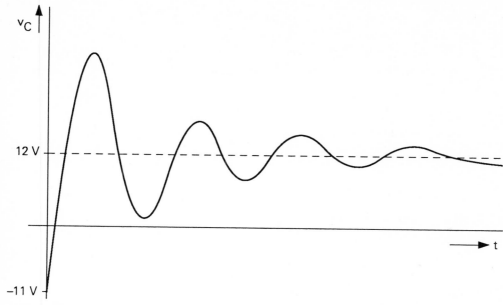

Figure 9.16

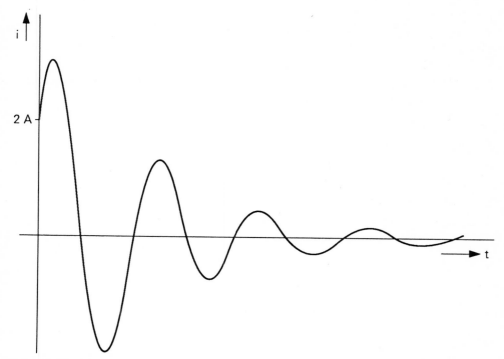

Figure 9.17

a. Find v and i for t → 0

Solution: For $t \to 0$ we have the steady state.

Introduce complex quantities: add the complex voltage $V_1 = 7$ V to the source voltage. The complex current is then:

$$I = \frac{7}{2 + 2j} = \frac{7}{4}(1 - j) \text{ A.}$$

So the time function is

$$i = \frac{7}{4}(\cos 2t + \sin 2t) \text{ A} \qquad\qquad t \to 0.$$

We further have $v = 7 \cos 2t$ V $\qquad\qquad t \to 0.$

b. Find v(0⁻) and i(0⁻)

Solution: Substitute $t = 0^-$:

$$v(0^-) = 7 \text{ V} \quad \text{and} \quad i(0^-) = \frac{7}{4} \text{ A.}$$

c. Determine the order for t > 0

Solution: For a chosen source intensity of zero there is no loop of capacitors. The only capacitor adds one to the order. For a chosen source intensity of zero there is no cut-set of inductors. The only inductor adds one to the order.

Result: the order is two.

d. Check whether v and i are continuous at t = 0

Solution: For a chosen source intensity of zero there is no loop of capacitors and make contacts only. So v is continuous.

For a chosen source intensity of zero there is no cut-set of inductors and break contacts only. So i is continuous.

In other words:

$$v(0^+) = v(0^-) = 7 \text{ A,}$$

$$i(0^+) = i(0^-) = \frac{7}{4} \text{ A.}$$

e. Find the steady state of v

Solution: Use complex quantities, this time for the network with opened switch.

$$V_p = 7 \cdot \frac{\dfrac{\frac{1}{2j} \cdot (2j + 2)}{\frac{1}{2j} + 2j + 2}}{\dfrac{\frac{1}{2j} \cdot (2j + 2)}{\frac{1}{2j} + 2j + 2} + \frac{5}{4}}$$

consequently $V_p = \frac{8}{17}(3 - 5j)$ V.

(p stands for particular solution).

The corresponding time function is

$$v_p = 1.41 \cos 2t + 2.35 \sin 2t \text{ V} \qquad t \to 0.$$

f. Find the characteristic equation

Solution: The two-port has the impedance

$$Z = \frac{5}{4} + \frac{(\lambda + 2) \cdot \frac{1}{\lambda}}{\lambda + 2 + \frac{1}{\lambda}} = \frac{5\lambda^2 + 14\lambda + 13}{4(\lambda^2 + 2\lambda + 1)}.$$

The numerator set to zero is the characteristic equation:

$$5\lambda^2 + 14\lambda + 13 = 0.$$

g. Find the homogenous solution of v

Solution: The roots of the characteristic equation are $\lambda_{1,2} = -1.4 \pm 0.8j$.
So

$$v_h = A_1 e^{(-1.4 + 0.8j)t} + A_2 e^{(-1.4 - 0.8j)t} \text{ V}$$

(h stands for homogenous solution).

It can also be written as

$$v_h = e^{-1.4t} (F_1 \cos 0.8t + F_2 \sin 0.8t) \text{ V}.$$

h. Find $\left(\dfrac{dv}{dt}\right)_{t=0^+}$

Solution: Consider the network for $t = 0^+$ (see Figure 9.19).

We find $i_C(0^+) = -\frac{7}{4}$ A, so $\left(\dfrac{dv}{dt}\right)_{t=0^+} = -\frac{7}{4}$ (C = 1 F).

i. Give the total solution

Solution: $v = e^{-1.4t} (F_1 \cos 0.8t + F_2 \sin 0.8t) + 1.41 \cos 2t + 2.35 \sin 2t \text{ V}.$

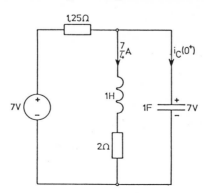

Figure 9.19

$t = 0^+ \Rightarrow v(0^+) = 7 = F_1 + 1.41$ subsequently $F_1 = 5.59$.

Further

$$\frac{dv}{dt} = -1.4e^{-1.4t} (F_1 \cos 0.8t + F_2 \sin 0.8t)$$

$$+ e^{-1.4t} (-0.8 F_1 \sin 0.8t + 0.8 F_2 \cos 0.8t)$$

$$- 2.82 \sin 2t + 4.71 \cos 2t.$$

$t = 0^+ \Rightarrow (\frac{dv}{dt})_{t=0^+} = -1.75 = -1.4 F_1 + 0.8 F_2 + 4.71$

subsequently $F_2 = 1.71$.

So the solution is:

$$v = e^{-1.4t} (5.59 \cos 0.8t + 1.71 \sin 0.8t) + 1.41 \cos 2t - 2.35 \sin 2t \text{ V}   t \geq 0.$$

j. Check whether the initial conditions suffice

Solution: $t = 0^+ \Rightarrow v(0^+) = 5.59 + 1.41 = 7$ V; this suffices.

Further

$$\frac{dv}{dt} = e^{-1.4t}(-6.46 \cos 0.8t - 6.87 \sin 0.8t) + 4.71 \cos 2t - 2.82 \sin 2t \text{ A}$$

$$t \geq 0.$$

$t = 0^+ \Rightarrow i(0^+) = -6.46 + 4.71 = -1.75$ A; this also suffices.

9.10 Conclusion

We have met the following problems in this chapter:
– Continuity and discontinuity and related to that, the impulse function.
– The question of the initial conditions.
– The calculation of a particular solution.
The charm of the *Laplace's method* is that all these problems are reduced to quite simple calculations.

However, due to the strictly mathematical calculation a disadvantage may be that the physical understanding might be lost.

Extension of the physical understanding is provided by the graph theory and the theory of state equations.

These subjects, however, are outside the scope of this book.

9.11 Problems

Notes

a. *The networks hold for all t < 0.*

b. *At t = 0 the switch is or switches are operated simultaneously.*

c. *The time of switching is zero.*

d. *If not otherwise stated the source intensities are constant.*

e. *If not otherwise stated the solution for t ≥ 0 is asked.*

9.1 $v_C(0^-) = 7$ V.

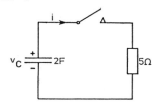

a. Find v_C.
b. Find i.
c. Sketch these functions.
d. Calculate $v_C(t)$ for
   t = 0, 1, 2, 3, 4, 5, 10, 20 and
   30 s to 4 decimal places.

9.2 $i_L(0^-) = 7$ A.

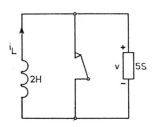

a. Find i_L.
b. Find v.
c. Sketch these functions.

9.3 $v_C(0^-) = 0$ V.
a. Find v_C.
b. Find i.

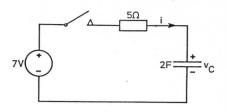

c. Sketch these functions.
d. Calculate $v_C(t)$ for t = 1, 2, 3, 4
   and 5 s to 4 decimal places.

9.4 Same questions as in Problem 9.3a,
but now with $v_C(0^-) = 3$ V.

9.5 Same questions as in Problem 9.3a,
but now with $v_C(0^-) = 7$ V.

9.6

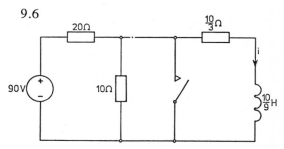

a. Find $i(0^-)$.
b. Check whether i is continuous at
   t = 0.
c. Find i and sketch this function
   for all t.

9.7 In the network the conductors are
indicated in Siemens.
a. Find the voltage $v(0^-)$.
b. Check whether the capacitor
   voltage is continuous at t = 0.
c. Find v(t) and give a sketch of this
   function for all t.

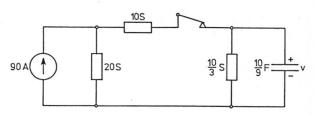

9.8 S is a switch of which the isolation resistance is R (as drawn in the circuit).

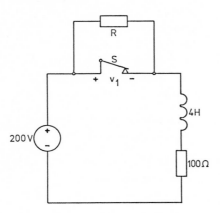

a. Find v_1 as a function of t and of R.
b. Sketch v_1 for a (positive) value of R.
c. Calculate the greatest value of v_1 if R = 1 MΩ.

9.9 v = 25 cos 2t V.

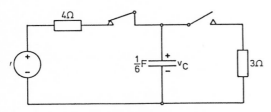

a. Find v_C for t < 0.
b. Find $v_C(0^-)$ and $v_C(0^+)$.
c. Find v_C for t ≥ 0.

9.10 $v_0 = 5 \cos 3t$ V $v_C(0^-) = 0$ V.
a. Find v_C.
b. Find v_C for $t = \frac{1}{20}$ s
and for $t = \frac{1}{10}$ s.

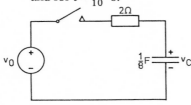

9.11 $v_1 = 3 \cos 2t$ V.
Find i.

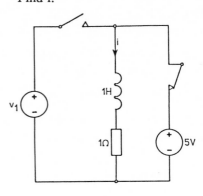

9.12 $v_1 = 50 \cos 2t$ V.

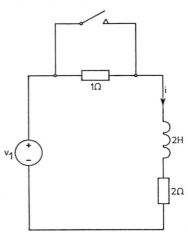

a. Find i for t < 0.
b. Find $i(0^-)$ and $i(0^+)$.
c. Find i for t → ∞.
d. Find i for t ≥ 0.

9.13

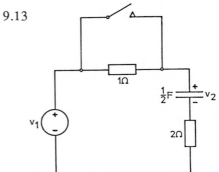

$v_1 = 5 \sin 2t$ V.
a. Find v_2 for $t < 0$.
b. Find $v_2(0^-)$ and $v_2(0^+)$.
c. Find v_2 for $t \to \infty$.
d. Find v_2 for $t \geq 0$.

9.14 $v = 4$ V.
$v_C(0^-) = 0$ V.
$L = 1$ H, $R = 1\,\Omega$, $C = 2$ F.
Find v_C.

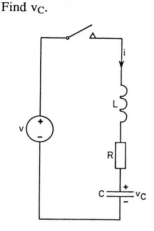

9.15 The same questions as in Problem 9.14, but now for
$L = \frac{3}{2}$ H, $R = 2\,\Omega$, $C = 2$ F.

9.16 The same questions as in Problem 9.14, but now for
$L = 2$ H, $R = 2\,\Omega$, $C = 2$ F.

9.17 The same questions as in Problem 9.14, but now for
$v_C(0^-) = 1$ V.

9.18 The same questions as in Problem 9.14, but now for
$v_C(0^-) = 1$ V and $i(0^-) = 2$ A.

9.19 The same questions as in Problem 9.14, but now for
$v = 4 \cos t$ V.

9.20 $v = 10 \sin 2t$ V.
Find i.

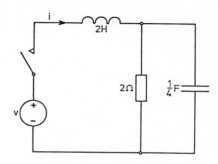

9.21 $v_1 = 5 \cos \frac{1}{2} t$ V.
a. Find $v_C(0^-)$ and $i(0^-)$.
b. Find v_2.
c. Find $v_2 (0.1)$.

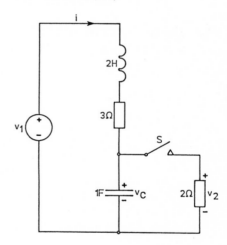

9.22 $v_1 = 12$ V.

Given: $L = 2$ H

Find v_2.

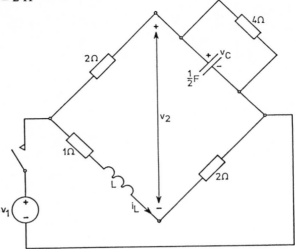

9.23   The same question as in Problem 9.22, but now for $L = 3$ H.

9.24 $v_0 = 6$ V.

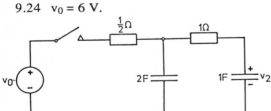

a. Find v_2.
b. Find v_2 for $t = 0.1$ s and for $t = 0.2$ s.
c. Sketch $v_2(t)$

9.25   The same questions as in Problem 9.24.a, but now for $v_0 = 5 \cos t$ V.

9.26 $v = 26$ V.

Find v_C.

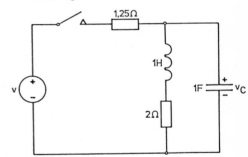

9.27   The same question as in Problem 9.26, but now for $v = 7 \cos 2t$ V.

10

Computer aided analysis*

10.1 Introduction

As a consequence of the increasing complexity of electronic circuits, in particular the *integrated circuits*, the necessity arose to use the *digital computer* in analysing networks. Besides resistors and capacitors, transistors and *diodes* are frequently used. Out of these elements, which are mostly nonlinear, *models* are made which are used in the computing process.

We shall introduce some basic concepts in this chapter, and assume that the conversion to the model has already occurred. A very simple model of a transistor has been given in Section 1.16.

More intricate models also contain diodes.

10.2 The NA-matrix

The NA-matrix (*nodal analysis*) is the matrix created in the node method (see Section 1.8). If the network does not contain voltage sources (but does contain current sources), we have

$$\mathcal{Y} V = I. \tag{10.1}$$

The matrix

$$NA = [\mathcal{Y} \| I] \tag{10.2}$$

is called the NA-matrix. The $\mathcal{Y}$-matrix is put to the left of the double line, the current source intensities known to the right. The voltages are indicated with respect to a reference point, called *datum* (which is mostly earthed).

If the network does contain voltage sources with a series impedance one can rebuild this network to the former type by transforming them into the Norton equivalence.

Example (see Figure 10.1).
If we introduce the Norton equivalences, the network of Figure 10.2 arises.
We find

$$\begin{bmatrix} 4+2j+j & -4-j \\ -4-j & 4+1+j \end{bmatrix} \begin{bmatrix} V_1 \\ V_2 \end{bmatrix} = \begin{bmatrix} -16+2 \\ 16-26 \end{bmatrix}.$$

So the NA-matrix is

---

* The author owes much to C.W. Ho, A.E. Ruehli & P.A. Brennan: The modified Nodal Approach to Network Analysis, IEEE Transactions on Circuits and Systems, June 1975.

$$NA = \begin{bmatrix} 4+2j+j & -4-j \\ -4-j & 4+1+j \end{bmatrix} \begin{vmatrix} -16+2 \\ 16-26 \end{vmatrix}.$$

We have mentioned the value of each network element in the NA-matrix separately on purpose.

10.3 The concept 'stamp'

Each network element influences the elements of the NA-matrix. The conductor of 4 S in the previous example occurs four times in the NA-matrix. This conductance prints its *stamp* on the NA-matrix of which the form is

	1	2
1	4	-4
2	-4	4

Not every element has a stamp of four places. The capacitor with the admittance of 2j S only takes one place in the NA-matrix. The reason for this is that one terminal of this capacitor has been connected to the datum.

The current sources also have their stamps and these occur in the right-hand side.

The sign of the stamp elements in a matrix is found as follows: In the left-hand side of the NA-matrix the positive current directions are chosen as *moving away* from the node, in the right-hand side as *moving towards* the node. In this way positive numbers arise at the *main diagonal* in the left-hand part.

In computer applications one starts with an 'empty' NA-matrix. The network elements are offered one by one to the matrix while each element prints its stamp on the NA-matrix (on the correct position and with the correct sign).

10.4 The MNA-matrix

The above-mentioned method fails if the network contains a voltage source without a series impedance. Also three of the four transactors (the 'vit' can be used in the NA-method) cause difficulties. In these cases one uses the MNA-matrix (*modified nodal analysis*) instead of the NA-matrix.

In this method one or more unknown currents are also solved in addition to the node voltages. These currents are therefore also written in the left-hand part of the matrix. The source current I_x in each voltage source without series impedance is chosen (the direction may be chosen at random). It is considered as an unknown quantity, while the voltage source intensity results in a known relation in the potentials of the nodes of the source. So we get

$$\begin{bmatrix} \mathcal{A} & \mathcal{B} \\ \mathcal{C} & \mathcal{D} \end{bmatrix} \cdot \begin{bmatrix} \mathcal{V} \\ I_x \end{bmatrix} = \begin{bmatrix} I_b \\ \mathcal{V}_b \end{bmatrix}. \tag{10.3}$$

V is the column vector of the node voltages, I_x are the unknown voltage source currents, I_b are the known current source intensities and V_b the known voltage source intensities.

The matrix $\begin{bmatrix} \mathcal{A} & \mathcal{B} \\ C & \mathcal{D} \end{bmatrix}$ is the left-hand part of the MNA-matrix, the vector $\begin{bmatrix} I_b \\ V_b \end{bmatrix}$ the right-hand part:

$$\text{MNA} = \begin{bmatrix} \mathcal{A} & \mathcal{B} & \| & I_b \\ C & \mathcal{D} & \| & V_b \end{bmatrix}. \tag{10.4}$$

Also if the network contains one or more of the transactors ivt, vvt and iit we have something similar. Then the vector I_x also contains transactor currents. The voltage-current-transactor can be written in the NA-matrix.

In the following outline all these cases will be discussed and it will become clear what the submatrices $\mathcal{A}$, $\mathcal{B}$, C and $\mathcal{D}$ look like.

10.5 Outline of the stamps

We shall now derive the stamps of the different network elements (and parts of networks). We shall again separate the left- and right-hand sides of the equations in the matrix notation by means of a double line.

1. The admittance

(See Figure 10.3).

The equations are (leaving currents positive)

node a: $YV_a - YV_b = 0$,

node b: $-YV_a + YV_b = 0$.

Figure 10.3

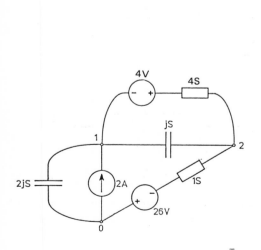

Figure 10.1

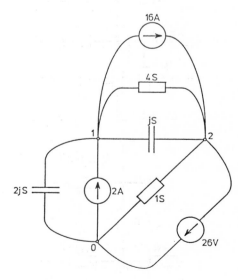

Figure 10.2

The nodes a and b are open, so the current entering is zero.
The stamp is

	a	b	
a	Y	−Y	0
b	−Y	Y	0

If Y = 0 this means open terminals, which result in the *zero stamp*. It can be omitted without difficulty.

2. The current source
(See Figure 10.4).
The equations are (entering currents positive)

node a: −I = 0,

Figure 10.4

node b: I = 0.

The stamp is

	a	b	
a	0	0	−I
b	0	0	I

If I = 0 (open terminals) the zero stamp results once more.

3. The voltage source
(See Figure 10.5)
Now it is necessary to introduce an unknown current I_x, which will be written in the part to be solved (the left-hand side). One may choose the direction at random.

node a: $I_x = 0$,

node b: $−I_x = 0$,

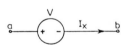

extra equation: $V_a − V_b = V$.

Figure 10.5

We shall indicate the extra row in the stamp with *.
The stamp is

	a	b	I_x	
a	0	0	1	0
b	0	0	−1	0
*	1	−1	0	V

The terms in the extra equation are expressed in volts, while the terms in both other equations are expressed in amperes.

If the voltage source intensity is $V = 0$ (which means a short circuit), then the right-hand side contains zeros only.

4. The voltage source with series impedance

(See Figure 10.6).

It is possible (see Section 10.2) to use the Norton equivalence. We shall not do so here but use an introduced current. The equations are:

node a: $I_x = 0$,

node b: $-I_x = 0$,

extra equation: $V_a - V_b = V + ZI_x$.

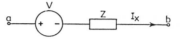

Figure 10.6

Consequently we get the stamp:

	a	b	I_x	
a	0	0	1	0
b	0	0	-1	0
*	1	-1	-Z	V

If $Z = 0$ the stamp of the voltage source results. If $V = 0$ we do not immediately have the stamp of the admittance $1/Z$, because I_x is present as a variable quantity. Elimination of I_x does result in the stamp of $1/Z$.

5. The voltage-current transactor

(See Figure 10.7).
The equations are:

node c: $G(V_a - V_b) = 0$,

node d: $G(-V_a + V_b) = 0$.

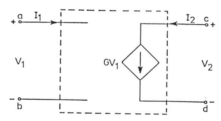

Figure 10.7

The nodes a and b give zeros which we shall neglect.
We get

	a	b	
c	G	-G	0
d	-G	G	0

This is the only transactor for which no current in the left-hand side need be introduced.

6. The current-current transactor

(See Figure 10.8).
The current I_1 is introduced as an unknown current to be solved.
The equations are:

node a: $I_1 = 0$,

node b: $-I_1 = 0$,

node c: $\alpha I_1 = 0$,

node d: $-\alpha I_1 = 0$,

extra equation $V_a - V_b = 0$.

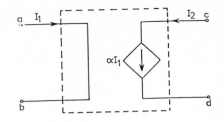

Figure 10.8

The stamp becomes:

	a	b	I_1	
a	0	0	1	0
b	0	0	-1	0
c	0	0	α	0
d	0	0	$-\alpha$	0
*	1	-1	0	0

7. The voltage-voltage transactor

(See Figure 10.9).

The current I_2 has to be introduced as an unknown quantity. The nodes a and b result in zeros. It further holds that

node c: $I_2 = 0$,

node d: $-I_2 = 0$,

extra equation: $V_c - V_d = \mu(V_a - V_b)$.

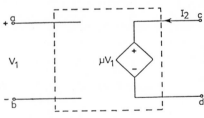

Figure 10.9

The stamp becomes:

	a	b	c	d	I_2	
c	0	0	0	0	1	0
d	0	0	0	0	-1	0
*	μ	$-\mu$	-1	1	0	0

8. The current-voltage transactor

(See Figure 10.10).

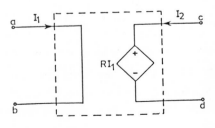

Figure 10.10

Both currents I_1 and I_2 have to be introduced as unknown quantities. We find

node a: $I_1 = 0$,

node b: $-I_1 = 0$,

node c: $I_2 = 0$,

node d: $-I_2 = 0$,

extra 1: $V_a - V_b = 0$,

extra 2: $V_c - V_d = RI_1$.

The stamp becomes:

	a	b	c	d	I_1	I_2	
a	0	0	0	0	1	0	0
b	0	0	0	0	-1	0	0
c	0	0	0	0	0	1	0
d	0	0	0	0	0	-1	0
*	1	-1	0	0	0	0	0
**	0	0	1	-1	-R	0	0

9. The gyrator

(See Figure 10.11).
The equations are

$$V_1 = -RI_2,$$

$$V_2 = RI_1.$$

In other words

$$I_1 = G(V_c - V_d),$$

$$I_2 = -G(V_a - V_b).$$

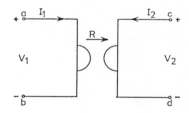

Figure 10.11

So

$$GV_C - GV_d = 0$$

and $\qquad -GV_a + GV_b = 0.$

(Both right-hand sides are zero because of the open terminals)
The stamp is

	a	b	c	d	
a	0	0	G	−G	0
b	0	0	−G	G	0
c	−G	G	0	0	0
d	G	−G	0	0	0

10. Magnetic coupled inductors

(See Figure 10.12).

The basic equations are

$$V_1 = \lambda L_1 I_1 + \lambda M I_2,$$

$$V_2 = \lambda M I_1 + \lambda L_2 I_2.$$

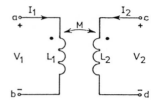

Figure 10.12

If we solve I_1 and I_2, with $V_1 = V_a − V_b$ and $V_2 = V_c − V_d$ we find:

$$I_1 = \frac{L_2}{\lambda D} V_a - \frac{L_2}{\lambda D} V_b - \frac{M}{\lambda D} V_c + \frac{M}{\lambda D} V_d,$$

$$I_2 = -\frac{M}{\lambda D} V_a + \frac{M}{\lambda D} V_b + \frac{L_1}{\lambda D} V_c - \frac{L_1}{\lambda D} V_d$$

with $D = L_1 L_2 - M^2 \neq 0$.

So the stamp is (omitting the zeros):

	a	b	c	d	
a	$\dfrac{L_2}{\lambda D}$	$-\dfrac{L_2}{\lambda D}$	$-\dfrac{M}{\lambda D}$	$\dfrac{M}{\lambda D}$	
b	$-\dfrac{L_2}{\lambda D}$	$\dfrac{L_2}{\lambda D}$	$\dfrac{M}{\lambda D}$	$-\dfrac{M}{\lambda D}$	
c	$-\dfrac{M}{\lambda D}$	$\dfrac{M}{\lambda D}$	$\dfrac{L_1}{\lambda D}$	$-\dfrac{L_1}{\lambda D}$	
d	$\dfrac{M}{\lambda D}$	$-\dfrac{M}{\lambda D}$	$-\dfrac{L_1}{\lambda D}$	$\dfrac{L_1}{\lambda D}$	

If $D = 0$, i.e. for complete coupling (see Section 4.4) we act as follows.

If we consider I_1 and I_2 as unknown quantities and take the basic equations as extra relations, we then find:

node a: $I_1 = 0$,

node b: $-I_1 = 0$,

node c: $I_2 = 0$,

node d: $-I_2 = 0$,

extra 1: $V_a - V_b = \lambda L_1 I_1 + \lambda M I_2$,

extra 2: $V_c - V_d = \lambda M I_1 + \lambda L_2 I_2$.

So the stamp becomes:

	a	b	c	d	I_1	I_2
a					1	
b					-1	
c						1
d						-1
*	1	-1			$-\lambda L_1$	$-\lambda M$
**			1	-1	$-\lambda M$	$-\lambda L_2$

11. The transformer
(See Figure 10.13).
The basic equations are

$$V_1 = \frac{V_2}{n},$$

$$I_1 + nI_2 = 0.$$

Thus we find

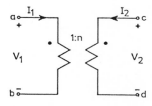

Figure 10.13

node a: $I_1 = 0$,

node b: $-I_1 = 0$,

node c: $I_2 = 0$, therefore $-\dfrac{I_1}{n} = 0$,

node d: $\dfrac{I_1}{n} = 0$,

extra: $V_a - V_b = \dfrac{V_c - V_d}{n}$.

So the stamp is

	a	b	c	d	I_1
a					1
b					-1
c					$-\dfrac{1}{n}$
d					$\dfrac{1}{n}$
*	1	-1	$-\dfrac{1}{n}$	$\dfrac{1}{n}$	

12. The NII

(See Figure 10.14).

The basic equations are

$$I_1 = G_1 V_2,$$

$$I_2 = G_2 V_1,$$

so that the following stamp results

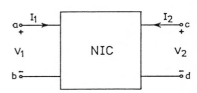

V_1 NII V_2

Figure 10.14

	a	b	c	d
a			G_1	$-G_1$
b			$-G_1$	G_1
c	G_2	$-G_2$		
d	$-G_2$	G_2		

13. The NIC

(See Figure 10.15).

The basic equations are

$$V_1 = \mu V_2,$$

$$I_2 = \alpha I_1.$$

Thus

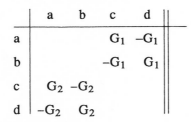

I_1 V_1 NIC V_2 I_2

Figure 10.15

node a: $I_1 = 0$,

node b: $-1_1 = 0$,

node c: $\alpha I_1 = 0$,

node d: $-\alpha I_1 = 0$,

extra: $V_a - V_b = \mu(V_c - V_d)$,

with which we find the following stamp:

	a	b	c	d	I_1
a					1
b					−1
c					α
d					$-\alpha$
*	1	−1	$-\mu$	μ	

14. The transistor

We use a greatly simplified model (see Figure 10.16).
We find

$$\text{node b: } I_b = GV_b - GV_e = 0,$$

$$\text{node e: } I_e = GV_e - GV_b - G_1(V_b - V_e) = 0,$$

$$\text{node c: } I_c = G_1(V_b - V_e) = 0,$$

so that the stamp becomes:

	b	e	c
b	G	$-G$	0
e	$-G-G_1$	$G+G_1$	0
c	G_1	$-G_1$	0

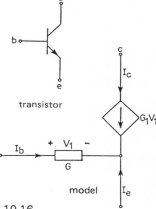

transistor

model

Figure 10.16

This stamp is the sum of the stamps of the conductor G and the voltage-current transactor.

15. The operational amplifier

For this stamp one takes that of the voltage-voltage transactor with a very large amplifying factor μ (e.g. 10^5).

10.6 Earthing a terminal

An element or a set of elements of which none of the nodes have been earthed are given. The question is how the stamp changes if one node is earthed. The potential of that node becomes zero so that the column concerned in the stamp can be left out. The row can also be left out because the concerning node equation is not relevant.

Result: *Earthing a terminal means leaving out the corresponding row and column.*

10.7 Connecting two terminals

Consider the following stamp:

	1	2	3	
1	a	b	c	I_1
2	d	e	f	I_2
3	g	h	i	I_3

The corresponding equations are

$$1) \ aV_1 + bV_2 + cV_3 = I_1,$$

$$2) \ dV_1 + eV_2 + fV_3 = I_2,$$

$$3) \ gV_1 + hV_2 + iV_3 = I_3.$$

Connect nodes 2 and 3. We shall call the newly formed node k. This results in the following equations:

$$V_2 = V_3 = V_k$$

and $\qquad I_k = I_2 + I_3.$

So we get

$$1) \ aV_1 + (b + c)V_k = I_1,$$

$$2) \ dV_1 + (e + f)V_k = I_2,$$

$$3) \ gV_1 + (h + i)V_k = I_3.$$

Adding both last equations results in:

$$1) \ aV_1 + (b + c)V_k = I_1,$$

$$k) \ (d + g)V_1 + (e + f + h + i)V_k = I_2 + I_3 = I_k.$$

with the stamp

	1	k	
1	a	b+c	I_1
k	d+g	e+f+h+i	I_k

If we compare this stamp with the original we can conclude that *the interconnection of two nodes comprises adding the corresponding rows and columns.*

10.8 A terminal becomes an internal node

The stamp of a network with k terminals means k node equations with k currents (sometimes an equation has been reduced to $I_m = 0$). If we do not consider a terminal as an external terminal the node voltage concerned must be solved from the equation of that node current set to zero and be substituted into the other node equations.

Example

Suppose the stamp is

	1	2	3	
1	a	b	c	I_1
2	d	e	f	I_2
3	g	h	i	I_3

and suppose we do not want to regard node 3 as an external terminal anymore.
We find

$$aV_1 + bV_2 + cV_3 = I_1,$$

$$dV_1 + eV_2 + fV_3 = I_2,$$

$$gV_1 + hV_2 + iV_3 = I_3.$$

It further holds that $I_3 = 0$.
So

$$V_3 = -\frac{g}{i} V_1 - \frac{h}{i} V_2.$$

So

$$aV_1 + bV_2 - \frac{cg}{i} V_1 - \frac{ch}{i} V_2 = I_1,$$

$$dV_1 + eV_2 - \frac{fg}{i} V_1 - \frac{fh}{i} V_2 = I_2.$$

So

$$I_1 = (a - \frac{cg}{i}) V_1 + (b - \frac{ch}{i}) V_2,$$

$$I_2 = (d - \frac{fg}{i}) V_1 + (e - \frac{fh}{i}) V_2,$$

with which we find the following stamp

	1	2	
1	$\dfrac{ai - cg}{i}$	$\dfrac{bi - cg}{i}$	I_1
2	$\dfrac{di - fg}{i}$	$\dfrac{ei - fh}{i}$	I_2

$i \neq 0.$

If $i = 0$ terminal 3 cannot be regarded as an internal node.

Conclusion: *If a terminal becomes an internal node we find the new stamp by means of an elimination process.*

10.9 Examples

Example 1 (Figure 10.17).

The stamps of the various network elements together give the MNA-matrix:

	1	2	I_x	
1	2+4	−2	1	0
2	−2	2	−1	8
*	1	−1	−3	4

Solution gives $V_1 = 2$ V, $V_2 = \frac{34}{7}$ V and $I_x = -\frac{16}{7}$ A.

Example 2 (Figure 10.18).

For the transistor we use a model that contains one more resistor than the model of Figure 10.16 (see Figure 10.19).

The source v_1 represents the input signal. The supply source V is a short circuit for a.c. components. If we replace v_1 and R_1 by their Norton equivalence, the circuit of Figure 10.20 results.

The NA-matrix becomes

	b	a	e	c	
b	$G_1+G_2+G_3+G_b$	$-G_b$	0	0	$\frac{v_1}{R_1}$
a	$-G_b$	G_b+G_a	$-G_a$	0	0
e	0	$-G-G_a$	G_a+G_4+G	0	0
c	0	G	$-G$	G_5	0

Solution gives, among other things, the voltage v_c as a function of t for a given $v_1(t)$, i.e. the amplification.

10.10 The solution of the matrix equation

The solution of a set of equations can be done in several ways. A method that is no longer used is *Cramer's rule.*

Matrix inversion is sometimes applied, but only for small matrices. The method used often nowadays is *L-U-factorisation.* It can be done manually small matrices ($< 5\times5$), for larger matrices the computer is used. In this method the matrix concerned is written as the product of the matrix L and the matrix U.

The L-matrix has zeros in the upper right-hand triangle, the U-matrix has zeros in the lower left-hand triangle. Moreover, the main diagonal of $\mathcal{U}$ contains elements of the value one only.

L stands for *lower*, $\mathcal{U}$ for *upper*. Treatment of this method is outside the scope of this book.

10.11 Numerical integration

So far we have only discussed the steady state in this chapter. We shall now consider networks with transient response. We start with a very simple network (see Figure 10.21).

Assume that the capacitor voltage has the value $v(0) \neq 0$ at $t = 0$. We now want to find

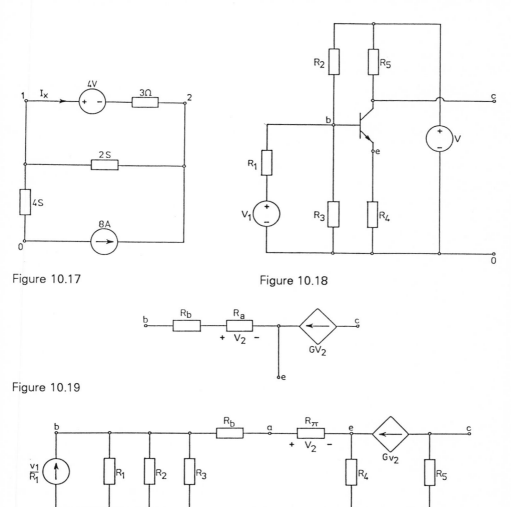

Figure 10.17

Figure 10.18

Figure 10.19

Figure 10.20

v(t) for t > 0. In order to be able to solve this problem with a digital computer we have to *discretesize* the function of v as a function of time, i.e. we have to choose a finite number of moments at the continuous time axis. In most cases equal intervals are chosen. We then speak of *equidistant* moments of time, which shall be indicated with the integer k. The distance between two sequential moments of time is called h. Let the number k = 0 coincide with the moment t = 0 so that

$$t = h \cdot k.$$

For the network of Figure 10.21:

$$v = -Ri,$$

$$i = C \frac{dv}{dt}.$$

If v(k), i.e. the voltage belonging to the number k, has been found then v(k + 1) can be found as follows. See Figure 10.22 in which a random function v(t) has been drawn.
The real value for k + 1 is v_r while the *approximate* value can be found from the value v(k) increased with the *difference*

$$\Delta = h \cdot \left(\frac{dv(t)}{dt}\right)_{t=k}. \tag{10.5}$$

We use the differential quotient at the moment k. An error arises which is proportionally smaller as h is chosen smaller. With a smaller value of h, however, the time of computing becomes greater so that a compromise must be found.
This method is known as the *Forward Euler Integration* method (FE)
We shall now solve the problem with this method.
We have

$$v(k+1) = v(k) + h\left(\frac{dv}{dt}\right)_{t=k}, \tag{10.6}$$

so

$$v(k+1) = v(k) + \frac{h}{C} i(k). \tag{10.7}$$

With

$$i(k) = -\frac{1}{R} v(k) \tag{10.8}$$

this results in

$$v(k+1) = v(k) \left(1 - \frac{h}{RC}\right). \tag{10.9}$$

So each value of v for a certain value of k can be found from the preceding value. We choose R = 1 Ω, C = 1 F, h = 0.2 s and v(0) = 1 V.
Then v(k+1) = 0.8v(k).
Consequently

v(1) = 0.8 V,

v(2) = 0.8² V,

v(3) = 0.8³ V,   etc.

which can easily be done on a pocket calculator. The real solution is

$$v(t) = e^{-\frac{t}{RC}} = e^{-t} \text{ V,}$$

and it is interesting to compare the computed numerical values with the exact values. The following table gives the result. The columns BE and TR will be discussed later.

time	k	exact(V)	FE(V)	BE(V)	TR(V)
0 s	0	1	1	1	1
0.2 s	1	0.82	0.80	0.83	0.82
0.4 s	2	0.67	0.64	0.69	0.67
0.6 s	3	0.55	0.51	0.58	0.55
0.8 s	4	0.45	0.41	0.48	0.45
1 s	5	0.37	0.33	0.40	0.37
2 s	10	0.14	0.11	0.16	0.13
5 s	25	0.007	0.004	0.010	0.006
∞	∞	0	0	0	0

In spite of the rather large step h (5 *samples* in the RC-time) the difference is quite small. This difference can be made smaller still by using the *Backward Euler* method (BE) instead of the FE. In this method we use the derivative at the following moment of time:

$$\Delta = h \cdot \left(\frac{dv}{dt}\right)_{t=k+1}. \tag{10.10}$$

For the considered network we now get:

$$v(k+1) = v(k) + h\left(\frac{dv}{dt}\right)_{t=k+1}, \tag{10.11}$$

so

$$v(k+1) = v(k) + \frac{h}{C} i \ (k+1). \tag{10.12}$$

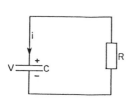

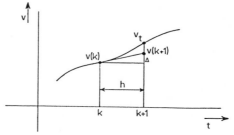

Figure 10.21   Figure 10.22

With

$$i(k+1) = -\frac{1}{R} v(k+1)$$

it follows that

$$v(k+1) = v(k) - \frac{h}{RC} v(k+1),$$

so that

$$v(k+1) = \frac{v(k)}{1 + \dfrac{h}{RC}} . \qquad (10.13)$$

For the chosen values of R, C and h we now get:

$$v(k+1) = \frac{v(k)}{1.2} ,$$

so that

$$v(1) = \frac{1}{1.2} V,$$

$$v(2) = \frac{1}{1.2^2} V, \quad \text{etc.}$$

These values are indicated in the BE column of the table. Note that FE gives too small a value and BE too large a value. So it is obvious to try a method which is the average of both the preceding methods. This is the *trapezoidal integration method* (TR).
For this it holds that:

$$v(k+1) = v(k) + \frac{1}{2} h \{(\frac{dv}{dt})_{t=k} + (\frac{dv}{dt})_{t=k+1}\}, \qquad (10.14)$$

so

$$v(k+1) = v(k) + \frac{h}{2C} \{i(k) + i(k+1)\}. \qquad (10.15)$$

With

$$i(k) = -\frac{1}{R} v(k) \text{ and } i(k+1) = -\frac{1}{R} v(k+1)$$

this results in

$$v(k+1) = v(k) - \frac{h}{2RC} \{v(k) + v(k+1)\}.$$

Consequently

$$v(k+1) (1 + \frac{h}{2RC}) = v(k)(1 - \frac{h}{2RC}),$$

so

$$v(k+1) = \frac{1 - h/2RC}{1 + h/2RC}\ v(k),$$

(10.16)

so that with the given values of R, C and h:

$$v(k+1) = \frac{1 - 0.1}{1 + 0.1}\ v(k) = \frac{0.9}{1.1}\ v(k),$$

therefore

$$v(1) = 0.82 \text{ V}, \ v(2) = 0.67 \text{ V, etc.}$$

These values are included in the table in the column TR. The difference is pleasantly small; for small values of k one can only notice the difference in the third decimal (not indicated).
We shall now analyse the problem once more, but now for the step value h = 3 s. We find:
FE

$$v(k+1) = v(k)(1-3) = -2\ v(k).$$

So $v(1) = -2$ V,

  $v(2) = \ \ 4$ V,

  $v(3) = -8$ V, etc.

thus $|v(\infty)| = \infty.$

We find values that differ much from the exact values; even the row is divergent!
Now the BE:

$$v(k+1) = v(k)\ \frac{1}{1 + 3}\ = \frac{1}{4}\ v(k),$$

so that

  $v(1) = 0.25$ V,

  $v(2) = 0.06$ V, etc.

  $v(\infty) = 0$ V.

These values differ greatly, but the function does approach zero for k → ∞.
Finally the TR:

$$v(k+1) = \frac{1 - 3/2}{1 + 3/2}\ v(k) = -\frac{1}{5}\ v(k),$$

thus

$$v(1) = -0.2 \text{ V},$$

$$v(2) = 0.04 \text{ V},$$

$$v(\infty) = 0 \text{ V}.$$

This row is convergent, too. We can see that the forward Euler method may cause errors. That is why one always uses the backward Euler method or still better the trapezoidal method.

10.12 Equivalent circuits

It appears to be possible to derive equivalent circuits for a capacitor, of which the basic equation has been discretesised. First consider the backward Euler method.
For the capacitor we have found (10.12):

$$v(k+1) = v(k) + \frac{h}{C} i(k+1).$$

Because the terms of this sequation are all voltages this equation represents a series circuit according to Kirchhoff's voltage law (see Figure 10.23).
Starting from (10.12) we can also derive a parallel circuit by multiplying both sides of the equation with C/h and by putting i(k+1) in the left-hand side:

$$i(k+1) = \frac{C}{h} v(k+1) - \frac{C}{h} v(k). \tag{10.17}$$

Now the terms are currents (see Figure 10.24).
Note that we now have a conductance.
We can also derive two equivalent circuits for an inductor. We have

$$v = L \frac{di}{dt},$$

$$i(k+1) = i(k) + h\left(\frac{di}{dt}\right)_{t=k+1},$$

so

$$i(k+1) = i(k) + \frac{h}{L} v(k+1). \tag{10.18}$$

We now find the circuit of Figure 10.25.
When we take v(k+1) to the left-hand side we get

$$v(k+1) = \frac{L}{h} i(k+1) - \frac{L}{h} i(k). \tag{10.19}$$

(See Figure 10.26).
We now have derived equivalent circuits for the *dynamic* elements inductor and capacitor (the name reactive elements is used in the theory of complex quantities). We can derive stamps from these circuits which can be used in the MNA-matrix.

If we call the upper terminal a and the lower one b, then for Figure 10.23 (the *series version* of the capacitor) follows:

	a	b	i		
a	0	0	1		0
b	0	0	-1		0
*	1	-1	$-\dfrac{h}{C}$		$v(k)$

and for the *parallel version*:

	a	b		
a	$\dfrac{C}{h}$	$-\dfrac{C}{h}$		$\dfrac{C}{h}\,v(k)$
b	$-\dfrac{C}{h}$	$\dfrac{C}{h}$		$-\dfrac{C}{h}\,v(k)$

For the inductor we find for the parallel version:

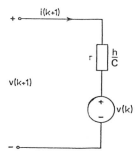

Figure 10.23

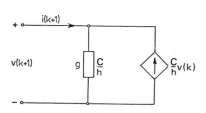

Figure 10.24

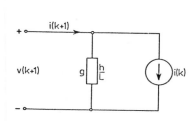

Figure 10.25

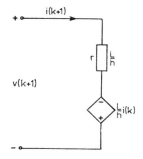

Figure 10.26

	a	b	
a	$\dfrac{h}{L}$	$-\dfrac{h}{L}$	$-i(k)$
b	$-\dfrac{h}{L}$	$\dfrac{h}{L}$	$i(k)$

and for the series version:

	a	b	i	
a	0	0	1	0
b	0	0	-1	0
*	1	-1	$-\dfrac{L}{h}$	$-\dfrac{L}{h}\, i(k)$

In the left-hand part of these stamps the quantities are indicated at the moment k + 1, in the right-hand part at the moment k. If the right-hand part is known at t = k, the left-hand part can be solved at t = k + 1.

Owing to the influence exercised by the rest of the network via the MNA-matrix, the right-hand side will be known at t = k + 1, after which the left-hand part can be solved at t = k + 2, etc. An exception is the parallel version of the inductor, which shall be discussed later.

We shall first show some examples.

Example 1

Consider the same example as in the beginning of the preceding section (see Figure 10.21). Using the parallel equivalence for the capacitor we find the network of Figure 10.27.

The NA-matrix is very simple:

	1	
1	$\dfrac{1}{R} + \dfrac{C}{h}$	$\dfrac{C}{h}\, v_1(k)$

We find $(\dfrac{1}{R} + \dfrac{C}{h})\, v_1(k+1) = \dfrac{C}{h}\, v_1(k)$, which is the same formula as (10.13). (We have called the voltage of node 1 v_1 here, instead of v.)

Example 2 (see Figure 10.28).

Find the equivalent network of Figure 10.29.

Choose h = 0.1 s.

We find

	1	
1	$10 + 1$	$\dfrac{v_b(k+1)}{2} + 10\, v_1(k)$

so that

$$11 \ v_1(k+1) = \frac{v_b(k+1)}{2} + 10 \ v_1(k).$$

Note that the known source voltage v_b is present in the right-hand side at $t = k + 1$.
For $k = 0$ we find:

$$11 \ v_1(1) = \frac{v_b(1)}{2} + 10 \ v_1(0).$$

Because $v_1(0)$ must be given (initial condition) we can calculate $v_1(1)$. For $k = 1$ this results in

$$11 \ v_1(2) = \frac{v_b(2)}{2} + 10 \ v_1(1).$$

After that $v_1(2)$ can be found, etc.
This can be done automatically, quickly and accurately by computer.

Example 3

Consider the example of Figure 9.15. The equivalent network (parallel version of the capacitor and series version of the inductor) is shown in Figure 10.30.
Choose $h = 0.01$ s.
We find the following MNA-matrix:

	1	2	i	
1	$\frac{1}{2}$	0	1	6
2	0	2	−1	$2 \ v_2(k)$
*	1	−1	−100	$-100 \ i(k)$

$$v_C = v_2$$

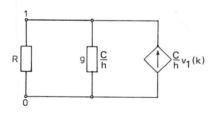

Figure 10.27

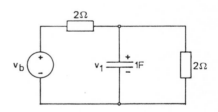

Figure 10.28

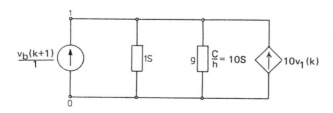

Figure 10.29

The initial conditions are given: $v_2(0) = -11$ V and $i(0) = 2$ A. With that $v_1(1)$, $v_2(1)$ and $i(1)$ can be found:

$$v_1(1) = 7.64 \text{ V}; \quad v_2(1) = -9.91 \text{ V}; \quad i(1) = 2.18 \text{ A}.$$

Substitute $v_2(1)$ and $i(1)$ in the right-hand side and calculate $v_2(2)$ and $i(2)$, etc. Computer calculations, in which k ranges from 1 to 350, result in the dotted plot in Figures 10.31 and 10.32. In these plots the drawn line is the exact function (according to the calculations in Chapter 9).

The computer program of this example is as follows:

```
Begin
k   = 1
m = 350
Read the left-hand side of MNA
Read the initial conditions (v₂(0) = -11 and i(0) = 2).
Substitute these into the right-hand part of MNA
While k < m repeat:
        Solve the set, i.e. find all voltages and
        currents (here v₁, v₂ and i) for k.
        Substitute that in the right-hand part of MNA.
        k : = k + 1
End.
```

Result: All node voltages (and the voltage source currents) for k = 1,2,...,m.

With this process we have found the solution of a network without using the method described in Chapter 9. Its advantage is that source intensities with a particular function of time can also be admitted. A list of source intensities with rising values of k suffices to find the solution.

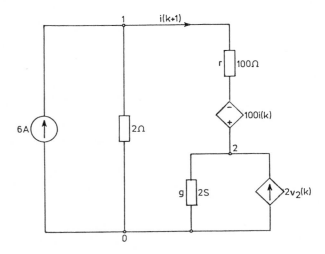

Figure 10.30

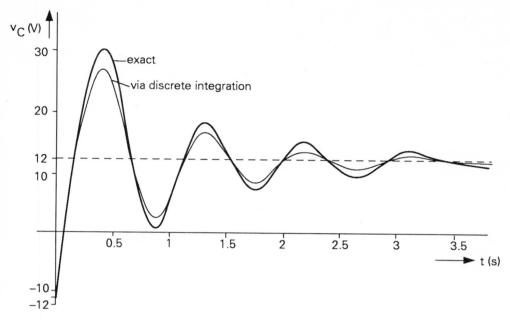

Figure 10.31

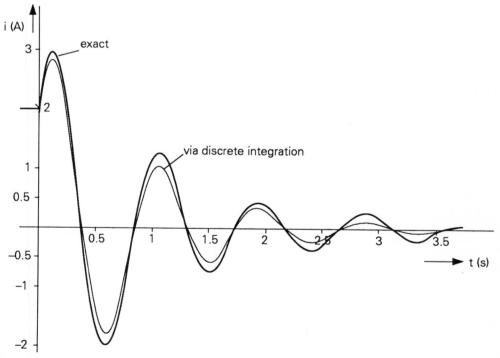

Figure 10.32

Before using the trapezoidal rule in networks, we must go back to the parallel version of the inductor (Figure 10.25). The stamp derived contains the current in the right-hand part of the MNA-matrix but not in the left-hand part. This is due to the fact that the stamp derived from Figure 10.25 is in fact a reduction of the real stamp. We shall investigate this further. In Figure 10.33 the equivalent network which starts with (10.18) has been drawn again.

The stamp including node c is:

	a	b	c	i	
a				1	
b		$\frac{h}{L}$	$-\frac{h}{L}$		i(k)
c		$-\frac{h}{L}$	$\frac{h}{L}$	-1	$-i(k)$
*	1		-1		

This represents four relations. Connecting the nodes a and c to one node a (as described in Section 9.7) results in the stamp found previously:

	a	b	
a	$\frac{h}{L}$	$-\frac{h}{L}$	$-i(k)$
b	$-\frac{h}{L}$	$\frac{h}{L}$	$i(k)$

This stamp means two equations, while the equation of node c – which just happens to be (10.18) – has been eliminated (the equation $v_a = v_c$ has also disappeared, but this is not important).

The conclusion is that we can use the reduced stamp of the parallel version of the inductor, provided we add (10.18) in the calculations.

Example 4 (Figure 10.34).
Find i(k) given i(0).

Solution
With the parallel version of the inductor we find Figure 10.35.
The (reduced) stamp of the inductor results in:

$$\text{MNA} = \begin{array}{c|c|c} & 1 & \\ \hline 1 & G + \dfrac{h}{L} & -i(k) \end{array}$$

Together with (10.18)

$$i(k+1) = i(k) + \frac{h}{L}\, v_1(k+1)$$

and the initial condition i(0) for the inductor this results in i(k) for all $k \geq 0$.

After all, with the given $i(0)$, $v_1(1)$ follows from the MNA-matrix. Subsequently (10.18) gives $i(1)$ which, substituted in the MNA-matrix gives $v_1(2)$, etc.

We further note that the 4×4-stamp for the inductor also leads to the solution, with the disadvantage, however, of an unnecessarily large MNA-matrix (more computer time and memory places).

We shall now discuss the numerical analysis of networks with the trapezoidal rule. Start with the capacitor. We have

$$i = C \frac{dv}{dt} .$$

The trapezoidal rule gives

$$v(k+1) = v(k) + \frac{h}{2} \{ (\frac{dv}{dt})_{t=k} + (\frac{dv}{dt})_{t=k+1} \},$$

so that we get

$$v(k+1) = v(k) + \frac{h}{2C} i(k) + \frac{h}{2C} i(k+1). \tag{10.20}$$

This results in the circuit of Figure 10.36.
With that the stamp becomes:

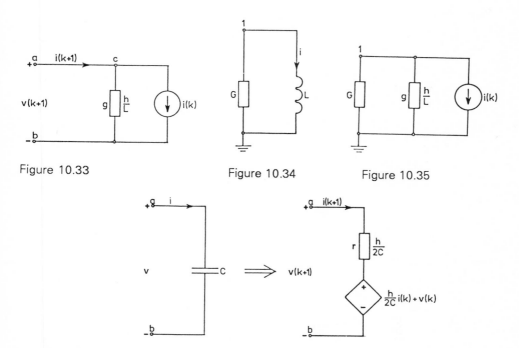

Figure 10.33

Figure 10.34

Figure 10.35

Figure 10.36

	a	b	i		
a			i		
b			-1		
*	1	-1	$-\dfrac{h}{2C}$		$\dfrac{h}{2C}\,i(k) + v(k)$

The disadvantage of this stamp is that besides the capacitor voltage at $k = 0$ we have to know the capacitor current at $k = 0$. This disadvantage always arises with the trapezoidal rule.

Moreover, the stamp is large, but its size can be reduced by using the parallel version. To that end we bring the current $i(k+1)$ to the left-hand side in (10.20):

$$i(k+1) = \frac{2C}{h}\,v(k+1) - \frac{2C}{h}\,v(k) - i(k). \tag{10.21}$$

The circuit becomes as Figure 10.37.
So the stamp is

	a	b		
a	$\dfrac{2C}{h}$	$-\dfrac{2C}{h}$		$\dfrac{2C}{h}\,v(k) + i(k)$
b	$-\dfrac{2C}{h}$	$\dfrac{2C}{h}$		$-\dfrac{2C}{h}\,v(k) - i(k)$

Here $v = v_a - v_b$.

As an extra relation we use (10.21), as discussed above.

Note the similarity of this stamp and that found in the backward Euler method.

Now consider the inductor. We have

$$v = L\,\frac{di}{dt}\,.$$

The trapezoidal rules gives:

$$i(k+1) = i(k) + \frac{h}{2}\,\left\{\left(\frac{di}{dt}\right)_{t=k} + \left(\frac{di}{dt}\right)_{t=k+1}\right\},$$

so that

$$i(k+1) = i(k) + \frac{h}{2L}\,v(k) + \frac{h}{2L}\,v(k+1) \tag{10.22}$$

arises. The circuit becomes as Figure 10.38.
The stamp is:

	a	b		
a	$\dfrac{h}{2L}$	$-\dfrac{h}{2L}$		$-i(k) - \dfrac{h}{2L}\,v(k)$
b	$-\dfrac{h}{2L}$	$\dfrac{h}{2L}$		$i(k) + \dfrac{h}{2L}\,v(k)$

Here $v = v_a - v_b$. The extra relation is (10.22)

For completeness' sake we shall derive the series version. Bring the voltage $v(k+1)$ to the left-hand side in (10.22):

$$v(k+1) = \frac{2L}{h} i(k+1) - \frac{2L}{h} i(k) - v(k).$$

(10.23)

This results in the circuit of Figure 10.39.
The stamp is:

	a	b	i	
a			1	
b			−1	
*	1	−1	$-\frac{2L}{h}$	$-\frac{2L}{h} i(k) - v(k)$

Figure 10.37

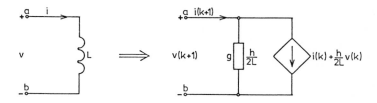

Figure 10.38

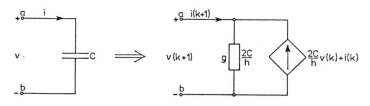

Figure 10.39

To conclude this section we shall once more discuss Example 3, but this time with the trapezoidal rule (see Figure 10.40).
We find:

$$
\text{MNA} = \begin{array}{c|cc|c}
 & 1 & 2 & \\
\hline
1 & G + \dfrac{h}{2L} & -\dfrac{h}{2L} & I_1 - i - \dfrac{h}{2L} v_1 + \dfrac{h}{2L} v_2 \\[2ex]
2 & -\dfrac{h}{2L} & \dfrac{h}{2L} + \dfrac{2C}{h} & 2i + \dfrac{h}{2L} v_1 - \dfrac{h}{2L} v_2 + \dfrac{2C}{h} v_2
\end{array}
$$

with the extra relation:

$$i(k+1) = \frac{2C}{h} v_2(k+1) - \frac{2C}{h} v_2(k) - i(k).$$

The other extra relation:

$$i(k+1) = i(k) + \frac{h}{2L} v_{12}(k) + \frac{h}{2L} v_{12}(k+1),$$

is not necessary, because the inductor current is similar to the capacitor current here.
The initial conditions were: $i(0) = 2$ A, $v_C(0) = v_2(0) = -11$ V. Together with $V_b = 12$ V (see Figure 9.15) this results in the value of v_1 at $t = 0$, viz. $v_1(0) = 8$ V. The source current $I_1 = 6$ A again.
After substituting the remaining values and after computer calculation it turns out that the difference with the exact values is so small that both plots nearly coincide.
In the table on the next pages the computer output of the capacitor voltage $v_2(k)$ for k, increasing from 0 to 357, is given, in which $t = 100k$.
The first column gives k, the second one gives the exact value of v_C according to the formula

$$v_C = e^{-t} (-23 \cos 7t + 11 \sin 7t) + 12 \text{ V},$$

while in the last column the capacitor voltage for numerical integration according to the

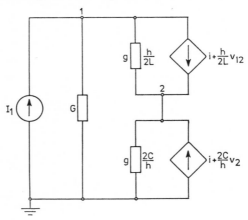

Figure 10.40

trapezoidal rules is given, and this only at those moments that the value differs from the value in the second column.

This clearly shows the great accuracy of the trapezoidal rule.

k	v_C (V) exact	v_C (V) with TR	k	v_C (V) exact	v_C (V) with TR	k	v_C (V) exact	v_C (V) with TR
0	−11,00		45	26,6		90	2,73	2,70
1	− 9,95		46	25,9		91	3,16	3,14
2	− 8,82		47	25,2		92	3,63	3,61
3	− 7,60	− 7,61	48	24,4		93	4,13	4,11
4	− 6,32		49	23,6		94	4,67	4,64
5	− 4,96	− 4,97	50	22,7		95	5,22	5,20
6	− 3,55	− 3,56	51	21,8		96	5,80	5,78
7	− 2,09	− 2,10	52	20,9		97	6,40	6,37
8	− 0,595	− 0,600	53	19,9		98	7,02	6,99
9	0,938	0,932	54	18,9	19,0	99	7,64	7,62
10	2,49		55	17,9	18,0	100	8,28	8,25
11	4,07	4,06	56	16,9	17,0	101	8,92	8,89
12	5,65	5,64	57	15,9	16,0	102	9,57	9,54
13	7,23	7,22	58	14,9		103	10,2	
14	8,80		59	13,9		104	10,9	10,8
15	10,4		60	12,9		105	11,5	
16	11,9		61	11,9		106	12,1	
17	13,4		62	11,0		107	12,7	
18	14,9		63	10,1		108	13,3	
19	16,3		64	9,15	9,17	109	13,9	
20	17,7		65	8,27	8,29	110	14,4	
21	19,0		66	7,44	7,45	111	15,0	
22	20,3	20,2	67	6,64	6,65	112	15,5	
23	21,4		68	5,88	5,89	113	16,0	15,9
24	22,6		69	5,17	5,18	114	16,4	
25	23,6		70	4,50	4,52	115	16,8	
26	24,6		71	3,89	3,90	116	17,2	
27	25,5		72	3,33	3,34	117	17,6	
28	26,3		73	2,82	2,83	118	17,9	
29	27,0		74	2,37		119	18,2	
30	27,6		75	1,97	1,98	120	18,4	
31	28,2		76	1,64		121	18,6	
32	28,6		77	1,36		122	18,8	
33	29,0		78	1,13		123	19,0	
34	29,3		79	0,97		124	19,1	
35	29,4		80	0,87	0,86	125	19,1	
36	29,5		81	0,82	0,81	126	19,1	
37	29,5		82	0,82		127	19,1	
38	29,4		83	0,89	0,88	128	19,1	
39	29,3		84	1,00	0,99	129	19,0	
40	29,0		85	1,17	1,16	130	18,9	
41	28,7		86	1,39	1,38	131	18,8	
42	28,3		87	1,66	1,64	132	18,6	
43	27,8		88	1,97	1,95	133	18,4	
44	27,2		89	2,33	2,31	134	18,1	18,2

k	v_C exact	with TR	k	v_C exact	with TR	k	v_C exact	with TR
135	17,9		184	9,06	9,04	233	13,2	13,1
136	17,6		185	9,29	9,27	234	13,1	
137	17,3		186	9,53	9,51	235	12,9	
138	17,0		187	9,78	9,76	236	12,7	12,8
139	16,6	16,7	188	10,0		237	12,6	
140	16,3		189	10,3		238	12,4	
141	15,9		190	10,5		239	12,2	12,3
142	15,5	15,6	191	10,8		240	12,1	
143	15,1	15,2	192	11,1		241	11,9	
144	14,7	14,8	193	11,3		242	11,8	
145	14,3	14,4	194	11,6		243	11,6	
146	13,9		195	11,9	11,8	244	11,5	
147	13,5		196	12,1		245	11,3	
148	13,1		197	12,4	12,3	246	11,2	
149	12,7		198	12,6		247	11,0	11,1
150	12,3		199	12,8		248	10,9	
151	11,9		200	13,0		249	10,8	
152	11,5		201	13,3	13,2	250	10,7	
153	11,1		202	13,5		251	10,6	
154	10,8		203	13,7	13,6	252	10,5	
155	10,4		204	13,8		253	10,4	
156	10,1		205	14,0		254	10,4	
157	9,74	9,75	206	14,2		255	10,3	
158	9,43	9,45	207	14,3		256	10,3	
159	9,15	9,16	208	14,4		257	10,2	
160	8,88	8,89	209	14,5		258	10,2	
161	8,64	8,65	210	14,6		259	10,2	
162	8,41	8,42	211	14,7		260	10,1	
163	8,21	8,22	212	14,8		261	10,1	
164	8,03	8,04	213	14,8		262	10,1	
165	7,88		214	14,9		263	10,2	
166	7,75		215	14,9		264	10,2	
167	7,64		216	14,9		265	10,2	
168	7,55		217	14,9		266	10,3	10,2
169	7,49		218	14,9		267	10,3	
170	7,45		219	14,8	14,90	268	10,4	
171	7,44	7,43	220	14,8		269	10,4	
172	7,45	7,44	221	14,7		270	10,5	
173	7,48	7,47	222	14,7		271	10,6	
174	7,53	7,52	223	14,6		272	10,6	
175	7,61	7,59	224	14,5		273	10,7	
176	7,70	7,69	225	14,4		274	10,8	
177	7,81	7,80	226	14,3		275	10,9	
178	7,95	7,93	227	14,1		276	11,0	
179	8,10	8,08	228	14,0		277	11,1	
180	8,26	8,24	229	13,9		278	11,2	
181	8,44	8,42	230	13,7		279	11,3	
182	8,64	8,62	231	13,6		280	11,4	
183	8,85	8,83	232	13,4		281	11,5	

k	v_C		k	v_C		k	v_C	
	exact	with TR		exact	with TR		exact	with TR
282	11,6		308	13,2		333	11,8	
283	11,8	11,7	309	13,2		334	11,8	
284	11,9	11,8	310	13,1		335	11,7	
285	12,0		311	13,1		336	11,7	
286	12,1		312	13,1		337	11,6	
287	12,2		313	13,0		338	11,6	
288	12,3		314	13,0		339	11,5	
289	12,4	12,3	315	13,0		340	11,5	
290	12,4		316	12,9		341	11,4	
291	12,5		317	12,9		342	11,4	
292	12,6		318	12,8		343	11,4	
293	12,7		319	12,7	12,8	344	11,3	
294	12,8		320	12,7		345	11,3	
295	12,8		321	12,6		346	11,3	
296	12,9		322	12,6		347	11,3	
297	13,0	12,9	323	12,5		348	11,3	
298	13,0		324	12,5		349	11,2	
299	13,0		325	12,4		350	11,2	
300	13,1		326	12,3		351	11,2	
301	13,1		327	12,2		352	11,2	
302	13,1		328	12,1	12,2	353	11,3	11,2
303	13,2		329	12,1		354	11,3	
304	13,2		330	12,0		355	11,3	
305	13,2		331	11,9	12,0	356	11,3	
306	13,2		332	11,9		357	11,3	
307	13,2							

10.13 Nonlinear elements

So far we have assumed all elements considered to be linear. In practice, however, one hardly ever comes across linear behaviour. For instance, a resistor does not have the same resistance for each current, because the resistance depends on the temperature, which itself is a function of the dissipation, and which in turn is determined by the current. This can be indicated by the formula

$$V = R(I). \tag{10.24}$$

R has to be regarded as a *function* here. The current-voltage plot will not be a straight line then.

A nonlinear resistance often met in practice is the diode. This can be a single network element but it is also present in more advanced equivalent circuits of a transistor. In Figure 10.41 its symbol has been drawn.

The relation between V and I is shown in Figure 10.42.

Another nonlinear element is the inductor with iron core. The magnetic flux is a nonlinear function of the current:

$$\Phi = L(I). \tag{10.25}$$

The capacitor with a dielectric has nonlinear behaviour as well. The charge is a function of the voltage:

$$Q = C(V). \tag{10.26}$$

A network with nonlinear elements can also be analysed with the aid of a computer.

Example (Figure 10.43).
V_b is a d.c. voltage source, R_1 is a linear resistance; R is a nonlinear resistance with the function:

$$I = G(V). \tag{a}$$

Find the current I.
In Figure 10.44 the plot of the nonlinear resistance together with the plot of the formula

$$V_b = R_1 I + V, \tag{b}$$

called *load line*, has been drawn.
Intersection S gives the solution. We shall now try to find this intersection with the help of a computer. We take an arbitrary value V(0), calculate the corresponding value I(0) of the current with (a) and determine at which point the tangent intersects the load line. The intersection is called S'(1). In general, S'(1) will not coincide with S.
The voltage V(1) belongs to S'(1) and this voltage determines the current I'(1) of the load line and the current I(1) of the nonlinear function. This results in point S(1). Determine the tangent in S(1) and intersect it with the load line. The intersection is called S'(2), where S(2) belongs to the nonlinear plot with the voltage V(2) and the current I(2), etc. The successive values of the voltages V(0), V(1), V(2), etc. are called *iterations*. We finish the calculation if the difference between two successive iterations has become smaller than a prescribed (very small) number ε. The tangent for the starting value is called

$$g(0) = \left(\frac{dI}{dV}\right)_{V=V(0)} = \left\{\frac{d}{dv} G(V)\right\}_{V=V(0)}. \tag{c}$$

From the plot we see that

$$I'(1) = I(0) + g(0) \{V(1) - V(0)\}. \tag{d}$$

It further holds that

$$V_b = R_1 I'(1) + V(1), \tag{e}$$

so that we get

$$\frac{V_b - V(1)}{R_1} = I(0) + g(0) \{V(1) - V(0)\}. \tag{f}$$

From this and with

$$I(0) = G\{V(0)\} \tag{g}$$

follows:

$$V(1) = \frac{R_1}{1 + g(0) R_1} \left[\frac{V_b}{R_1} - G\{V(0)\} + g(0)V(0) \right]. \tag{h}$$

With this we are able to calculate the value of V(1) from V(0). By repetition from (h) we find

$$V(m+1) = \frac{R_1}{1 + g(m) R_1} \left[\frac{V_b}{R_1} - G\{V(m)\} + g(m)V(m) \right], \tag{i}$$

with, see (c)

$$g(m) = \left(\frac{dI}{dV} \right)_{V=V(m)}. \tag{j}$$

We now solve the problem with $V_b = 12$ V; $R_1 = 2\ \Omega$; $I = 0.1\ V^2$; $\varepsilon = 0.1$. By the latter we mean that

$$|V(m+1) - V(m)| < 0.1.$$

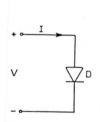

Figure 10.41

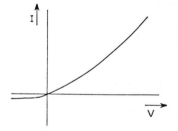

Figure 10.42

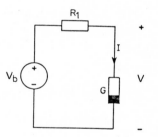

Figure 10.43

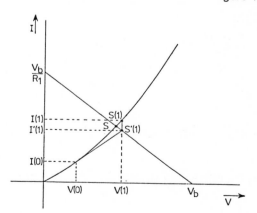

Figure 10.44

Choose the starting value V(0) = 1. We then find

$$g = (\frac{dI}{dV}) = 0.2V.$$

So

$$g(0) = 0.2 \ V(0) = 0.2$$

and

$$I(0) = 0.1.$$

From (h):

$$V(1) = \frac{2}{1 + 0.2 \cdot 2} \ (\frac{12}{2} - 0.1 + 0.2 \cdot 1) = 8.7.$$

We further find

$$g(1) = 0.2 \cdot V(1) = 1.74; \quad I(1) = 0.1 \cdot 8.7^2 = 7.59.$$

From (i):

$$V(2) = \frac{2}{1 + 1.74 \cdot 2} \ (6 - 7.59 + 1.74 \cdot 8.7) = 6.05.$$

We further find

$$g(2) = 0.2 \cdot 6.05 = 1.21; \quad I(2) = 0.1 \cdot 6.05^2 = 3.66.$$

So

$$V(3) = \frac{2}{1 + 1.21 \cdot 2} \ (6 - 3.66 + 1.21 \cdot 6.05) = 5.65.$$

To continue

$$g(3) = 0.2 \cdot 5.65 = 1.13, \quad I(3) = 0.1 \cdot 5,65^2 = 3.19,$$

$$V(4) = \frac{2}{1 + 1.13 \cdot 2} \ (6 - 3.19 + 1.13 \cdot 5.65) = 5.64.$$

We are now finished, because $|V(3) - V(4)| = 0.01 \ll \varepsilon$.
So the solution is V = 5.6 V and the current is I = 3.2 A, both on the load line and on the nonlinear characteristic.
Note how quickly the method converges.

We may also determine an equivalent circuit of a nonlinear element and a corresponding stamp. From (d):

$$I'(m+1) = I(m) + g(m)V(m+1) - g(m)V(m). \tag{10.27}$$

The quantity g(m) can be regarded as a conductance (see Figure 10.45).
The stamp is

	a	b	
a	g(m)	−g(m)	−I(m) + g(m)V(m)
b	−g(m)	g(m)	I(m) − g(m)V(m)

We shall solve the above example with the MNA-method. In Figure 10.46 the equivalent circuit has been drawn.

The MNA-matrix is

	1	
1	$\frac{1}{R_1}$ + g(m)	$\frac{V_b}{R_1}$ + g(m)V(m) − I(m)

So

$$\{\frac{1}{R_1} + g(m)\}\, V_1(m+1) = \frac{V_b}{R_1} + g(m)V(m) - I(m),$$

which results in formula (i) found above, if we set $I(m) = G\{V(m)\}$. So it is indeed possible to solve networks with nonlinear elements systematically with a computer. It even turns out that this method can also be used with *nonlinear dynamic elements*. However, we shall not discuss this in more detail.

We shall finally derive a computer program based on a somewhat larger network (Figure 10.47).

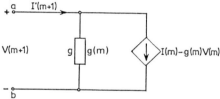

Figure 10.45

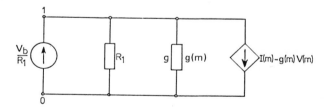

Figure 10.46

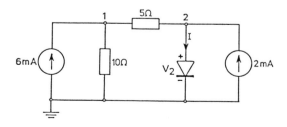

Figure 10.47

Given that the current source intensities are constant.

It further holds that $I = 10^{-3}(e^{V_2/0.025} - 1)$

Determine V_2.

Solution

The equivalent circuit is (see Figure 10.48).

The MNA-matrix is

	1	2		
1	0.3	−0.2		0.006
2	−0.2	0.2 + g(m)		$g(m)V_2(m) - I(m) + 0.002$

It further holds that $g = \dfrac{dI}{dV_2} = \dfrac{1}{25} \cdot e^{V_2/0.025}$.

We start with $V_2(0) = 10 \, mV$.

In that case it holds that

$$g(0) = \frac{1}{25} \cdot e^{10/25} = 0.059673$$

and $I(0) = 0.491852 \cdot 10^{-3} \, A.$

We substitute this in the MNA-matrix and solve the set. We find:

$$V_1(1) = 0.052214 \, V,$$

$$V_2(1) = 0.048321 \, V.$$

With that we calculate the new values of g and I:

$$g(1) = 0.276364,$$

$$I(1) = 5.909104 \cdot 10^{-3} \, A.$$

Solve →

$$V_1(2) = 0.046130 \, V,$$

$$V_2(2) = 0.039195 \, V.$$

The difference between $V_2(1)$ and $V_2(2)$ is still too great, that is why we continue

$$g(2) = 0.191843,$$

$$I(2) = 3.796085 \cdot 10^{-3} \, A.$$

Solve →

$$V_1(3) = 0.045075 \, V,$$

$$V_2(3) = 0.037613 \, V.$$

The difference becomes smaller. We continue:

$g(3) = 0.180080,$

$I(3) = 3.501992 \cdot 10^{-3}$ A.

Solve →

$V_1(4) = 0.045050$ V,

$V_2(4) = 0.037574$ V.

The difference has become smaller than 0.1 mV!
We do one more step:

$g(4) = 0.179799,$

$I(4) = 3.494975 \cdot 10^{-3}$ A.

Solve →

$V_1(5) = 0.045050$ V,

$V_2(5) = 0.037574$ V.

The difference to the earlier values cannot even be seen in the sixth decimal, so that the solution is: $V_2 = 37.6$ mV.
The above derivation leads to the following computer program:

```
Begin

    m = 1 (counter)
    V₂(0) = 1 (starting value)
    eps = 0.01 (accuracy desired)
    mmax = 10 (largest number of steps)
    Read the constant elements of MNA
    Calculate the tangent g(0) and the diode current I_D(0)
    Add the variables to MNA
    While m < mmax do:
```

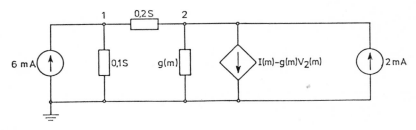

Figure 10.48

```
Solve the set, i.e. find V₂(1)

If |V₂(1) - V₂(0)| < eps

then:                              else:

write:                             calculate g(1) and I_D(1).

V₂(1) and call                     g(0) : = g(1);

it 'the solution'                  I_D(0) : = I_D(1)

                                   V₂(0) : = V₂(1);

                                   Add that to MNA:

                                   m : = m + 1

                                   If m = mmax then write

                                   convergence fault
```

End.

10.14 Problems

10.1 a. Give the NA-matrix.
      b. Find the solution.

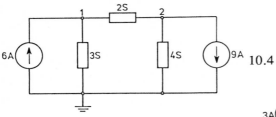

a. Give the MNA-matrix.
b. Alter the network so that an NA-matrix is possible. Give the matrix.
c. Give the solution.

10.2 Give the stamp of an impedance Z, if
      a. none of the terminals of Z is earthed,
      b. one of the terminals of Z is earthed.

10.4

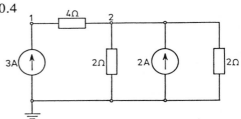

Find the NA-matrix, solve this set and calculate the mesh currents with the solution.

10.3

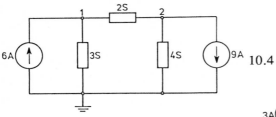

10.5

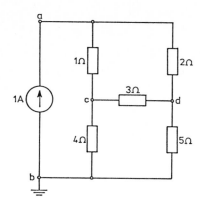

a. Give the NA-matrix.
b. Find the solution.

10.6

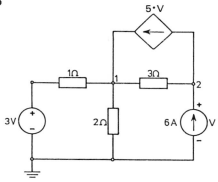

a. Give the NA-matrix.
b. Find the solution.

10.7

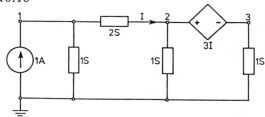

a. Give the NA-matrix.
b. Find the solution.

10.8

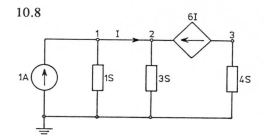

a. Give the MNA-matrix.
b. Find the solution.

10.9

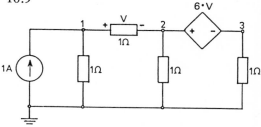

a. Give the MNA-matrix.
b. Find the solution.

10.10

a. Give the MNA-matrix.
b. Find the solution.

10.11 Find the stamp.

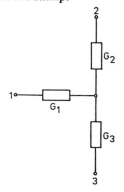

10.12 Find the stamp
  a. directly,
  b. by setting $G_3 \to \infty$ in the
     preceding problem.

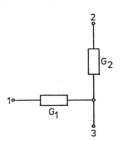

10.13 Give the stamp.

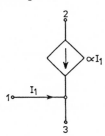

10.14 Give the stamp of this transistor
  equivalence
  a. directly,
  b. as the sum of two stamps.

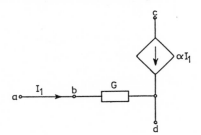

10.15 Find the stamp of this transistor
  equivalence.

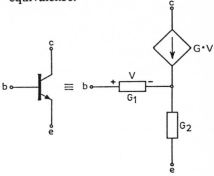

10.16 Find the stamp
  a. directly,
  b. by using a limit in Problem
     10.15.

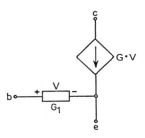

10.17 A transistor amplifier is given by
  the equivalence drawn.

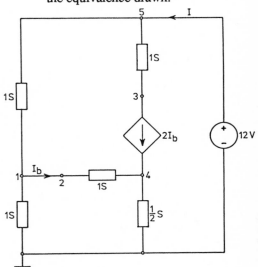

a. Give the MNA-matrix.
Change the current-current transactor into a voltage-current transactor and combine the terminals 1 and 2 into one terminal 1.

  b. Give the MNA-matrix of the network thus created.

  c. Find the solution.

10.18 This three-terminal network is part of a larger network. Give the stamp.

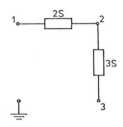

10.19 Node 3 in the network of Problem 10.18 is earthed. Find the stamp.

10.20 What will the stamp of 10.18 become if node 2 is earthed?

10.21 What will the stamp of 10.18 become if node 2 is no longer considered to be an external terminal?

10.22 The stamp of a three-terminal network is

	1	2	3
1	a	b	c
2	d	e	f
3	g	h	i

Terminal 2 is earthed. Give the stamp.

10.23 One connects the nodes 1 and 2 in the network of Problem 10.22. Call

it node k. Give the stamp.

10.24 One does not want to regard node 2 of the network of Problem 10.22 as an external terminal anymore. Give the stamp.

10.25 Given is $v_C(0) = 7$ V.

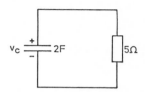

a. Give the equivalence for the numerical solution with the backward Euler method. Choose the parallel equivalence for the capacitor. Set $h = 1$ s.

  b. Find $v_C(t)$ for $t = 1, 2, 3, 4$ and 5 s.

  c. Prove that the solution for all t is

$$v_C(t) = 7 \left(\frac{1}{1.1}\right)^t \text{ V.}$$

  d. Find $v_C(t)$ for $t = 10, 20$ and 30 s.

10.26 The same questions as in Problem 10.25, but now with the series version for the capacitor.

10.27 The same questions as in Problem 10.25, but now for $h = \frac{1}{2}$ s.

(The solution for all t now is $v_C(k) = 7 \left(\frac{4}{4.2}\right)^k$ V with $k = 2$ t.)

10.28 The same questions as in Problem 10.25, but now with the trapezoidal rule.

(The solution for all t now is $v_C(k) = 7 \left(\frac{3.8}{4.2}\right)^k$ V.)

10.29 Given: $v_C(0) = 0$ V. Choose $h = 1$ s and BE (backward Euler).

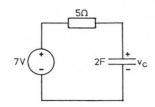

a. Express $v_C(t+1)$ into $v_C(t)$.
b. Find $v_C(t)$ for
   $t = 1, 2, 3, 4$ and 5 s.

10.30 As Problem 10.29, but now with
       TR, $h = 1$ s again.

10.31 Given: $v_0 = 5 \cos 3t$ V;
       $v_C(0) = 0$ V; $h = 0.05$ s.
       Find $v_C(t)$ for $t = 0.05$ s and for
       $t = 0.1$ s. Use BE.

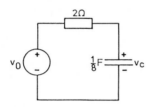

10.32 Given $v_0 = 6$ V;
       $v_1(0) = v_2(0) = 0$ V and $h = 0.1$ s.
       a. Find $v_1(t)$ and $v_2(t)$ for $t = 0.1$ s
          and for $t = 0.2$ s. Use BE.
       b. Explain the large difference of
          v_2 between these calculated
          values and the exact values (see
          Problem 9.24).
       c. Find $v_2(t)$ for $t = 0.1$ s and for $t
          = 0.2$ s. Use TR (choose $h =$
          0.1 s).

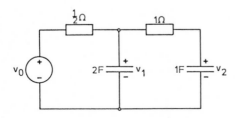

10.33 Choose BE and $h = 0.1$ s.
       $v_b = 5 \cos \frac{1}{2}t$ V; $v_{20}(0) = 1$ V;
       $i(0) = 1.5$ A.

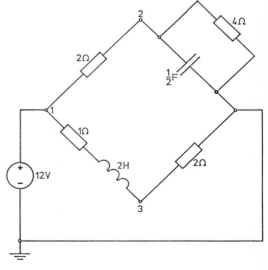

a. Give the MNA-matrix.
b. Find $v_1(t)$ and $v_2(t)$ for
   $t = 0.1$ s.

10.34 Choose BE and $h = 0.1$ s.
       Give the MNA-matrix.

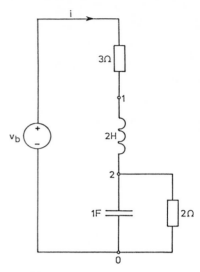

10.35 Give the MNA-matrix of the
       network in Problem 9.26.

10.36 Write a computer program for Problem 10.33.

10.37 For the nonlinear resistance holds

$$I = 0.1 \, V^2.$$

Find V
a. by direct calculation,
b. graphically,
c. by means of iterations (choose the starting value V = 1 V and stop if the difference between two iterations is smaller than 0.1).

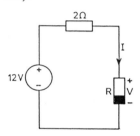

10.38 For the diode it holds that

$$I = 10^{-4}(e^{V/0.026} - 1) \, A.$$

Find V
a. graphically,
b. by means of iterations via the equivalent circuit (use the Norton equivalence for the source and series resistance).

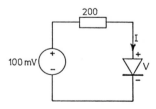

10.39 For the diode

$$I = I_S(e^{\frac{V}{V_r}} - 1) \, A.$$

Give the equivalent circuit and find the stamp.

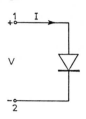

10.40 For the diode holds

$$I_D = 4(e^{\frac{V_D}{4}} - 1) \, A.$$

a. Give the equivalent circuit.
b. Find the MNA-matrix.

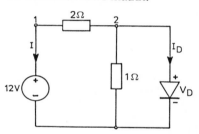

10.41 Write a computer program to determine the voltage V_D in Problem 10.40 with an accuracy of 1 mV.

10.42 Give the MNA-matrix.
The left-hand side must contain the voltages V_a, V_b, V_c and V_e with the argument m+1, the right-hand side voltages and currents with the argument m.

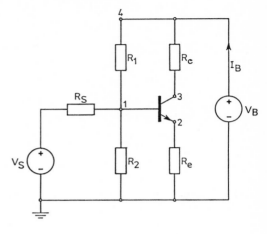

10.43 Give the equivalent circuit and the MNA-matrix for this transistor equivalence.

10.44 Give the MNA-matrix of this transistor amplifier.
Use the transistor equivalence of Problem 10.43.

10.45 a. Give the stamp of a voltage-voltage transactor with amplifying factor μ.
   b. Derive the parallel equivalence of a capacitor C by using the backward Euler method. The step is h. Give the stamp.
   c. Give the MNA-matrix of this network. Change the opamp into a voltage-voltage transactor with $\mu = 100$. Set h = 1 s.

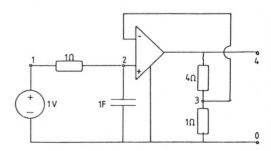

10.46 a. Derive the equivalent series circuit for an inductor L, using the backward Euler method. The step is h.

Next consider the network below. The source is a d.c. voltage source.

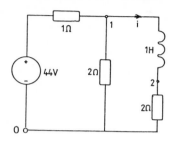

b. Change the voltage source and the series resistance into the Norton equivalence and give the MNA-matrix. The step is h = 0.1 s.

c. Find i(t) for t = 0.1 s if i(0) = 12 A.

10.47 The source is a d.c. voltage source. The initial condition is
$v_{12}(0) = 1.6$ V.
One wants to find the node voltages with the backward Euler method. Choose h = 1 s.

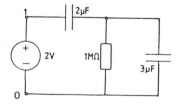

a. First derive the parallel equivalence for a capacitor C and give the stamp (the step is h).

b. Give the total equivalent circuit for the network and give the MNA-matrix.

c. Find $v_{20}(1)$, i.e. the voltage of node 2 with respect to zero at the moment 1 s.
The exact solution is
$v_{20} = 0.4\ e^{-0.2t}$ V; $t \geq 0$.

d. Determine the difference of the calculation in percents.

10.48 $v_1 = 10 \cos 2t$ V; $i(0) = 0$ A; h = 0.05 s.
Use the backward Euler method.

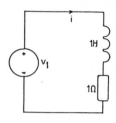

a. Give the MNA-matrix.

b. Derive the current i for k = 1 and compare it with the exact solution:
$i(t) = -2e^{-t} + 2 \cos 2t + 4 \sin 2t$ A.

10.49 The source is a d.c. current source.
$v_1(0) = 0$ V; $i(0) = 0$ A; h = 0.02 s.
Use the backward Euler method.

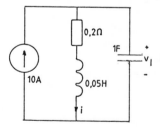

a. Give the MNA-matrix.

b. Find $v_1(k)$ for k = 1 and for k = 2.

10.50 Compile a list of the errors you have found in this book and send it to the publisher.

Appendix I

Linearity and superposition

Consider the network of Figure 1.23 with the excitations V_1 and V_2 and the response I. This network is shown again in Figure I.1.

We now bring this network in two different states, which can be realised by giving both excitations different values.

The first state is given one prime, the second state two.

So, from the equation, derived in Section 1.10

$$I = \frac{1}{24} V_1 + \frac{1}{12} V_2 \tag{A}$$

we get

$$I' = \frac{1}{24} V_1' + \frac{1}{12} V_2' \tag{B}$$

and

$$I'' = \frac{1}{24} V_1'' + \frac{1}{12} V_2''. \tag{C}$$

If α and β are two arbitrary constants, then

$$\alpha I' + \beta I'' = \alpha(\frac{1}{24} V_1' + \frac{1}{12} V_2') + \beta(\frac{1}{24} V_1'' + \frac{1}{12} V_2'')$$

so

$$\alpha I' + \beta I'' = \frac{1}{24} (\alpha V_1' + \beta V_1'') + \frac{1}{12} (\alpha V_2' + \beta V_2''). \tag{D}$$

In general: If the excitation (V_1, V_2) gives the response I, then the excitation $(\alpha V_1' + \beta V_1'', \alpha V_2' + \beta V_2'')$ will give the response $(\alpha I' + \beta I'')$.

This is a general characteristic of linear systems.

As an example, try to find out whether

$$P = VI \tag{E}$$

is a linear function.

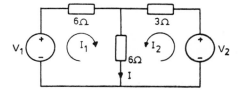

Figure I.1

Of course, this is not the case here because we have a product. We shall, however, determine it formally.

First state: $P' = V'I'$,

Second state: $P'' = V''I''$.

The excitation $(\alpha V' + \beta V'', \alpha I' + \beta I'')$ gives the response

$$(\alpha V' + \beta V'')\cdot(\alpha I' + \beta I''), \tag{F}$$

while the response $\alpha P' + \beta P''$ equals

$$\alpha V'I' + \beta V''I''. \tag{G}$$

The expressions (F) and (G) are not equal, in other words the function is not linear.

As a second example, take

$$v = L\frac{di}{dt}. \tag{H}$$

This is the formula for the inductor which was introduced in Section 2.12.

Consider (H) as a system with i as excitation and v as response.
The excitation $\alpha i' + \beta i''$ gives the response

$$L\frac{d}{dt}(\alpha i' + \beta i''), \tag{J}$$

while the response $\alpha v' + \beta v''$ equals

$$\alpha L\frac{di'}{dt} + \beta L\frac{di''}{dt}. \tag{K}$$

(J) equals (K), so the system is linear.

The mesh method and the node method give linear equations. To find the solution of a network, we only use a linear process: We multiply by constants and add or subtract equations. The result is always a linear equation.

For a network with two sources formula (1.24) turns out to be linear in this way. (The excitations may, of course, be two current sources or one voltage source and one current source).

We now choose $\alpha = \beta = 1$ and find: The excitation $(V_1' + V_1'', V_2' + V_2'')$ gives the response $(I' + I'')$.

Next take $V_2' = 0$, i.e. the source V_2 becomes a short circuit in the first state while we also make $V_1'' = 0$, which means that the source V_1 becomes a short circuit in the second state.

We thus get: The excitation $(V_1' + 0, 0 + V_2'')$ gives the response $(I' + I'')$ where I' follows from V_1' only and I'' from V_2'' only. Which can also be written as

$$(V_1',0) \Rightarrow I'$$

$$(0,V_2'') \Rightarrow I''$$

$$I = I' + I''$$

For our examples of Figure 1.20 this means the state as drawn in Figure I.2. Calculation results in

$$I' = \frac{1}{24} V_1' \qquad I'' = \frac{1}{12} V_2''.$$

So $$I = I' + I'' = \frac{1}{24} V_1' + \frac{1}{12} V_2''.$$

In general

$$I = \frac{1}{24} V_1 + \frac{1}{12} V_2.$$

Compare this result with the result found earlier. Addition like this is called (simple) superposition and we say: *For a linear system the superposition law holds.*

For three excitations (we take three voltage sources) one can use *compound superposition.*
We have

$$(V_1, V_2, V_3) \Rightarrow I$$

First state: $(V_1,0,0) \Rightarrow I'$

Second state: $(0,V_2,V_3) \Rightarrow I''$

$$I = I' + I''$$

The second state can be regarded as the result of two states:

$$(0,V_2,0) \Rightarrow I'''$$

$$(0,0,V_3) \Rightarrow I''''$$

$$I'' = I''' + I''''$$

The result thus is:

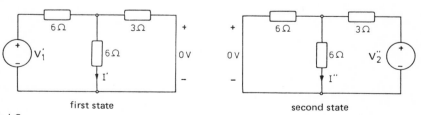

first state   second state

Figure I.2

$$(V_1,0,0) \Rightarrow I'$$

$$(0,V_2,0) \Rightarrow I'''$$

$$(0,0,V_3) \Rightarrow I''''$$

$$I = I' + I''' + I''''$$

This is the principle of superposition.

We finally determine whether the equation

$$\frac{dv}{dt} = L\frac{d^2i}{dt^2} + R\frac{di}{dt} + \frac{1}{C}i$$

is linear. v is considered as excitation and i as response.
If the excitation is $\alpha v'$, the response is $\alpha i'$ according to

$$\alpha\frac{dv'}{dt} = \alpha(L\frac{d^2i'}{dt^2} + R\frac{di'}{dt} + \frac{1}{C}i'). \tag{L}$$

If the excitation is $\beta v''$ the response is $\beta i''$ according to

$$\beta\frac{dv''}{dt} = \beta(L\frac{d^2i''}{dt^2} + R\frac{di''}{dt} + \frac{1}{C}i''). \tag{M}$$

Now assume the response to be $\alpha i' + \beta i''$, will the excitation be $\alpha v' + \beta v''$?
For the response:

$$L\frac{d^2}{dt^2}(\alpha i' + \beta i'') + R\frac{d}{dt}(\alpha i' + \beta i'') + \frac{1}{C}(\alpha i' + \beta i'')$$

$$= \alpha(L\frac{d^2i'}{dt^2} + R\frac{di'}{dt} + \frac{1}{C}i') + \beta(L\frac{d^2i''}{dt^2} + R\frac{di''}{dt} + \frac{1}{C}i'').$$

According to (L) and (M) this equals

$$\alpha\frac{dv'}{dt} + \beta\frac{dv''}{dt} \,,$$

while this can also be written as:

$$\frac{d}{dt}(\alpha v' + \beta v'')$$

The excitation is indeed $\alpha v' + \beta v''$, therefore this system is linear.

Appendix II

Tellegen's theorem

Consider a network containing sources and resistors.

First consider two nodes a and b in the network, between which there are p branches in parallel with currents $I_{ab}{}^{(1)}$, $I_{ab}{}^{(2)}$, $I_{ab}{}^{(3)}$, ..., $I_{ab}{}^{(p)}$. We shall indicate the sum of these currents by

$$\tilde{I}_{ab} = I_{ab}{}^{(1)} + I_{ab}{}^{(2)} + I_{ab}{}^{(3)} + \ldots + I_{ab}{}^{(p)}. \tag{A}$$

The number of terms in the right-hand side can also be one. If that is the case there is only one branch between a and b. The number of terms can also be zero, then there is no branch between a and b at all. So we have

$$\tilde{I}_{ba} = -\tilde{I}_{ab}. \tag{B}$$

Thus for all b branches of the network with k nodes we find:

$$\sum_{n=1}^{b} V_n I_n = V_{12}\tilde{I}_{12} + V_{13}\tilde{I}_{13} + V_{14}\tilde{I}_{14} + \ldots + V_{1k}\tilde{I}_{1k}$$

$$+ V_{23}\tilde{I}_{23} + V_{24}\tilde{I}_{24} + \ldots + V_{2k}\tilde{I}_{2k}$$

$$+ V_{34}\tilde{I}_{34} + \ldots + V_{3k}\tilde{I}_{3k}$$

$$+ \ldots \ldots \ldots \ldots \ldots + V_{k-1,k}\tilde{I}_{k-1,k}. \tag{C}$$

Now, in general, for two nodes r and s holds:

$$V_{rs} = V_r - V_s. \tag{D}$$

We substitute (D) in (C) and use (B):

$$\sum_{n=1}^{b} V_n I_n = 0 + V_1\tilde{I}_{12} + V_1\tilde{I}_{13} + \ldots + V_1\tilde{I}_{1k}$$

$$+ V_2\tilde{I}_{21} + 0 + V_2\tilde{I}_{23} + \ldots + V_2\tilde{I}_{2k}$$

$$+ V_3\tilde{I}_{31} + V_3\tilde{I}_{32} + 0 + \ldots + V_3\tilde{I}_{3k} \tag{E}$$

$$+$$

$$\cdot$$

$$\cdot$$

$$+ \ldots \ldots \ldots \ldots \ldots + 0.$$

In this expression we have added zeros in order to obtain a symmetrical notation. Each row in (E) contains the sum of k terms, which is zero according to the current law. The right-hand side of (E) is therefore zero, and so we find:

$$\sum_{n=1}^{b} V_n I_n = 0,$$

where b is the number of branches in a network.

We have only used Kirchhoff's laws and not Ohm's law. So the theorem holds for nonlinear networks. *It even holds for two networks with the same graph, in which the branch voltages are used in one network and the branch currents in the other.*

Appendix III

Reciprocity

Consider a two-port consisting of resistors in a certain configuration and bring it into two different states (see Figure III.1).

We emphasise that the two-port does *not* contain sources. The voltages and currents in the first state are given one prime and those in the second state two. Now we first apply Tellegen's theorem for the voltages in state 1 and the currents in state 2.
We find

$$V_1' I_1'' + V_2' I_2'' = - \sum_{int} V_i' I_i''. \tag{a}$$

The sum is to be taken over all internal branches $i = 1,2,3,\dots,b$.
We once more apply Tellegen's theorem for the voltages in state 2 and the currents in state 1:

$$V_1'' I_1' + V_2'' I_2' = - \sum_{int} V_i'' I_i'. \tag{b}$$

For each internal branch holds

$$V_1' = R_i I_1' \quad \text{and} \quad V_1'' = R_i I_1''. \tag{c}$$

So

$$V_i' I_i'' = R_i I_i' I_i'' = R_i I_i'' I_i' = V_i'' I_i', \tag{d}$$

together with (a) results in

$$V_1' I_1'' + V_2' I_2'' = V_1'' I_1' + V_2'' I_2'. \tag{III.1}$$

We shall call this the *general reciprocity theorem*.
If all current directions of the ports are changed, all terms get a negative sign and therefore (III.1) remains valid.
Now consider the resistance matrix of a two-port

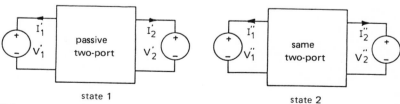

state 1   state 2

Figure III.1

$$\mathcal{R} = \begin{bmatrix} R_{11} & R_{12} \\ R_{21} & R_{22} \end{bmatrix}.$$

We have

$$V_1 = R_{11}I_1 + R_{12}I_2. \tag{e}$$

$$V_2 = R_{21}I_1 + R_{22}I_2. \tag{f}$$

Again consider two states

$$V_1' = R_{11}I_1' + R_{12}I_2' \qquad\qquad V_1'' = R_{11}I_1'' + R_{12}I_2'',$$

$$V_2' = R_{21}I_1' + R_{22}I_2' \qquad\qquad V_2'' = R_{21}I_1'' + R_{22}I_2''.$$

If we use this in the general reciprocity theorem we find

$$R_{11}I_1'I_1'' + R_{12}I_2'I_1'' + R_{21}I_1'I_2'' + R_{22}I_2'I_2'' =$$

$$= R_{11}I_1''I_1' + R_{12}I_2''I_1' + R_{21}I_1''I_2' + R_{22}I_2''I_2'.$$

So

$$R_{12}I_2'I_1'' + R_{21}I_1'I_2'' = R_{12}I_2''I_1' + R_{21}I_1''I_2'.$$

Therefore

$$I_2'I_1''(R_{12} - R_{21}) = I_1'I_2''(R_{12} - R_{21}).$$

This equation must hold for all currents, so

$$R_{21} = R_{12}. \tag{III.2}$$

We shall call this the *derived reciprocity theorem*.

This theorem says that the resistance matrix of a passive two-port is symmetrical with respect to the main diagonal.

We now return to the change of short circuit and voltage source. In the first state $V_1' = V$ and $V_2' = 0$. In the second state we have $V_2'' = V$ and $V_1'' = 0$. This substituted in (III.1) gives

$$VI_1'' = VI_2'.$$

So

$$I_1'' = I_1'.$$

We do indeed find the same current in both short circuits.

We again consider (e) and (f). From these we solve I_1 and I_2. We find

$$I_1 = G_{11}V_1 + G_{12}V_2, \tag{g}$$

$$I_2 = G_{21}V_1 + G_{22}V_2, \tag{h}$$

with the 'conductance' matrix

$$G = \begin{bmatrix} G_{11} & G_{12} \\ G_{21} & G_{22} \end{bmatrix},$$

in which G is the reverse of $\mathcal{R}$:

$$G = \mathcal{R}^{-1}. \tag{III.3}$$

We can also find

$$G_{21} = G_{12}. \tag{III.4}$$

(Do not confuse this with (8.20)).
The conductance matrix is symmetric as well.

From (h) follows

$$G_{21} = \left(\frac{I_2'}{V_1}\right)_{V_2'=0}.$$

The condition $V_2' = 0$ means a short circuit of the output (in the first state) (see Figure III.2(a)).
From (g) it follows that

$$G_{12} = \left(\frac{I_1''}{V_2''}\right)_{V_1''=0},$$

belonging to Figure III.2(b).
With $G_{21} = G_{12}$ and $V_1' = V_2'' = V$:

$$I_1'' = I_2',$$

a result we earlier found directly from (III.1).

Finally we give an example of a non-reciprocal two-port (see Figure III.3).
This is the *gyrator*. See also Sections 1.18 and 8.4.
The basic formulas are:

$$V_1 = -RI_2,$$

$$V_2 = RI_1.$$

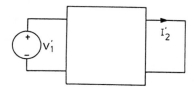 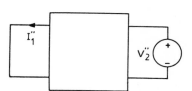

Figure III.2   a   b

So the resistance matrix is

$$\mathcal{R} = \begin{bmatrix} 0 & -R \\ R & 0 \end{bmatrix},$$

with $R_{21} \neq R_{12}$, so the two-port is not reciprocal.

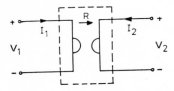

Figure III.3

Appendix IV

Thévenin's and Norton's theorems

Consider the network of Figure IV.1, which is the same as Figure 1.35.
We want to find the current I. We bring the resistor R outside the remaining part of the network (see Figure IV.2).
Using the mesh method in Figure IV.1 we find:

$$V_1 = 6I_1 + V, \tag{a}$$

$$V_2 = V + 3I_2, \tag{b}$$

$$I_1 + I_2 = I. \tag{c}$$

We have not used the resistor R.
Elimination of I_1 and I_2 (these are internal quantities) gives:

$$3V + 6I = V_1 + 2V_2$$

or in general

$$AV + BI = C. \tag{IV.1}$$

We find a linear relation for each network consisting of sources and resistors, because we have used Kirchhoff's laws and Ohm's law and because we have used linear processes only. A, B and C are constants, which depend on the values of the elements and on the

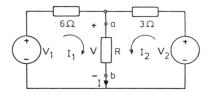

Figure IV.1

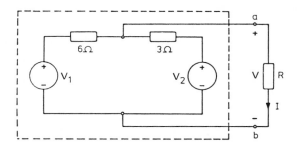

Figure IV.2

graph of the network. Moreover, the relation is homogenous, i.e. there is no constant term, while C is linear dependent on the internal source intensities.

We now solve the port voltage:

$$V = \frac{C}{A} - \frac{B}{A} I.$$

For the time being we use $A \neq 0$.

The corresponding diagram is shown in Figure IV.3(a).

This is called the *Thévenin equivalence*. If we make $\frac{C}{A} = V_T$ the Thévenin voltage and $\frac{B}{A} = R_T$ the Thévenin resistor, the network of Figure IV.3(b) results.

If we remove R, open nodes are created and the current I becomes zero. From (IV.1) then follows:

$$AV = C.$$

So

$$\frac{C}{A} = V = V_T.$$

This means that the voltage arising on the open terminals a and b, called *open voltage* (which we shall indicate by $\overset{\circ}{V}_{ab}$), equals the Thévenin voltage:

$$V_T = \overset{\circ}{V}_{ab}. \tag{IV.2}$$

So (IV.2) is a way to find the Thévenin voltage of a network.

If we short-circuit the terminals a and b then $V = 0$. The current then appearing is called the *short circuit current* I_s.

From (IV.1):

$$BI_s = C.$$

So

$$I_s = \frac{C}{B} .$$

For the time being we exclude $B = 0$.

We find

$$\frac{B}{A} = \frac{C/A}{C/B} .$$

So

$$R_T = \frac{V_T}{I_s} , \tag{IV.3}$$

with which we have found a way to determine the Thévenin resistance.

The second way to find R_T is the following.

If we make all source intensities zero, then $C = 0$, because C is linear dependent on the source intensities. From (IV.1):

$$AV + BI = 0.$$

So

$$\frac{V}{-I} = \frac{B}{A}. \qquad \text{(IV.4)}$$

The left-hand side is the quotient of port voltage and port current (note the sign) and this is the resistance we measure between the terminals a and b. We shall indicate this resistance by R_{ab}.

Consider the terminals on the right of the network, so we measure to the left, as it were. All source intensities should be made zero.

So

$$R_{ab} = R_T. \qquad \text{(IV.5)}$$

We shall now discuss the exceptions.
If $A = 0$ then from (IV.1):

$$BI = C.$$

The port current then is $I = \frac{C}{B}$, i.e. the equivalence is a current source, see Figure IV.4. If $B = 0$ we have

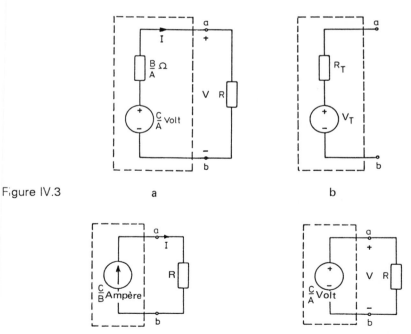

Figure IV.3

a

b

Figure IV.4

Figure IV.5

$$AV = C.$$

Then the port voltage is $V = \frac{C}{A}$, which means that the equivalence is a voltage source, see Figure IV.5.

With (IV.1) we now solve I:

$$I = \frac{C}{B} - \frac{A}{B} V \qquad \text{with } B \neq 0.$$

We have already discussed $B = 0$.

The equivalence is as shown in Figure IV.6.

The dimension of $\frac{A}{B}$ is that of conductance. With $\frac{C}{B} = I_N$ (the *Norton current*) and $G_N = \frac{A}{B}$ (the *Norton conductance*) we find the so-called *Norton equivalence* (see Figure IV.7).

The short circuit current is the Norton current:

$$I_s = I_N. \tag{IV.6}$$

The open port voltage is $\overset{\circ}{V}_{ab} = \frac{I_N}{G_N}$.

So

$$V_T = \frac{I_s}{G_N}.$$

Consequently

$$G_N = \frac{1}{R_T}. \tag{IV.7}$$

Finally, on the analogy of (IV.5) we find:

$$G_{ab} = G_N. \tag{IV.8}$$

If the network does contain transactors we also find equation IV.1:

$$AV + BI = C,$$

in which A, B and C, however, will be functions of the transactor constants α, μ, R and G. This will be demonstrated with the network of Figure IV.8.

$$V_1 = 2I_1 - I$$

$$-V = -I_1 + 2I + \alpha I_1.$$

Elimination of I_1 gives

$$2V + (3 + \alpha)I = (1 - \alpha)V_1.$$

In this example B and C are functions of the transactor constant α.

Here, too, equation IV.4 holds, so that resistance measured at the terminals a and b depends on the transactor quantities, which implies that the controlled source intensities must not be made zero.

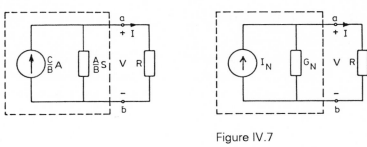

Figure IV.6   Figure IV.7

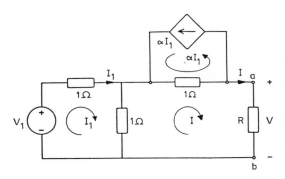

Figure IV.8

Appendix V

Star-delta transformation

Consider the networks of Figure V.1.

Both *three-terminal resistor networks* delta and star much behave the same way, which means that the voltages between the nodes 1, 2 and 3 must be the same in the delta circuit as the corresponding voltages in the star circuit for arbitrary terminal currents I_a, I_b and I_c.

First choose $I_a \neq 0$ and $I_b = I_c = 0$. This state gets one prime. For the delta circuit:

$$V'_{12\Delta} = \frac{A(B + C)}{A + B + C} I_a$$

and for the star circuit:

$$V'_{12\lambda} = (D + F)I_a.$$

If we subsequently only take $I_b \neq 0$ we get (two primes):

$$V''_{12\Delta} = A \cdot - \frac{B}{A + B + C} \cdot I_b \quad \text{and} \quad V''_{12\lambda} = -DI_b.$$

Finally only $I_c \neq 0$ (three primes):

$$V'''_{12\Delta} = A \cdot - \frac{C}{A + B + C} \cdot I_c \quad \text{and} \quad V'''_{12\lambda} = -FI_c.$$

Superposed and equalised:

$V_{12\Delta} = V_{12\lambda}$ gives: $\dfrac{A(B + C)I_a - ABI_b - ACI_c}{A + B + C} = (D + F)I_a - DI_b - FI_c.$

This must hold for all I_a, I_b and I_c, so

$$D = \frac{AB}{A + B + C} \tag{V.1.a}$$

and

$$F = \frac{AC}{A + B + C}. \tag{V.1.b}$$

So the 'star resistance' equals the product of both 'delta resistances' that are connected to the corresponding terminal, divided by the sum of the three delta resistances. We thus find

$$E = \frac{BC}{A + B + C}. \tag{V.1.c}$$

Each of the formulas (V.1) results in both others by cyclic changing: The cycle (A-B-C) gives the cycle (D-E-F), i.e. replace A by B, B by C and C by A in (V.1.a), then this means replacing D by E, E by F and F by D. In this way (V.1.c) is created. One step further in both cycles gives (V.1.b).

Next we shall pose the problem the other way round: The star circuit is given, the equivalent delta circuit is required. Now D, E and F are the quantities known and A, B and C the quantities desired. One can try to solve the set (V.1).

Calculation, however, requires a lot of writing.

Easier and more interesting is to use dual reasoning. Consider Figure V.2 in which the elements are *conductors* now.

The terminal currents must be equal for all voltage source intensities in the left- and in the right-hand circuits. We first choose $V_a \neq 0$ and $V_b = V_c = 0$. We then have

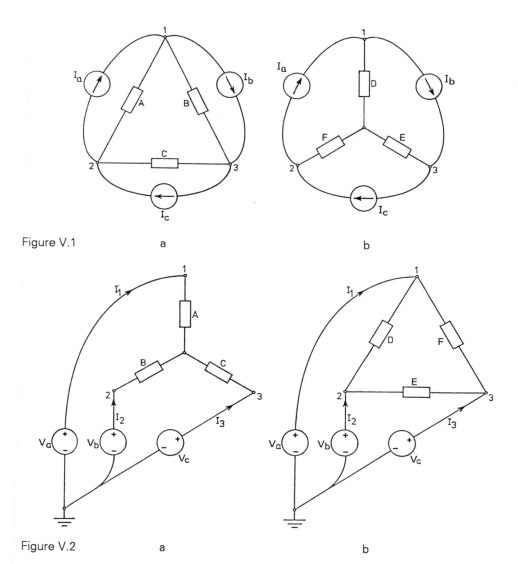

Figure V.1   a   b

Figure V.2   a   b

$$I'_{1\lambda} = \frac{V_a}{\dfrac{1}{A} + \dfrac{1}{B + C}} = \frac{A(B + C)}{A + B + C} V_a$$

and $I'_{1\Delta} = (D + F)V_a.$

Now choose $V_b \neq 0$ and $V_a = V_c = 0$:

$$I''_{1\lambda} = -\frac{V_b}{\dfrac{1}{A + C} + \dfrac{1}{B}} \cdot \frac{A}{A + C} = -\frac{AV_b}{1 + \dfrac{A + C}{B}} = \frac{-ABV_b}{A + B + C}$$

and $I''_{1\Delta} = -DV_b.$

Finally, if only $V_c \neq 0$:

$$I'''_{1\lambda} = -\frac{V_c}{\dfrac{1}{C} + \dfrac{1}{A + B}} \cdot \frac{A}{A + B} = \frac{-AV_c}{1 + \dfrac{A + B}{C}} = \frac{-ACV_c}{A + B + C}$$

and $I'''_{1\Delta} = -FV_c.$

Superposition and equalising result in:

$$\frac{A(B + C)V_a - ABV_b - ACV_c}{A + B + C} = (D + F)V_a - DV_b - FV_c.$$

This must holds for all V_a, V_b and V_c, so

$$D = \frac{AB}{A + B + C},$$

which is the same formula as (V.1.a).

The formulas (V.1.b) and (V.1.c) also result, so that we conclude:

The formulas (V.1) give the transformation from delta to star for resistances and from star to delta for conductances.

Appendix VI

Foster's theorem

For a loss-less one-port it holds that

$$\frac{dX}{d\omega} \geq 0.$$

We shall prove this theorem by making some propositions:

a. *If a sinusoidal a.c. current with amplitude |I| flows through an inductor the average magnetic energy stored is*

$$W_L = \tfrac{1}{4} L|I|^2 \tag{a.1}$$

Proof: We have, cf. (2.21);

$$w_L = \tfrac{1}{2} Li^2.$$

So the magnetic energy is a function of the current. The average energy over one period T thus is

$$W_L = \frac{1}{T} \int_0^T w_L dt = \frac{L}{2T} \int_0^T i^2 dt.$$

This equals $\tfrac{1}{2} L I_{eff}^2$ according to (2.8) and this equals $\tfrac{1}{4} L|I|^2$ according to (2.15), with which this proposition has been proved.

In a dual manner one can prove that the average electric energy stored in a capacitor equals

$$W_C = \tfrac{1}{4} C|V|^2, \tag{a.2}$$

where |V| is the amplitude of the sinusoidal capacitor voltage.

We must emphasise that we are concerned here with a harmonic voltage and current respectively, i.e. with a constant amplitude, and that it concerns the *average* energy. This energy is caused by the reactive power: $Q = 2\omega (W_m - W_e)$ supplied by the source as will become clear from proposition e below. The average real power supplied by the source is zero! Therefore it is not possible, for example, to convert this stored energy into heat in resistors (which can, however, be done with a constant energy in a capacitor or an inductor) with the aid of switches. Switching off the system disturbs the harmonic character and the average real power supplied may then not be zero. If that is the case we speak of *transient response*, see Chapter 9.

b. *Kirchhoff's voltage law also holds if we take the derivatives to ω of the voltages instead of the voltages themselves:*

$$\sum_{m=1}^{l} \frac{dV_m}{d\omega} = 0. \tag{b.1}$$

In this formula V_m are the l complex branch voltages in a loop, which are functions of ω.

Proof: Start with the voltage law

$$\sum_{m=1}^{l} V_m(\omega) = 0.$$

First take the angular frequency ω_1 and subsequently ω_2 with $\omega_2 > \omega_1$.
We find

$$\sum_{m=1}^{l} \{V_m(\omega_2) - V_m(\omega_1)\} = 0.$$

We divide both sides by $\omega_2 - \omega_1$:

$$\sum_{m=1}^{l} \frac{V_m(\omega_2) - V_m(\omega_1)}{\omega_2 - \omega_1} = 0.$$

With $\Delta V_m(\omega) = V_m(\omega_2) - V_m(\omega_1)$ and $\Delta\omega = \omega_2 - \omega_1$:

$$\sum_{m=1}^{l} \frac{\Delta V_m(\omega)}{\Delta\omega} = 0,$$

subsequent to which the proof is completed after the limit transition.
Similarly from Kirchhoff's current law one can derive

$$\sum_{n=1}^{b} \frac{dI_n}{d\omega} = 0. \tag{b.2}$$

c. *From the above, together with Tellegen's theorem:*

$$\sum VI = 0,$$

where the sum must be taken over all branches of the network;

$$\sum \frac{dV}{d\omega} I^* = 0 \tag{c.1}$$

and

$$\sum V^* \frac{dI}{d\omega} = 0. \tag{c.2}$$

d. *For the reactance $X(\omega)$ of a loss-less one-port it holds that:*

$$\frac{dX(\omega)}{d\omega} = \frac{4}{|I|^2} (W_m + W_e), \tag{d}$$

where |I| is the amplitude of the sinusoidal port current and W_m and W_e respectively are the total magnetic and electric energies stored in the one-port.

For the proof consider Figure VI.1.

Set the complex source current I to real for all ω so that $\dfrac{dI}{d\omega} = 0$.

With

$$V = ZI,$$

$$Z = jX$$

we find

$$V = jXI.$$

With that we find

$$\frac{dV}{d\omega} = j\frac{dX}{d\omega}I + jX\frac{dI}{d\omega},$$

so

$$\frac{dV}{d\omega} = j\frac{dX}{d\omega}I.$$

Applying proposition (c.1) we get

$$-j\frac{dX}{d\omega}II^* + \sum_L \frac{dV_L}{d\omega}I_L^* + \sum_C \frac{dV_C}{d\omega}I_C^* = 0.$$

The minus sign arises because the source voltage and the source current do not belong together. The summation must be executed over all inductors and capacitors respectively. Subsequently

$$j\frac{dX}{d\omega}|I|^2 = \sum_L \frac{dV_L}{d\omega}I_L^* + \sum_C \frac{dV_C}{d\omega}I_C^*.$$

Now $V_L = j\omega L I_L$, so $\dfrac{dV_L}{d\omega} = jLI_L + j\omega L\dfrac{dI_L}{d\omega}$

and $V_C = \dfrac{1}{j\omega C}I_C$, so $\dfrac{dV_C}{d\omega} = -\dfrac{1}{j\omega^2 C}I_C + \dfrac{1}{j\omega C}\dfrac{dI_C}{d\omega}$.

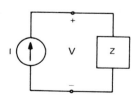

Figure VI.1

With that we get

$$j|I|^2 \frac{dX}{d\omega} = \sum_L jL|I_L|^2 + \sum_L j\omega L \frac{dI_L}{d\omega} I_L^* + \sum_C -\frac{1}{j\omega^2 C}|I_C|^2 + \sum_C \frac{1}{j\omega C}\frac{dI_C}{d\omega}I_C^*.$$

Next consider the second and fourth sums in the right-hand side.
With

$$V_L^* = -j\omega L I_L^*$$

$$V_C^* = -\frac{I_C^*}{j\omega C}$$

for these sums we find:

$$-\sum_L V_L^* \frac{dI_L}{d\omega} - \sum_C V_C^* \frac{dI_C}{d\omega}.$$

While according to proposition (c.2) it holds that

$$\sum_L V_L^* \frac{dI_L}{d\omega} + \sum_C V_C^* \frac{dI_C}{d\omega} - V^* \frac{dI}{d\omega} = 0.$$

The last term in the left-hand side of this equation is zero.
Consequently the second and the fourth sums in the total expression are zero.
We therefore get

$$j|I|^2 \frac{dX}{d\omega} = \sum_L jL|I_L|^2 + \sum_C -\frac{1}{j\omega^2 C}|I_C|^2.$$

With $V_C = \frac{1}{j\omega C} I_C$ and therefore with $V_C^* = -\frac{1}{j\omega C} I_C^*$ follows

$$|V_C|^2 = \frac{1}{\omega^2 C^2}|I_C|^2,$$

with which the last term in the total expression becomes

$$\sum_C jC |V_C|^2.$$

So

$$j|I|^2 \frac{dX}{d\omega} = \sum_L jL |I_L|^2 + \sum_C jC |V_C|^2.$$

Thus

$$|I|^2 \frac{dX}{d\omega} = 4(W_m + W_e)$$

according to proposition (a).

With this the proof of proposition (d) has been completed.

e. *The reactance $X(\omega)$ of a loss-less one-port is*

$$X(\omega) = \frac{4\omega}{|I|^2} (W_m - W_e),$$ (e)

in which $|I|$ is the amplitude of the port current and W_m and W_e respectively is the total magnetic and electric energy respectively stored in the network.

Proof: The complex power supplied by the source (see Figure VI.1) is

$$S = \frac{1}{2} VI^*.$$

This equals the total complex power consumed by the inductors and the capacitors:

$$S = \frac{1}{2} \sum_L V_L I_L^* + \frac{1}{2} \sum_C V_C I_C^*.$$

With $V_L = j\omega L I_L$ and $I_C = j\omega C V_C$:

$$S = \frac{1}{2} \sum_L j\omega L |I_L|^2 - \frac{1}{2} \sum_C j\omega C |V_C|^2.$$

But we have

$$S = P + jQ = jQ$$

because the real power consumed by an inductor or a capacitor is zero. So

$$Q = \frac{1}{2} \omega \left(\sum_L L |I_L|^2 - \sum_C C |V_C|^2 \right).$$

It also holds that

$$S = \frac{1}{2} VI^* = \frac{1}{2} jX |I|^2 = jQ.$$

With which

$$Q = \frac{1}{2} X |I|^2.$$

So

$$X = \frac{\omega}{|I|^2} \left(\sum_L L |I_L|^2 - \sum_C C |V_C|^2 \right),$$

with which the proof has been completed through proposition (a).

The proof of Foster's theorem now follows directly from proposition (d), because the energies are not negative, so

$$\frac{dX}{d\omega} \geq 0,$$

but the propositions (d) and (e) together result in a still *finer* theorem, one with more information.

We find

$$\frac{dX}{d\omega} - \frac{X}{\omega} = \frac{8}{|I|^2} W_e$$

and

$$\frac{dX}{d\omega} + \frac{X}{\omega} = \frac{8}{|I|^2} W_m.$$

The energies are not negative, so

$$\frac{dX}{d\omega} \geq \frac{X}{\omega} \qquad\qquad\qquad\qquad \text{(VI.1.a)}$$

and

$$\frac{dX}{d\omega} \geq -\frac{X}{\omega}. \qquad\qquad\qquad\qquad \text{(VI.1.b)}$$

If $W_e = 0$ we have the equal sign in (VI.1.a) and if $W_m = 0$ we have equal sign in (VI.1.b) The first case occurs with a single inductor: $X = \omega L$, so

$$\frac{dX}{d\omega} = \frac{X}{\omega} = L, \qquad\qquad\qquad\qquad \text{(VI.2.a)}$$

the second case with a single capacitor: $X = \dfrac{-1}{\omega C}$, so

$$\frac{dX}{d\omega} = -\frac{X}{\omega} = \frac{1}{\omega^2 C}. \qquad\qquad\qquad\qquad \text{(VI.2.b)}$$

From the above we can find the plot $X = f(\omega)$ with some more detail than with the rule that poles and zeros alternate.

Formula (VI.1.a) says that the tangent is always larger than the slope of the reactance of a single inductor. See point A in Figure VI.2.

Formula (VI.1.b) says that the tangent in each point is always larger than the slope of the reactance of a single capacitor. See point B. So the plots have the shape as drawn in all figures $X(\omega)$.

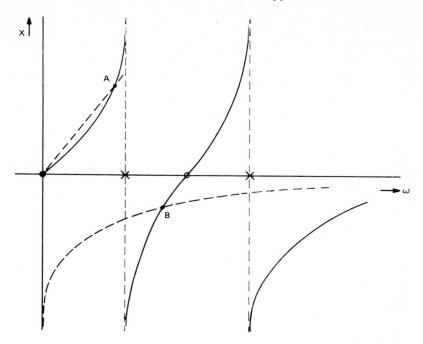

Figure VI.2

Answers to Problems

Problems Chapter 1

1.1. 11 A

1.2. 11 V

1.3 3 V

1.4 3 A

36 W, 18 W, 12 W, 1 W, 5 W

1.5 1 Ω

1.6 1 S

1.7 6 V

1.8 6 A

1.9 120 W

1.10 2 V

1.11 $P_I = 10$ W; $P_V = 6$ W; $P_R = 16$ W

1.12 4 A

1.13 4 A

1.14 $V_{ab} = \dfrac{61}{21}$ V; $R_{ab} = \dfrac{61}{21}$ Ω

1.15 3 A; $\dfrac{1}{2}$ A; $\dfrac{5}{2}$ A

1.16 23 V

1.17 $\dfrac{9}{13}$ V; $-\dfrac{33}{26}$ V

1.18 $I_1 = 2$ A, $I_2 = 6$ A, $I_3 = -2$ A
$V_1 = 2$ V, $V_2 = -2$ V, $V_3 = 18$ V,
$V_4 = 26$ V, $V_5 = 10$ V

1.19 −4 V, 20 V, 8 V

1.20 $-\dfrac{18}{11}$ A, $\dfrac{15}{11}$ A, $\dfrac{9}{11}$ A

1.21 not planar

1.22 I is indefinite

1.23 V is indefinite

1.24 9 A

1.25 a. $\dfrac{1}{4}$ V, $-\dfrac{7}{4}$ V, $\dfrac{3}{4}$ V

b.

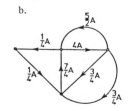

1.26 a. $R_{ab} = \dfrac{28 + 16\,R}{15 + 8\,R}$

b. $\lim\limits_{R\to 0} R_{ab} = \dfrac{28}{15}$ Ω en $\lim\limits_{R\to\infty} R_{ab} = 2\,\Omega$

1.27 a. Choose the current I_2 in the voltage source
of 2 V and the current I_{12} in the voltage
source of 12 V, both from plus to minus.
Then arises:
1) $I_2 - 4 - 6$ $= 3\,V_1$
2) $-I_2 - 3$ $= 14\,V_2 - 9\,V_5$
3) $6 + 3 - I_{12}$ $= 12\,V_3 - 12\,V_4$
4) $4 - 13$ $= -12\,V_3 + 14\,V_4 - 2\,V_5$
5) $15 + 13 + I_{12}$ $= -9\,V_2 - 2\,V_4 + 11\,V_5$
*) $V_2 - V_1 = 2$
**) $V_3 - V_5 = 12$

b. The mesh method is possible, because the
network is planar.

c. 28 A from node 5 to node 2.

1.28 The voltage source delivers 18 W, the current
source delivers 108 W. The resistors (counted
from left to right) dissipate 36, 18, 36 and
36 W respectively.

1.29 a. $V_1 = 2$ V, $V_2 = -1$ V.

b.

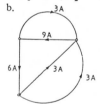

1.30 a. $V_1 = 0$ V, $V_2 = 3$ V, $V_3 = -3$ V.

b.

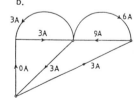

1.31 a. $V_1 = 0.8$ V, $V_2 = -0.2$ V, $V_3 = 0.4$ V.

b.

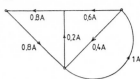

c. The voltage source delivers 0.8 W;
   the current source delivers 0.4 W.

1.32 $V_1 = 6$ V, $V_2 = 9$ V, $V_3 = -3$ V.

1.33 $I_x = I_y$.

1.34 a. $V_1 = 6$ V, $V_2 = -6$ V, $V_3 = 2$ V.

b.

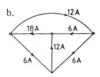

1.35 4 V

1.36 $\dfrac{11}{3}$ A

1.37 no

1.38 b. no

1.41 a. -1 V, $-\dfrac{7}{8}$ V, $-\dfrac{1}{8}$ V, -1 V

1.42 a. $I = -2 + \sqrt{4 + V}$

b. 1 A, 2 A
c. $-2 + \sqrt{21} = 2{,}58$ A
d. no

1.43 48 V, 24 V, 24 V, 24 V;
    9 A, -9 A, -3 A, -6 A

1.44 $V' = V'' = 8$ V

1.45 $V_T = 10$ V, $R_T = 10\,\Omega$, $I_N = 1$ A

1.46 $V_T = \dfrac{5}{8}$ V, $R_T = \dfrac{1}{8}\,\Omega$, $I_N = 5$ A

1.47 $V_T = 1.2$ V, $R_T = \dfrac{11}{15}\,\Omega$, $I_N = \dfrac{18}{11}$ A

1.48 $V_T = 60$ V, $R_T = \dfrac{130}{3}\,\Omega$, $I_N = \dfrac{18}{13}$ A

1.49 a. 32 W

b. $\dfrac{16}{3}$ W

c. $V_T = 8$ V, $R_T = 3\,\Omega$

d. $\dfrac{32}{3}$ W

e. $\dfrac{16}{3}$ W

f. There is only one equivalence for the external network (here R_b).

1.50 0 W, $5\dfrac{3}{25}$ W, $\dfrac{16}{3}$ W, $5\dfrac{3}{25}$ W, 0 W

1.51 a. 2 V

b. $\dfrac{1}{2}$ A

c. $4\,\Omega$, $\dfrac{1}{4}$ W

1.52 1 A

1.53 The power in R is zero for each value of R.

1.54 a. 3 A
     b. 108 W, 108 Ω

1.55 0.99 Ω

1.56 $0, \dfrac{1}{4}, \dfrac{1}{2}, \dfrac{3}{4}, 1, \dfrac{5}{4}, \dfrac{3}{2}, \dfrac{7}{4}$ V.

1.57 $I = 2$ A.

1.58 $V_T = \dfrac{4}{3}$ V, $R_T = 1.51\,\Omega$.

1.59 $I = 4.68$ A.

1.60 $V_{ab} = \dfrac{1}{12}$ V.

1.61 A voltage-voltage transactor

1.62 $V_1 = V_2$ and $I_1 = -I_2$

1.63 forbidden network

1.64 5 V

1.65 $\dfrac{V_2}{V_1} = -\dfrac{aR_3R_4}{R_2(R_3 + R_4) + \mu aR_3R_4}$

1.66 $\dfrac{3}{2}$ V

1.67 1 V, $\dfrac{7}{8}$ V, $\dfrac{1}{8}$ V

1.68 0.3 A

1.69 $\dfrac{2}{11}$ V

1.70 $\dfrac{2}{7}$ A

1.71 $V_T = -\dfrac{R_2GV_1}{1 - R_1G}$, $R_T = R_2$

1.72 $V_T = 7$ V, $R_T = \dfrac{1}{2}\,\Omega$

1.73 $V_T = 15$ V, $R_T = 30\,\Omega$

1.74 $V_T = \dfrac{AR_2}{R_1 + R_2 + AR_2}$, $R_T = \dfrac{R_1R_2}{R_1 + R_2 + AR_2}$

1.75 0.8 A

1.76 $\dfrac{3}{4}$

1.77 a. 4) $0 = (G_1 + G_2)V_4 - G_1V_1$
       5) $0 = (G_3 + G_4)V_5 - G_4V_3 - G_3V_2$

b. $V_3 = \dfrac{KG_1(G_3 + G_4)}{(G_1 + G_2)(G_3 + G_4 + KG_4)} V_1$

$\qquad - \dfrac{KG_3}{G_3 + G_4 + KG_4} V_2$

c. $\lim\limits_{K \to \infty} V_3 = \dfrac{G_1(G_3 + G_4)}{G_4(G_1 + G_2)} V_1 - \dfrac{G_3}{G_4} V_2$

1.78 $V_4 = -2\,V.$

1.79 $V_T = 0.5\,V \quad R_T = 0.5\,\Omega$

1.80 $V_{20} = 1\,V.$

1.81 The independent source delivers -4 W, the
dependent source delivers 60 W. The resistors
(counted from left to right) dissipate 4 W,
16 W, 36 W.

1.82 The independent source delivers -12 W, the
dependent source delivers 120 W. The resistors
(counted from left to right) dissipate 4 W,
32 W, 72 W.

1.83 $H = 1 + \dfrac{R_1}{R_2}.$

Problems Chapter 2

2.1 a. $I_{gem} = \dfrac{|I|}{\pi}\,A$

b. $I_{eff} = \dfrac{|I|}{2}\,A$

2.2 a. $V_{gem} = \dfrac{2|V|}{\pi}\,V$

b. $V_{eff} = \dfrac{|V|}{\sqrt{2}}\,V$

2.3 $V_{eff} = 1.5\sqrt{2}\,V,\ V_{gem} = 1.5\,V$

2.4 $V_{eff} = 5.03\,V,\ V_{gem} = 3.33\,V$

2.5 $I_{eff} = 4.08\,A$

2.6 a. $p = 12\,W,\ P = 12\,W$
b. $p = 24\cos^2 2t\,W,\ P = 12\,W$

c. $p = 24\cos\left(3t + \dfrac{\pi}{8}\right)\cos 3t\,W,\ P = 11.09\,W$

2.7 $5.43\,A$

2.8 a. $T = \dfrac{2\pi}{\omega}\,s$

b. $I_{eff} = \sqrt{\dfrac{|I_1|^2 + |I_2|^2}{2}}$

2.9 $10\,ms$

2.10 $6.47\cos 3t - 2.68\sin 3t\ V$

2.11 a. $\sqrt{29}\,\sin\left(4t + \arctan\dfrac{5}{2}\right)\,A$

b. $\sqrt{29}\,\cos\left(4t - \arctan\dfrac{2}{5}\right)\,A$

2.12 a.

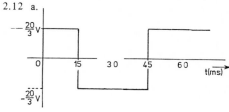

b.

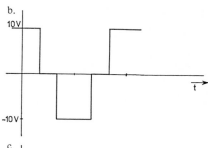

c.

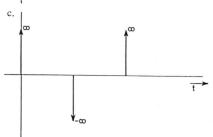

2.13 An inductor of $\dfrac{10}{377}$ H.

2.14 $v = \dfrac{1}{\omega C}\,|I|\sin\omega t\ V$

2.15 $i = -\dfrac{1}{\omega L}\,|U|\cos\omega t\ A$

2.16

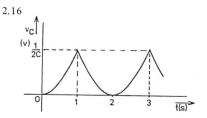

2.17

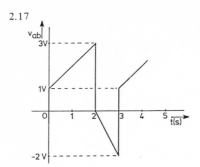

2.18   a & b.

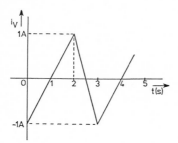

2.19 $v_{ab} = 10 \cos 2t$ V

2.20 $W = 166$ J

2.21 $i = 500 \dfrac{C_1 C_2}{C_1 + C_2} \cos 20t$ A

2.22 $L = \dfrac{L_1 L_2}{L_1 + L_2}$

2.23   a. $i = 3 + 4 \cos 5000t + 2 \sin 5000t$ A

    b. $I_{eff} = \sqrt{19}$ A

2.24   a. $v_{ab} = 24 \cos 3t + 8 \sin 3t$ V

    b. $i = 0.8 \sin 3t + 2.4 \cos 3t$ A

    c. A resistor of 10 Ω.

2.25   a. $i = 24 \cos 3t + 8 \sin 3t$ A

    b. $v_{ab} = 0.8 \sin 3t + 2.4 \cos 3t$ V

    c. A conductor of 10 S.

2.26   a. $i_3 = 13 \sin(\omega t + \arctan \dfrac{5}{12})$ A

    b. $v = 13 \sin(\omega t + \arctan \dfrac{5}{12})$ V

2.27

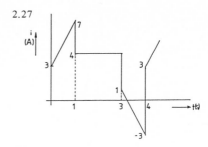

2.28   a. $v = \tfrac{3}{2} \cos 2t + 3 \sin 2t$ V,

        $i = \tfrac{5}{2} \sin 2t + \cos 2t$ A.

    b. $L = \dfrac{3}{58}$ H, $R = \dfrac{36}{29}$ Ω.

2.29   a. $15 - j$

    b. $-1 + 5j$

    c. $0.1 - 0.7j$

    d. $6 + 3j$

2.30   a. $\sqrt{53}\ e^{j \arctan \frac{7}{2}}$

    b. $e^{j \frac{\pi}{2}}$

    c. $\sqrt{10}\ e^{j \arctan 3}$

    d. $\sqrt{5}\ e^{j(\pi + \arctan \frac{1}{2})}$

2.31   a. 2

    b. $\dfrac{5}{2}$

    c. $\dfrac{3}{5}$

2.32   a. $A^* + B^*$

    b. A

    c. AB^*

    d. $\dfrac{A^*}{B^*}$

2.33   a. $\sqrt{26}$

    b. 3

    c. 1

2.34   a. $\arctan \dfrac{1}{5}$

    b. $\dfrac{\pi}{8}$

    c. $2 \arctan \dfrac{2}{5}$

2.35   a. j

    b. $-0.04 + j$

    c. $1.54 + 1.28j$

    d. 0.21

    e. $-1.07 + 4.05j$

2.37 a. no

b. yes

c. yes

d. yes

2.38 a. $4 \cos \omega t$

b. $4 \cos(\omega t + \frac{\pi}{4})$

c. $8.77 \cos \omega t$

2.39 a. $4e^{j\,5t}$

b. $4e^{j(5t + \frac{\pi}{4})}$

c. $4e^{j(5t - \frac{\pi}{2})}$

d. $4e^{j(5t - \frac{\pi}{4})}$

e. $4e^{(-i + 5j)t}$

f. $4e^{\{-t + j(5t - \frac{\pi}{2})\}}$

2.40 $Z = R + j(\omega L - \frac{1}{\omega C})$

$$Y = \cfrac{1}{R + j(\omega L - \frac{1}{\omega C})}$$

2.41 $Z = \dfrac{16j\omega}{8 - 2\omega^2} \ \Omega$

a. an inductor of $\frac{8}{3}$ H

b. open terminals

c. capacitor of 69.4 mF

2.42 Als $\omega = \dfrac{1}{\sqrt{LC}}$

2.43 a. $Z_1 Z_4 = Z_2 Z_3$

b. $R_1 R_4 = R_2 R_3$ en $L = R_1 R_4 C$

2.44 $R = \sqrt{\dfrac{L(1 - \omega^2 LC)}{C}}$

2.45 $\omega = \sqrt{\dfrac{1}{LC} - \dfrac{1}{R^2 C^2}}$

2.46 $\omega_1 = 0$, $\omega_2 = \dfrac{1}{\sqrt{LC}}$, $\omega_3 = \infty$, $L = R^2 C$

2.49 $Z = \dfrac{4}{5} - \dfrac{3}{5}j \ \Omega$

2.50 $I_C = \dfrac{3}{5} + \dfrac{1}{5}j$ A

2.51

branch	V	I*	VI*
1	2	$\dfrac{-1-3j}{5}$	$\dfrac{-2-6j}{5}$
2	$1+j$	$\dfrac{-2+4j}{5}$	$\dfrac{-6+2j}{5}$
3	$\dfrac{1-3j}{5}$	$\dfrac{3-j}{5}$	$-\dfrac{2j}{5}$
4	$\dfrac{9+3j}{5}$	$\dfrac{1+3j}{5}$	$\dfrac{6j}{5}$
5	$\dfrac{4+8j}{5}$	$\dfrac{2-4j}{5}$	$\dfrac{8}{5}$
			$\overline{\quad 0 \quad}$ +

2.52 a. $i = 7 \cos 4t$ A

b. $i = 3 \cos 4t - 8 \sin 4t$ A $=$
$= \sqrt{73} \cos(4t + \arctan \frac{8}{3})$ A

2.53 a. $i = 7 \sin 4t$ A

b. $i = 3 \sin 4t + 8 \cos 4t$ A $=$
$= \sqrt{73} \sin(4t + \arctan \frac{8}{3})$ A

2.54 a. $I = 4$ A

b. $I = 2(1+j)\sqrt{2}$ A

c. $I = 2j$ A

d. $I = \dfrac{9}{2}(1 + j\sqrt{3})$ A

e. $I = 15 + 14j$ A

2.55 a. $i = 2\sqrt{10} \cos(3t + \arctan \frac{1}{3})$ A

b. $i = 2\sqrt{10} \sin(3t + \arctan \frac{1}{3})$ A

c. $i = 2\sqrt{10} \sin(3t + \varphi + \arctan \frac{1}{3})$ A

2.56 a. $v_{24} = 20 \sin 10^3 t$ V

b. $L = 20$ H

2.57 $\arctan \frac{1}{2}$

2.58

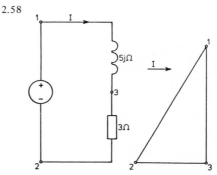

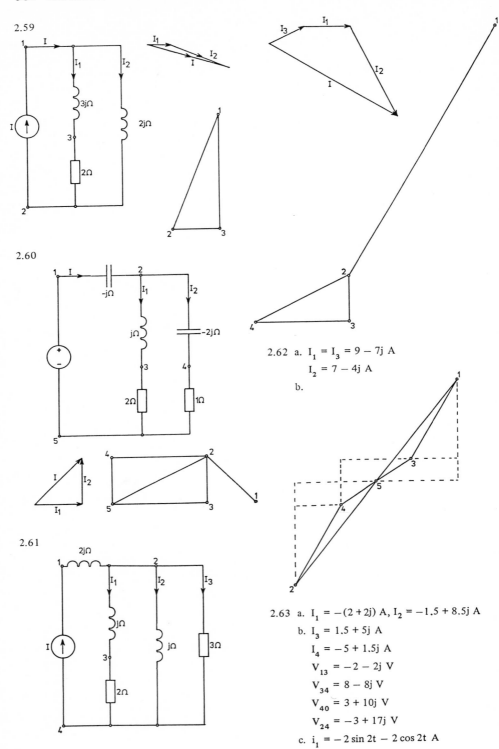

2.59

2.60

2.61

2.62 a. $I_1 = I_3 = 9 - 7j$ A
$I_2 = 7 - 4j$ A

b.

2.63 a. $I_1 = -(2 + 2j)$ A, $I_2 = -1.5 + 8.5j$ A
b. $I_3 = 1.5 + 5j$ A
$I_4 = -5 + 1.5j$ A
$V_{13} = -2 - 2j$ V
$V_{34} = 8 - 8j$ V
$V_{40} = 3 + 10j$ V
$V_{24} = -3 + 17j$ V
c. $i_1 = -2 \sin 2t - 2 \cos 2t$ A

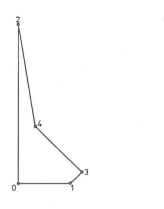

$$V_{13} = 8 + 14j \ V$$
$$V_{23} = 9 + 6j \ V$$
$$V_{24} = 10 - 2j \ V$$
$$V_{34} = 1 - 8j \ V$$

c.

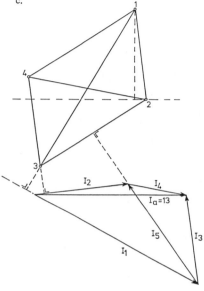

2.64 a.

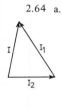

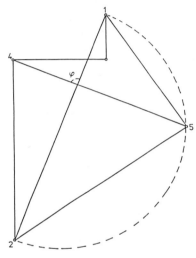

b. not exact $90°$, but $90.2°$.

2.65 a.

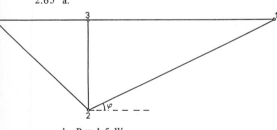

b. $P = 1.5 \ W$
c. $v_{12} = \sqrt{5} \sin (t + \arctan 0.5) \ V$

2.66 a. $I_1 = 14 - 8j \ A$, $I_2 = 8 + j \ A$
b. $I_3 = -1 + 8j \ A$, $I_4 = 5 - j \ A$
$I_5 = -6 + 9j \ A$
$V_{12} = -1 + 8j \ V$

d. $v_{23} = |V_{23}| \cos (\omega t + \arg V_{23}) \ V$
dus $v_{23} = \sqrt{117} \cos (300t + \arctan \frac{2}{3}) \ V$

2.67 b.

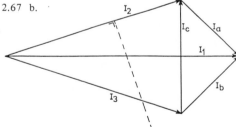

c.

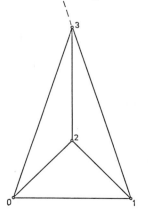

d. $v_{32}(t) = -5 \sin t$ V

2.68 a. $I = \dfrac{3 - 11j}{5}$ A

b. $S_1 = 2.2 + 1.4j$ VA

$S_2 = 0.4 - 1.2j$ VA

c. $Q_L = 3.4$ VA$_r$

$Q_C = -3.2$ VA$_r$

2.69 a. 0 en $\sqrt{\dfrac{5}{3}}$ rad/s

b. $Q_C = -0.1$ VA$_r$

2.70 a. $A = 1, B = -1$

b. $I_V = -\dfrac{2 + 6j}{5}$ A

c. $S_V = -\dfrac{1}{5} + \dfrac{3}{5}j$ VA

$S_I = \dfrac{1}{2} - \dfrac{1}{2}j$ VA

d. $P = 0.3$ W

2.71 a. $H = \dfrac{j\omega}{1 + 2j\omega}$

b. $\dfrac{3 + j}{4}$ Ω

c. $\dfrac{1}{3}$ W

2.72 $R = |Z_T|$

2.73 a. $1 + 3j$ V

b. $\dfrac{5}{2}j$ A

c. $1.2 + 0.4j$ Ω

$\dfrac{4}{3}$ H parallel met $\dfrac{4}{3}$ Ω

d. 1.26 Ω

2.74 a. $-7 + 11j$ ampère

b. $V_T = 1 + 2j$ volt

$Z_T = (1 + 2j)/2$ Ω

2.75 a. $I = 0.6 - 2.2j$ A.

b. $i = 2.28 \cos(5t - 1.3 \text{ rad})$ A.

2.76 a. $I_z = \dfrac{1.2(1 + 3j)}{0.8(3 - j) + Z}$ A.

b. $Z = 0.8(3 + j)$ Ω, $S = \dfrac{3 + j}{4}$ VA.

2.77 $R = \frac{1}{2}\Omega$.

Problems Chapter 3

3.1 a.

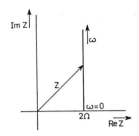

b.

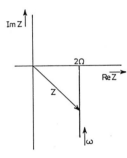

3.2 a.

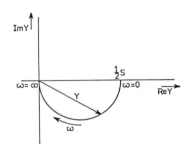

b.

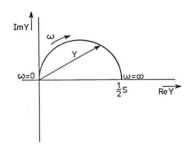

3.3 $I = \dfrac{12jC}{1 + 2jC}$

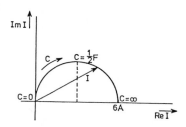

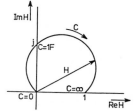

b.

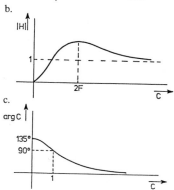

3.4 a. $H = \dfrac{1}{2} \cdot \dfrac{j\omega RC - 1}{j\omega RC + 1}$

b.

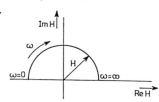

c.

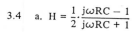

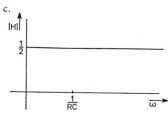

c.

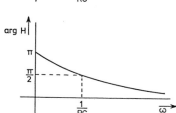

3.5 a.

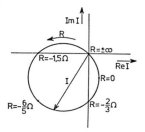

b. I is real if $R = -\dfrac{3}{2}\Omega$

I is imaginary if $R = -\dfrac{3}{2}\Omega$

3.6 a. $H = \dfrac{C}{C - 1 - j}$

3.7 a. $I = \dfrac{8 + 6j}{5R + 2 - 6j}$ A

b.

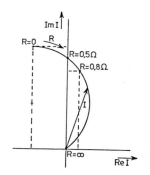

3.8 a. $I_1 = 1$ A

b. $I_1 = \dfrac{3 + 4j}{3 + jL}$ A

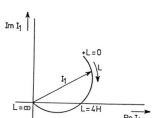

c.

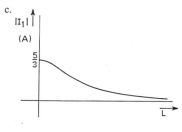

d.

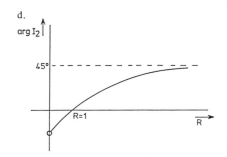

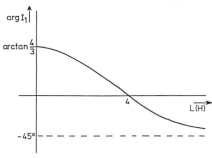

d. L = 0 H and L = 4 H.

3.9   a. $H = -j \dfrac{R+4}{4R - j(R+4)}$

b.

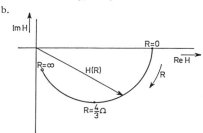

c. $v_2 = \dfrac{15}{41}(5 \cos \omega t + 4 \sin \omega t)$ V

3.10 a. $I_2 = \dfrac{1+j}{1+2j+R}$ A

b. R = 1 Ω

c.

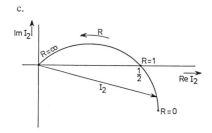

3.11

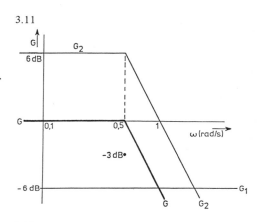

3.12

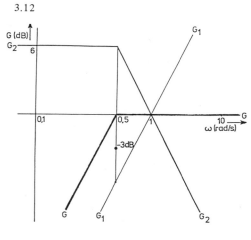

3.13

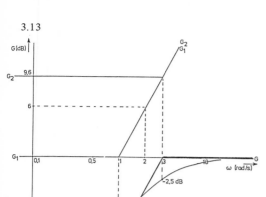

3.14

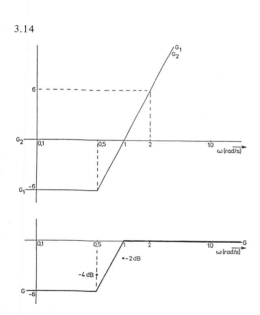

3.15

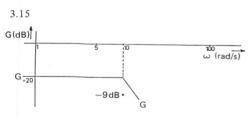

3.16

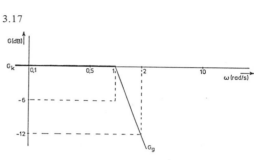

3.17

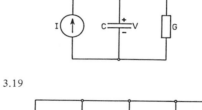

3.18

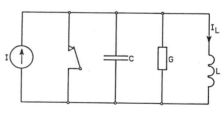

3.19

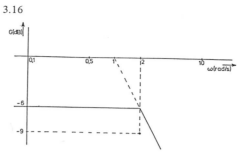

3.20

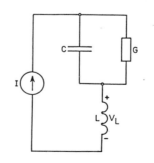

3.21 $\dfrac{1}{\sqrt{LC}}$

3.22 $\dfrac{1}{\sqrt{LC}}$

3.23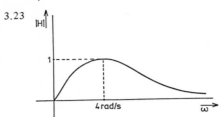

3.24   a. 43 rad/s
      b. 75 and 25 rad/s
      c. $Q_0 = 0.85$, $B = 50$ rad/s

3.25 $\dfrac{1}{\sqrt{LC}} \sqrt{\dfrac{R_1^2 C - L}{R_2^2 C - L}}$ and $\omega = 0$ and $\omega = \infty$.

3.26   All frequencies

3.27   No frequency at all

3.28 $\dfrac{1}{\sqrt{LC}} \sqrt{1 - \dfrac{R^2 C}{L}}$

3.29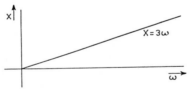

3.30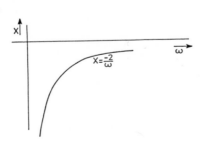

3.31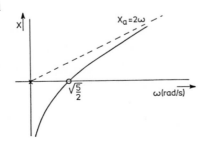

3.32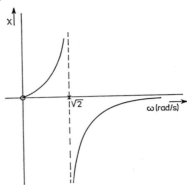

3.33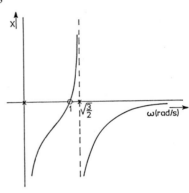

3.34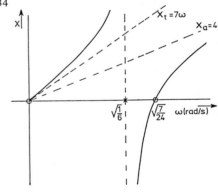

3.35

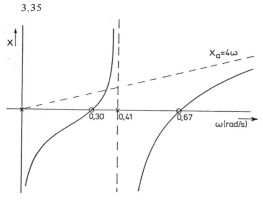

3.36 $z_1 = \omega_0$, $z_2 = 1.9\,\omega_0$, $z_3 = 0.55\,\omega_0$.
$p_1 = 0$, $p_2 = 0.7\,\omega_0$, $p_3 = 1.4\,\omega_0$,
with $\omega_0 = \dfrac{1}{\sqrt{LC}}$. Asymptote $\omega_a = \dfrac{1}{2}\omega L$.

3.37 a. $X = 3\omega\,\dfrac{3 - 2\omega^2}{6 - 5\omega^2}$

b. $z_1 = 0$, $z_2 = \sqrt{\dfrac{3}{2}}$, $p_1 = \sqrt{\dfrac{6}{5}}$

$X_r = \dfrac{3}{2}\omega$, $X_a = \dfrac{6}{5}\omega$

3.38 a. $X = \dfrac{2\omega^4 - 4\omega^2 + 1}{2\omega(\omega^2 - 1)}$

b. $z_1 = 0.54$, $z_2 = 1.3$, $p_1 = 0$, $p_2 = 1$
$X_a = \omega$

c.

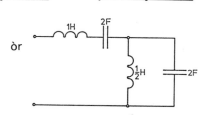

òr

3.39. a. $Z = jX$, $X = \omega\,\dfrac{5 - 14\omega^2}{1 - 6\omega^2}$

$z_1 = 0$, $z_2 = \sqrt{\dfrac{5}{14}}$, $p_1 = \sqrt{\dfrac{1}{6}}$

b. $X_r = 5\omega$, $X_a = \dfrac{7}{3}\omega$

c.

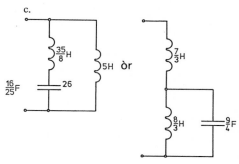

3.40 a. $X = \omega\,\dfrac{42\omega^2 - 15}{32\omega^2 - 8}\ \Omega$

b. $z_1 = 0$, $z_2 = \sqrt{\dfrac{5}{14}}$, $p_1 = \dfrac{1}{2}$

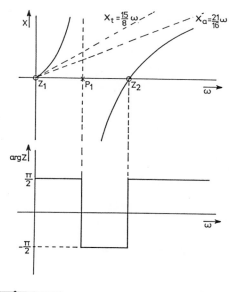

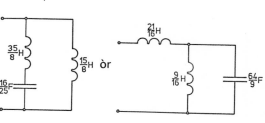

3.41 a. $Z = jX$, $X = \omega\,\dfrac{6 - 7\omega^2}{5 - 6\omega^2}\ \Omega$

b. $\lim\limits_{\omega \to 0}\ \dfrac{Z}{\omega} = \dfrac{6}{5}j$

$\lim\limits_{\omega \to \infty}\ \dfrac{Z}{\omega} = \dfrac{7}{6}j$

c. $z_1 = 0$, $z_2 = \sqrt{\dfrac{6}{7}}$, $p_1 = \sqrt{\dfrac{5}{6}}$

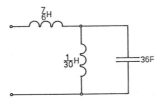

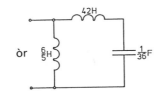

3.42 a. $X = \dfrac{\frac{5}{2}\omega}{1 - \omega^2}\ \Omega,$

zero $z_1 = 0$

pole $p_1 = 1$

b. $L = \dfrac{5}{2}$ H and $C = \dfrac{2}{5}$ F in parallel.

Problems Chapter 4

4.1 $Z_i = \dfrac{\omega^2(M^2 - L_1 L_2) + j\omega L_1 R}{R + j\omega L_2}$

4.2 The same formula as in Answer 4.1

4.3 $Z_i = \dfrac{1}{13}(1 + 21j)\ \Omega$

4.4 $Z_i = j\omega(L_1 + L_2 + 2M)$

4.5 $Z_i = j\omega(L_1 + L_2 + 2M_3) + \dfrac{\omega^2(M_1 + M_2)^2}{Z + j\omega L_3}$

4.6 $Z_i = j\omega\dfrac{L_1 L_2 - M^2}{L_1 + L_2 - 2M}$

4.7 As 4.6, but + in the denominator

4.8 $H = 2\dfrac{1 + jL}{2 + 5jL}$

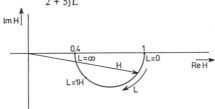

4.9   a. yes

b. $Z_i = \dfrac{j\omega LZ}{4j\omega L + Z}$ $\lim_{L \to \infty} Z_i = \dfrac{Z}{4}$

c. Ideal transformer with 1:n and with
n = 2.

4.10 $Z_i = \dfrac{Z}{4}$

4.11 $Z_i = \dfrac{Z}{4}$

4.12 n = ± 2

4.13 n = ± 5   P = 0.1 W

4.14 n = ± 200   C = 250 pF

4.15 $Z_i = \dfrac{Z}{(n - 1)^2}$

4.16 $Z_i = Z(n + 1)^2$

4.17 $Z_i = Z\dfrac{(n + 1)^2}{n^2}$

4.18 a. $V_T = \dfrac{1}{1 + 500jC}\ V_1$ $Z_T = \dfrac{500}{1 + 500jC}\ \Omega$

b. n = ± 5 $C = \dfrac{1}{250}$ F

c. $S = \dfrac{1}{200} \cdot \dfrac{9 + 8j}{145}$ VA

P = 0.31 mW

4.19

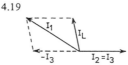

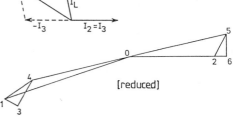

[reduced]

4.20 $H = \dfrac{1 - 3C}{4 + 2j - (6 - 2j)C}$

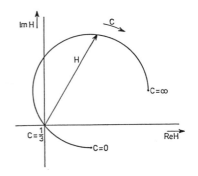

4.21 a. $X_{ab} = 4\omega \dfrac{1 - \omega^2}{1 - 2\omega^2}$

b. $z_1 = 0 \quad z_2 = 1 \quad p_1 = \dfrac{1}{2}\sqrt{2}$

c. $X_r = 4\omega$

4.22 a. $Z = jX \quad X = \dfrac{10\omega^4 - 8\omega^2 + 1}{\omega(6\omega^2 - 1)}$

b. $z_1 = 0.8 \quad z_2 = 0.39 \quad p_1 = 0 \quad p_2 = 0.41$

c. $X_a = \dfrac{5}{3}\omega$

4.23 a. $I = \dfrac{50}{5 + j(5 + M)}$

b.
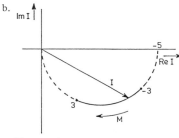

c. The part between M = −3 and M = +3
d. M = −2,1 H

4.24 $V_T = 3 + j$ V, $Z_T = 0,2 + 0,4j$ Ω.

4.25 $V_T = \dfrac{32 + 7j}{29}$ V, $Z_T = \dfrac{49 - 21j}{29}$ Ω.

Problems Chapter 5

5.1 a. 240 V
b. 415 V
c. 720 V

5.2 indefinite
5.3 a.

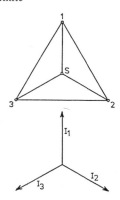

b. P = 17.3 kW

5.4 a.

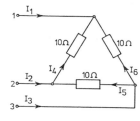

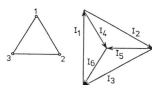

b. P = 51.7 kW

5.5 a.

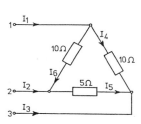

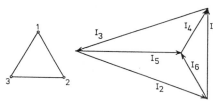

b. P = 23 kW

5.6 75.4 A
5.7 a. 71880 πC A
b. $\dfrac{1}{3}10^{-3}$ F

5.8

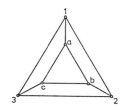

5.9 0 A
5.10 13.8 A

5.11

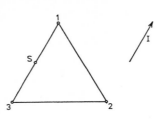

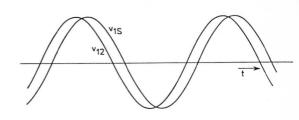

5.12

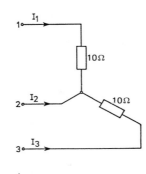

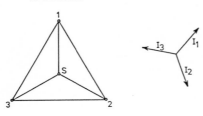

5.13

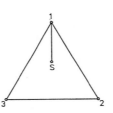

5.14   Choose line currents I_1, I_2 and I_3 directed
to the load.

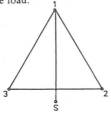

5.15   a.   Choose line currents I_1, I_2 and I_3 directed
to the load.

5.18 $|I| = 27.6$ A

5.19   a. $V_{ab} = \frac{1}{2} V_{13}$

b. $I_k = \frac{1}{2} V_{13}$

5.20   a. $I = \dfrac{100jC}{1 + jC}$ A

b.

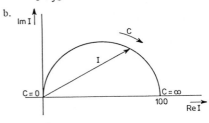

c. $C = \dfrac{1}{\sqrt{3}}$ F

5.21   a. $I = \dfrac{100}{1 + jL}$ A

b.

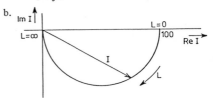

c. $L = \sqrt{3}$ H

5.22   a. $I_3 = 208$ A
b.   Choose line currents I_1, I_2 and I_3 directed
to the load.

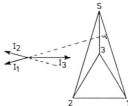

5.23   a. $I = \dfrac{138}{R - j}$ A

b.

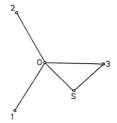

c.

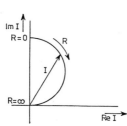

d. P = 9522 W

5.24 a. $V_{S1} = -\dfrac{100j\sqrt{3}}{2R+j} R$

b.

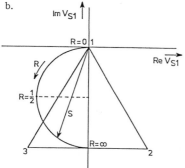

5.25 a. yes

    b. 86.4 kW

5.26 224 μF

5.27 a. $I_L = 0$ A

    b. $I_L = \dfrac{150(1+j\sqrt{3})}{3jL}$ A

    c. $L = \dfrac{1}{3}$ H

5.28 a. $n_1 = -1$ $n_2 = 1$

b.

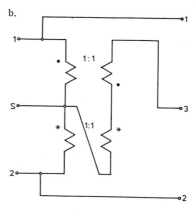

8.29 b.

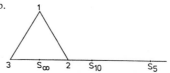

c. If $R \to \infty$ then S (S_∞) lies in the centre between 2 and 3. We apply superposition for R = 0: Choose first $V_{20} = V_{30} = 0$, then we have resonance with $Z_L = 10j$ Ω and $Z_C = -10j$ Ω in series. The current is infinite and thus S lies in infinity. If only the source V_{20} or the source V_{30} is working then the result is a limited current.

5.30 b. R = 0Ω $V_{S1} = 6(2+j)$ V

        R $\to \infty$ $V_{S1} = 15$ V

        R = 1Ω $V_{S1} = \dfrac{30}{13}(5+j)$ V

        $|V_{S1}|$ min = 11.6 V

        $|V_{S0}|$ min = 2.4 V

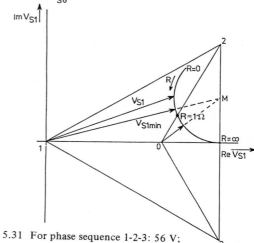

5.31 For phase sequence 1-2-3: 56 V;
For phase sequence 3-2-1: 207 V.

Problems Chapter 6

6.1 $f(t) = \dfrac{1}{2} A + \dfrac{2A}{\pi} (\sin \omega t + \dfrac{1}{3} \sin 3\omega t + \dfrac{1}{5} \sin 5\omega t + \dots) =$

$= \dfrac{1}{2} A + \dfrac{2A}{\pi} \sum\limits_{n=1}^{\infty} \dfrac{\sin (2n-1)\omega t}{2n-1}$ with $\omega = \dfrac{2\pi}{T}$.

6.2 $f(t) = \dfrac{1}{2} A + \dfrac{2A}{\pi} (\cos \omega t - \dfrac{1}{3} \cos 3\omega t + \dfrac{1}{5} \cos 5\omega t - \dots) =$

$= \dfrac{1}{2} A + \dfrac{2A}{\pi} \sum\limits_{n=1}^{\infty} (-1)^{n+1} \dfrac{\cos (2n-1)\omega t}{2n-1}$ with $\omega = \dfrac{2\pi}{T}$.

6.3 $f(t) = \dfrac{4A}{\pi} \sum\limits_{n=1}^{\infty} \dfrac{\sin (2n-1)\omega t}{2n-1}$ with $\omega = \dfrac{2\pi}{T}$.

6.4 $f(t) = \dfrac{8A}{\pi^2} \sum\limits_{n=0}^{\infty} (-1)^n \dfrac{\sin (2n+1)\omega t}{(2n+1)^2}$ with $\omega = \dfrac{2\pi}{T}$.

6.5 $f(t) = \dfrac{A}{2} - \dfrac{A}{\pi} \sum\limits_{n=1}^{\infty} \dfrac{\sin n\omega t}{n}$ with $\omega = \dfrac{2\pi}{T}$.

6.6 $f(t) = \dfrac{A}{4} - \dfrac{2A}{\pi^2} \sum\limits_{n=1}^{\infty} \dfrac{\cos (2n-1)\omega t}{(2n-1)^2} - \dfrac{A}{\pi} \sum\limits_{n=1}^{\infty} (-1)^n \dfrac{\sin n\omega t}{n}$ with $\omega = \dfrac{2\pi}{T}$.

6.7 $f(t) = \dfrac{2A}{\pi} - \dfrac{4A}{\pi} \sum\limits_{n=1}^{\infty} \dfrac{\cos n\omega t}{4n^2 - 1}$ with $\omega = \dfrac{2\pi}{T}$.

6.8 $f(t) = \dfrac{A}{\pi} + \dfrac{A}{2} \sin \omega t + \dfrac{2A}{\pi} \sum\limits_{n=1}^{\infty} \dfrac{1}{1 - 4n^2} \cos 2n\omega t$ with $\omega = \dfrac{2\pi}{T}$.

6.9 $f(t) = \dfrac{2A}{\pi} + \dfrac{4A}{\pi} \sum\limits_{n=1}^{\infty} (-1)^{n+1} \dfrac{\cos n\omega t}{4n^2 - 1}$ with $\omega = \dfrac{2\pi}{T}$

6.10 $f(t) = \dfrac{A}{\pi} + \dfrac{2A}{\pi} \sum\limits_{n=1}^{\infty} (-1)^{n+1} \dfrac{\cos 2n\omega t}{4n^2 - 1}$ with $\omega = \dfrac{2\pi}{T}$.

6.11 $f(t) = \dfrac{4B}{\pi a} \sum\limits_{n=0}^{\infty} \dfrac{\sin (2n+1)a \cdot \sin (2n+1)\omega t}{(2n+1)^2}$ with $\omega = \dfrac{2\pi}{T}$ and $a = \dfrac{2\pi t_1}{T}$

6.12 $f(t) = \dfrac{2B \sqrt{3}}{\pi} \{ \sin \omega t - \dfrac{1}{5} \sin 5\omega t - \dfrac{1}{7} \sin 7\omega t + \dfrac{1}{11} \sin 11\omega t + \dfrac{1}{13} \sin 13\omega t + \dots \}$

with $\omega = \dfrac{2\pi}{T}$.

6.13 T = 2π s.

6.14 $v = \dfrac{3}{2} \cos 98t + 8 \cos 100t + \dfrac{3}{2} \cos 102t$ V

6.15 a. Only odd sine terms.
b. Only odd terms.
c. d.c. term and cosine terms.
d. d.c. term and odd sine terms.
e. d.c. term and odd cosine terms.

6.16 a. 1.

b. Only sine terms.

c. $1 + \dfrac{10}{\pi} \sin \omega t - \dfrac{2}{\pi} \sin 2\omega t + \dfrac{10}{3\pi} \sin 3\omega t.$

6.17 a. $B_n = 0$, that means only cosine terms, thus even function symmetry. Example: half or full wave rectified cosine.

b. Now there must be only sine terms. Example: see 6.3 and 6.4.

6.18 $f(t) = \dfrac{2B\sqrt{3}}{\pi} \{\cos \omega t - \dfrac{1}{5} \cos 5\omega t$

$+ \dfrac{1}{7} \cos 7\omega t - \dfrac{1}{11} \cos 11\omega t + \dfrac{1}{13} \cos 13\omega t + \dots \}$

with $\omega = \dfrac{2\pi}{T}$.

Problems Chapter 7

7.1 a. $z_1 = -\dfrac{3}{2}$ no poles

b. $z_1 = -\dfrac{1}{12}$ $p_1 = 0$

c. $p_1 = -\dfrac{1}{10}$ no zeros.

7.2 Poles become zeros and visa versa.

7.3 $z_1 = 0$

$p_1 = -\dfrac{1}{2} + \dfrac{7}{2}j$

$p_2 = -\dfrac{1}{2} - \dfrac{7}{2}j$

$|I| = 0.23$ A

7.4 $z_1 = -1$

$z_2 = -0.33 + 0.85j$

$z_3 = -0.33 - 0.85j$

$p_1 = 0$

$p_2 = 1.22j$

$p_3 = -1.22j$

$|Z| = 1.17 \ \Omega$

arg $Z = 59°$

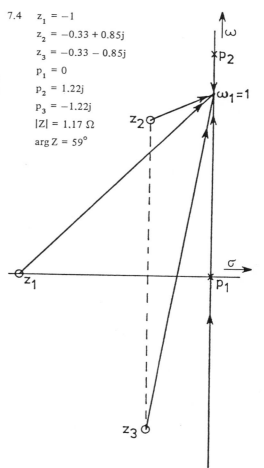

7.5 a.

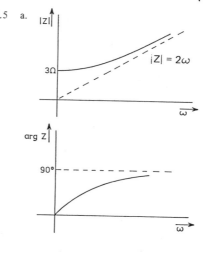

$|Z| = 2\omega$

3Ω

arg Z

$90°$

b.

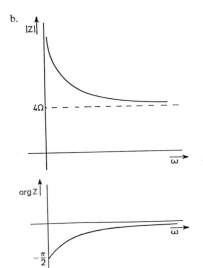

c.

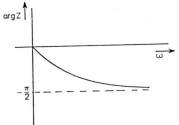

7.6   a. $z_1 = -\dfrac{3}{2}$ $p_1 = -\dfrac{1}{2}$ $p_2 = -1$

   b. $z_1 = -\dfrac{1}{5}$ $p_{1,2} = -\dfrac{1}{10} \pm \dfrac{3}{10} j$

   c. $z_1 = -2$ $p_{1,2} = -1$

7.7

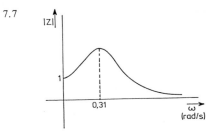

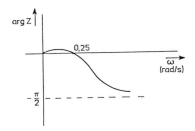

7.8   a. $z_1 = -\dfrac{3}{2}$ $z_2 = -4$ $p_{1,2} = -1 \pm j$

   b. both second order

7.9   a. $z_1 = -\dfrac{1}{2}$ $z_2 = -1$

      $p_1 = 0$ $p_2 = -\dfrac{2}{3}$

   b. both second order

7.10 $2\lambda^2 + 2\lambda + 25 = 0$

7.11   Short circuited terminals: $(2\lambda + 3)(\lambda + 4) = 0$
      Open terminals $: \lambda^2 + 2\lambda + 2 = 0$

7.12   a. $\lambda^2 + 3\lambda + 2 = 0$
      b. $(\lambda + 1)^2 = 0$
      c. $z_{1,2} = -1$ $p_1 = -1$ $p_2 = -2$

7.13 $z_{1,2} = -\dfrac{1}{6}$ $p_1 = 0$ $p_2 = -\dfrac{1}{6}$

7.14 $z_{1,2} = -2$ $p_1 = -2$ $p_2 = -3$

7.15 $z_{1,2} = p_{1,2} = -\dfrac{1}{6}$

7.16   a

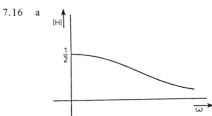

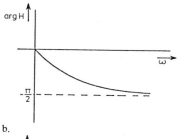

b.

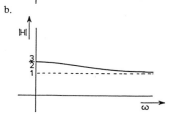

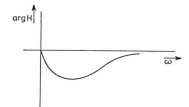

c.

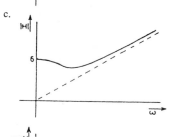

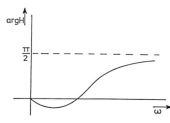

d.

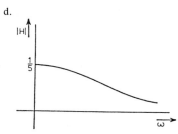

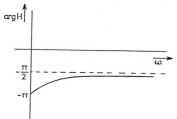

e.

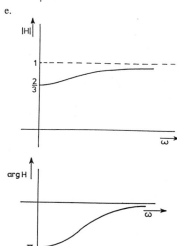

f.

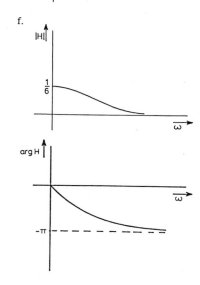

g.

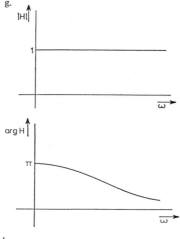

j. |H|

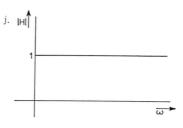

h.

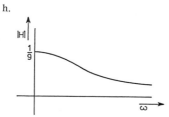

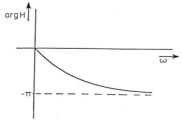

7.17   a, b, c

7.18 $z_1 = 0$ $z_{2,3} = \pm j\sqrt{\dfrac{7}{24}}$ $p_{1,2} = \pm j\sqrt{\dfrac{1}{6}}$

7.19   a. $z_{1,2} = \dfrac{1}{2} \pm \dfrac{1}{2}j\sqrt{7}$ $p_{1,2} = -\dfrac{1}{2} \pm \dfrac{1}{2}j\sqrt{7}$

b.

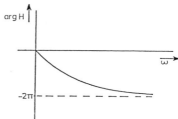

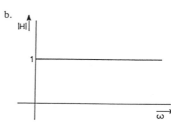

i.

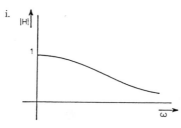

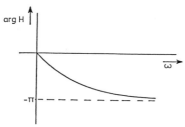

c. $0, -\pi, -2\pi$ rad

7.20   a. $H = \dfrac{\lambda^2\,RLC}{\lambda^2\,RLC + \lambda L + R}$

c.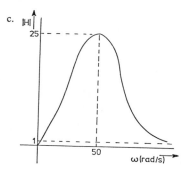

7.21 a. $H = \dfrac{6\lambda - 2}{6\lambda + 1}$

b.

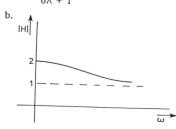

c.

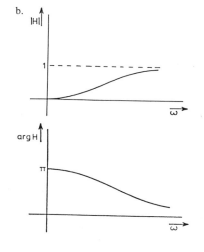

d. $\dfrac{1}{6}\sqrt{2}$ rad/s

7.22 a. $z_{1,2} = 0$ $p_1 = -0{,}38$ $p_2 = -2{,}62$

b.

7.23 a. $Z = 2\,\dfrac{2\lambda^2 + 13\lambda + 22}{(2\lambda + 5)(\lambda + 4)}$

b. $\lim\limits_{\lambda \to 0} Z = \dfrac{11}{5}\,\Omega$ $\lim\limits_{\lambda \to \infty} Z = 2\,\Omega$

c. $z_{1,2} = -3.25 \pm 0.66j$
   $p_1 = -2.5$ $p_2 = -4$

7.24 a. $Z = \dfrac{(\lambda + 1)(\lambda + 2)}{\lambda^2 + 2\lambda + 2}$

   $z_1 = -1$ $z_2 = -2$ $p_{1,2} = -1 \pm j$

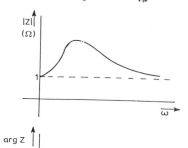

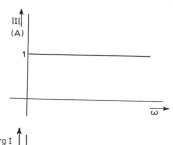

7.25 a. $I = \dfrac{\lambda - 1}{\lambda + 1}$

b. $z_1 = 1$ $p_1 = -1$

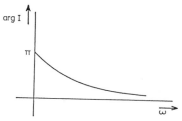

c.

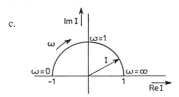

d. $i = 0,6 \cos 2t - 0,8 \sin 2t =$

$= \cos(2t + \arctan \frac{4}{3})$ A

7.26 a. $H = \dfrac{3\lambda}{3\lambda^3 + 7\lambda^2 + 4\lambda + 4}$

b. $z_1 = 0 \quad P_1 = -2 \quad P_{2,3} = -\dfrac{1}{6} \pm \dfrac{1}{6}j\sqrt{23}$

c.

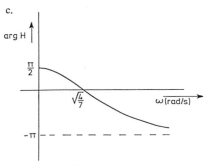

7.27 a. $H = \dfrac{2}{4\lambda^2 + 8\lambda + 5}$

b. no zeros

$P_{1,2} = -1 \pm \dfrac{1}{2}j$

$|H| = \dfrac{1}{4}\sqrt{2} \quad \arg H = -\dfrac{\pi}{4}$

7.28 a. $H = \dfrac{\lambda^3}{2\lambda^3 + 3\lambda^2 + 2\lambda + 1}$

$z_{1,2,3} = 0$

$P_1 = -1 \quad P_{2,3} = -0,25 \pm 0,66j$

b.

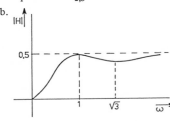

7.29 a. $V_2 = \dfrac{\lambda(\lambda - 2)}{(\lambda^2 + \lambda + 1)(3\lambda + 1)}$

$z_1 = 0 \quad z_2 = 2$

$P_{1,2} = -0,5 \pm 0,5j\sqrt{3} \quad P_3 = -\dfrac{1}{3}$

b.

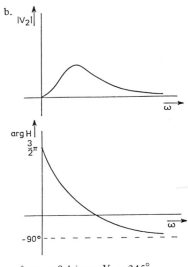

for $\omega = 0.1$ is arg $V_2 = 245°$

for $\omega = 2$ is arg $V_2 = -2°$

c. The natural oscillations are determined by the denominator of $V_2(\lambda)$.
The form of the natural oscillations is thus

$$A e^{P_1 t} + A^* e^{P_2 t} + B e^{P_3 t}$$

This must be a damped oscillation; so, poles in the left half-plane.
There are no conditions for the position of the zeros concerning stability.

7.30 a. $H = \dfrac{\lambda^3 + \lambda^2 - 2}{(\lambda + 2)(\lambda^2 + \lambda + 1)}$

b. $z_1 = 1 \quad z_{2,3} = -1 \pm j$

$P_1 = -2 \quad P_{2,3} = -\dfrac{1}{2} \pm \dfrac{1}{2}j\sqrt{3}$

c. $|H| = 1,41$

d. The position of the zeros does not cause instability.

7.31 b. $X = \dfrac{\omega^2 - 1}{\omega(2 - \omega^2)}$

c.

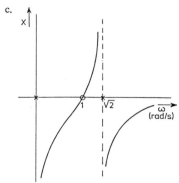

d.

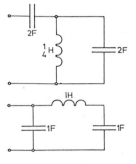

2F òr

$$7.32 \quad \text{a. } H = \frac{5\lambda^2 C + 1}{5\lambda^2 C + 5\lambda + 2}$$

b. $z_{1,2} = \pm j\sqrt{0.2}$ $P_{1,2} = -0.5 \pm 0.1j\sqrt{15}$

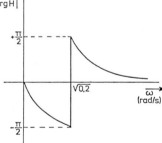

c.

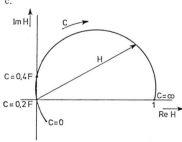

$$7.33 \quad \text{a. } H = \frac{\lambda^2 + 1}{(3\lambda + 1)(\lambda + 2)}$$

b. $z_{1,2} = \pm j$ $P_1 = -\frac{1}{3}$ $P_2 = -2$

c.

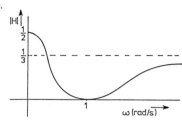

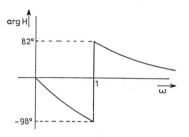

$$7.34 \quad \text{a. } H = \frac{4(\lambda + 2)}{5\lambda^2 + 14\lambda + 13}.$$

b. $|H| = 0.58.$

$$7.35 \quad \text{a. } H = \frac{2\lambda^3}{2\lambda^3 + 17\lambda^2 + 21\lambda + 6}.$$

b. Threefold zero $z_{1,2,3} = 0$;
poles $p_1 = -1$, $p_2 = -7.08$, $p_3 = -0.42$.

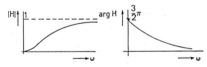

d. $\omega = 0.59$ rad/s.

Problems Chapter 8

8.1
$$\mathcal{Z} = \begin{bmatrix} Z_1 + Z_3 & Z_3 \\ Z_3 & Z_2 + Z_3 \end{bmatrix}$$

$$\mathcal{Y} = \frac{1}{Z_1 Z_2 + Z_2 Z_3 + Z_3 Z_1} \cdot \begin{bmatrix} Z_2 + Z_3 & -Z_3 \\ -Z_3 & Z_1 + Z_3 \end{bmatrix}$$

$$\mathcal{K} = \frac{1}{Z_3} \begin{bmatrix} Z_1 + Z_3 & Z_1 Z_2 + Z_2 Z_3 + Z_3 Z_1 \\ 1 & Z_2 + Z_3 \end{bmatrix}$$

8.2
$$\mathcal{Z} = \begin{bmatrix} Z & Z \\ Z & Z \end{bmatrix} \qquad \mathcal{Y} \text{ does not exist}$$

8.3
$$\mathcal{Y} = \frac{1}{Z} \begin{bmatrix} 1 & -1 \\ -1 & 1 \end{bmatrix} \qquad \mathcal{Z} \text{ does not exist}$$

8.4   a. $\mathcal{K} = 1_2$

b. $\mathcal{K} = \begin{bmatrix} -1 & 0 \\ 0 & -1 \end{bmatrix} = -1_2$

8.5 $\mathcal{Z} = \mathcal{O}$ $\mathcal{Y}$ and $\mathcal{K}$ do not exist

8.6 $\mathcal{Z} = \lambda \begin{bmatrix} L_1 & M \\ M & L_2 \end{bmatrix}$ no $\mathcal{Y}$ if $L_1 L_2 - M^2 = 0 \rightarrow$ ideal coupling.

8.7   Ideal transformer: a. ideal coupling → no $\mathcal{Y}$
b. $L_1 \rightarrow \infty$, $L_2 \rightarrow \infty$, $M \rightarrow \infty$.
No $\mathcal{Z}$.

8.8 $I_2' = \dfrac{-Z_{21} V}{\det \mathcal{Z}}$ $I_1'' = \dfrac{-Z_{12} V}{\det \mathcal{Z}}$

8.9 $V_T = \dfrac{V_1}{K_{11}}$ $Z_T = \dfrac{K_{12}}{K_{11}}$ $K_{11} \neq 0$

8.10 $\mathcal{Y} = \dfrac{1}{44} \begin{bmatrix} 31 & -23 \\ -23 & 27 \end{bmatrix}$

8.11 $\mathcal{Z} = \dfrac{1}{2} Z \begin{bmatrix} 2 & 1 \\ 1 & 2 \end{bmatrix}$

8.12   a. $\mathcal{Z}_a = \begin{bmatrix} 3 & 1 \\ 1 & 2 \end{bmatrix}$

b. $\mathcal{Z}_b = \begin{bmatrix} 2 & 1 \\ 1 & 3 \end{bmatrix}$

c. $\mathcal{Z} = \begin{bmatrix} 4 & 3 \\ 3 & 4 \end{bmatrix}$ $\mathcal{Z} \neq \mathcal{Z}_a + \mathcal{Z}_b$

8.13 $\mathcal{K}_1 = \begin{bmatrix} 3 & 2 \\ 1 & 1 \end{bmatrix}$ $\mathcal{K}_2 = \begin{bmatrix} 1 & 2 \\ 0 & 1 \end{bmatrix}$

$\mathcal{K} = \mathcal{K}_1 \mathcal{K}_2 = \begin{bmatrix} 3 & 8 \\ 1 & 3 \end{bmatrix}$

$\mathcal{K} = \mathcal{K}_2 \mathcal{K}_1 = \begin{bmatrix} 5 & 4 \\ 1 & 1 \end{bmatrix}$

8.14 $\mathcal{K} = \begin{bmatrix} \dfrac{1}{n} & 0 \\ 0 & n \end{bmatrix}$

8.15   a. $Z_i = \dfrac{AZ_0 + D}{CZ_0 + A}$

b. $Z_0 = \sqrt{\dfrac{D}{C}}$

8.16 $\mathcal{Z} = \dfrac{1}{2} \begin{bmatrix} Z_1 + Z_2 & Z_1 - Z_2 \\ Z_1 - Z_2 & Z_1 + Z_2 \end{bmatrix}$

8.17 $\mathcal{Y} = \dfrac{1}{\det \mathcal{K}} \begin{bmatrix} K_{22} & K_{12} \\ K_{21} & K_{11} \end{bmatrix}$

8.18   a, b, c, f, g, h, i, j

8.19   a. 34 V
b. 17,5 V

8.20 $\dfrac{1}{2} \begin{bmatrix} 5 & -3 \\ -3 & 5 \end{bmatrix}$

8.21   a. $\mathcal{K} = \begin{bmatrix} \dfrac{Z_1}{Z_2} + 1 & Z_1 \\ \dfrac{1}{Z_2} & 1 \end{bmatrix}$

b. $\mathcal{K} = -1$

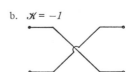

8.22 a. $\mathscr{K} = \dfrac{1}{4}\begin{bmatrix} 13 & 11 \\ 8 & 8 \end{bmatrix}$

b. $V_T = \dfrac{32}{13}$ V $\qquad R_T = \dfrac{11}{13}$ Ω

8.23 5 V

8.24 a. $V_2 = 0$ V

b. $\mathscr{Y} = \dfrac{1+j}{2}\begin{bmatrix} 1 & 0 \\ 0 & 1 \end{bmatrix}$

c. $\mathscr{K}_A = \begin{bmatrix} 1+2j & 4+4j \\ j & 1+2j \end{bmatrix}$

8.25 2 Ω

8.26 a. $Z_i = \dfrac{K_{11}}{K_{21}}$

b. $\mathscr{K}_4 = \dfrac{1}{16}\begin{bmatrix} 171 & 170 \\ 85 & 86 \end{bmatrix}$

c. $R_{n+1} = \dfrac{2+3R_n}{2+R_n}$

8.27 a. $H = \dfrac{1}{K_{11}}$

b. $\mathscr{K}(1) = \dfrac{1}{3}\begin{bmatrix} 5 & 16 \\ 1 & 5 \end{bmatrix}$ $\qquad H(1) = 0,6$

c. Only $\mathscr{K}_B$. Ideal transformer 3 : 1.

d. $H(n) = \dfrac{2}{3^n + (\frac{1}{3})^n}$

8.28 a. $\mathscr{K} = \begin{bmatrix} 11 & 8 \\ 4 & 3 \end{bmatrix}$ $\qquad \det \mathscr{K} = 1$

b. $\mathscr{K}_{total} = \begin{bmatrix} 9 & 6 \\ 3 & 2 \end{bmatrix}$ $\qquad \det \mathscr{K} \neq 1$

Not reciprocal

c. first case $\dfrac{1}{11}$
second case $\dfrac{1}{9}$

8.29 a. $H = \dfrac{-y_{21}}{1 + y_{11} + y_{22} + \det \mathscr{Y}}$

b. $\mathscr{Y}_1 = \dfrac{1}{3\lambda}\begin{bmatrix} 2 & -1 \\ -1 & 2 \end{bmatrix}$

$\mathscr{Y}_2 = \dfrac{1}{3}\lambda\begin{bmatrix} 1 & -1 \\ -1 & 1 \end{bmatrix}$

$\mathscr{Y} = \mathscr{Y}_1 + \mathscr{Y}_2$

c. $H = \dfrac{3\lambda(\lambda^2 + 1)}{6\lambda^3 + 11\lambda^2 + 12\lambda + 3}$

d. $z_1 = 0$ $\qquad z_{2,3} = \pm j$

$p_1 = -\dfrac{1}{3}$ $\qquad p_{2,3} = -\dfrac{3}{4} \pm \dfrac{1}{4}j\sqrt{15}$

e.

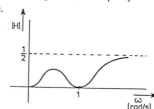

8.30 $V_T = \dfrac{5}{\lambda RC + 1}V_1$ $\qquad Z_T = 0$ Ω

8.31 a. $H = \dfrac{1}{(\lambda + 1)^2}$ $\quad$ poles $p_{1,2} = -1$
cut-off frequency 0.6 rad/s

b. $Z_i = \dfrac{\lambda + 1}{\lambda}$ Ω

c. the same as under a.

d. $v_4 = 0.32 \sin 3t - 0.24 \cos 3t$ V

8.32 $V_T = \dfrac{1}{\lambda^2 + \lambda + 1}$

$Z_T = \dfrac{-\lambda}{\lambda^2 + \lambda + 1}$

$Z_i = \dfrac{\lambda^2 + \lambda + 1}{2}$

8.33 $\dfrac{V_5}{V_1} =$

$\dfrac{1}{\lambda^3 R^3 C_1 C_2 C_3 + 2\lambda^2 R^2 C_3 (C_1 + C_2) + \lambda R(C_1 + C_3) + 1}$

8.34 $V_T = \dfrac{\lambda + 1}{\lambda + 3}$ $\quad , \quad Z_T = \dfrac{2}{\lambda + 3}.$

Problems Chapter 9

9.1 a. $v_C = 7e^{-0.1t}$ V

b. $i = 1.4e^{-0.1t}$ A

c.
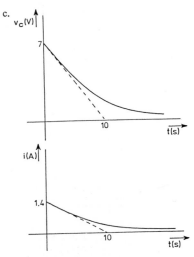

d. $v_C(0) = 7.0000$ V
$v_C(1) = 6.3339$ V
$v_C(2) = 5.7312$ V
$v_C(3) = 5.1857$ V
$v_C(4) = 4.6922$ V
$v_C(5) = 4.2457$ V
$v_C(10) = 2.5752$ V
$v_C(20) = 0.9473$ V
$v_C(30) = 0.3485$ V

9.2 dual of 9.1

9.3 a. $v_C = 7(1 - e^{-0.1t})$ V

b. $i = 1.4e^{-0.1t}$ A

c.
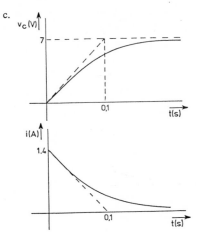

d. $v_C(0) = 0$ V
$v_C(1) = 0.6661$ V
$v_C(2) = 1.2689$ V
$v_C(3) = 1.8143$ V
$v_C(4) = 2.3078$ V
$v_C(5) = 2.7543$ V

9.4. $v_C = 7 - 4e^{-0.1t}$ V

9.5. $v_C = 7$ V

9.6. a. $i(0^-) = 3$ A
b. is continuous
c. $i = 3e^{-3t}$ A

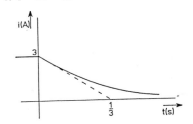

9.7. dual of 9.6

9.8. a. $v_1 = 2R(\dfrac{25}{a} + \dfrac{a - 25}{a}e^{-at})$ V

with $a = 25 + \dfrac{R}{4}$

b.
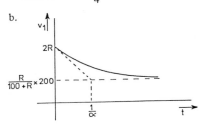

c. 2MV

9.9. a. $v_C = 9\cos 2t + 12\sin 2t$ V $t < 0$
b. $v_C(0^+) = v_C(0^-) = 9$ V
c. $v_C = 9e^{-2t}$ V $t \geqslant 0$

9.10 a. $v_C = -3.2 e^{-4t} + 3.2 \cos 3t + 2.4 \sin 3t$ V

b. $v_C(\frac{1}{20}) = 0.9028$ V

$v_C(\frac{1}{10}) = 1.6213$ V

9.11 $i = 4.4 e^{-t} + 0.6(\cos 2t + 2 \sin 2t)$ A

9.12 a. $i = 2(3 \cos 2t + 4 \sin 2t)$ A $t < 0$

b. 6 A

c. $i = 5(\cos 2t + 2 \sin 2t)$ A $t \to \infty$

d. $i = e^{-t} + 5(\cos 2t + 2 \sin 2t)$ A $t \geqslant 0$

9.13 a. $v_2 = \frac{1}{2} \sin 2t - \frac{3}{2} \cos 2t$ V $t < 0$

b. $v_2(0^+) = v_2(0^-) = -\frac{3}{2}$ V

c. $v_2 = \sin 2t - 2 \cos 2t$ V $t \to \infty$

d. $v_2 = \frac{1}{2} e^{-t} + \sin 2t - 2 \cos 2t$ V $t \geqslant 0$

9.14 $v_C = 4 - 4 e^{-\frac{1}{2}t}(\cos \frac{1}{2}t + \sin \frac{1}{2}t)$ V

9.15 $v_C = 4 + 2 e^{-t} - 6 e^{-t/3}$ V

9.16 $v_C = 4 - 4 e^{-\frac{1}{2}t} + 2t e^{-\frac{1}{2}t}$ V

9.17 $v_C = 4 - 3 e^{-\frac{1}{2}t}(\cos \frac{1}{2}t + \sin \frac{1}{2}t)$ V

9.18 $v_C = 4 - e^{-\frac{1}{2}t}(3 \cos \frac{1}{2}t + \sin \frac{1}{2}t)$ V

9.19 $v_C = 0.8(2 \sin t - \cos t)$

$+ e^{-\frac{1}{2}t}(0.8 \cos \frac{1}{2}t - 2.4 \sin \frac{1}{2}t)$ V

9.20 $i = -3 \cos 2t + \sin 2t + e^{-t}(3 \cos t + \sin t)$ A

9.21 a. $v_C(0^-) = 1$ V $i(0^-) = 1.5$ A

b. $v_2 = \frac{5}{4}(\cos \frac{1}{2}t + \sin \frac{1}{2}t)$

$+ \frac{1}{4} e^{-t}(\sin \frac{1}{2}t - \cos \frac{1}{2}t)$ V

c. $v_2(0.1) = 1.0963$ V

9.22 $v_2 = 0$ V

9.23 $v_2 = 8(e^{-t} - e^{-\frac{3}{2}t})$ V

9.24 a. $v_2 = 6 + 2 e^{-2t} - 8 e^{-\frac{1}{2}t}$ V

b. $v_2(0.1) = 0.0276$ V

$v_2(0.2) = 0.1019$ V

c.

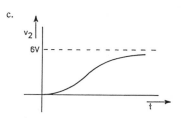

9.25 $v_2 = \frac{4}{3}(e^{-2t} - e^{-\frac{1}{2}t}) + 2 \sin t$ V $t \geqslant 0$

9.26 $v_C = 16 - e^{-1.4t}(16 \cos 0.8t + 2 \sin 0.8t)$ V

9.27 $v_C = e^{-1.4t}(-\frac{24}{17} \cos 0.8t - \frac{23}{17} \sin 0.8t) +$

$+ \frac{8}{17}(3 \cos 2t + 5 \sin 2t)$ V

Problems Chapter 10

10.1 a.

$$NA = \begin{array}{c|cc} & 1 & 2 \\ \hline & 5 & -2 \\ & -2 & 6 \end{array} \begin{array}{|c} \\ 6 \\ -9 \end{array}$$

b. $V_1 = \frac{9}{13}$ V; $V_2 = -\frac{33}{26}$ V

10.2 a.

$$\begin{array}{c|cc} & 1 & 2 \\ \hline & Y & -Y \\ & -Y & Y \end{array} \begin{array}{|c} \\ 0 \\ 0 \end{array} \quad \text{with } Y = \frac{1}{Z}$$

b. If terminal 2 is earthed

$$\begin{array}{c|c} & 1 \\ \hline & Y \end{array} \begin{array}{|c} \\ 0 \end{array} \quad \text{with } Y = \frac{1}{Z}$$

10.3 a. Choose a current I in the voltage source from plus to minus:

	1	2	3	I	
	3	-3	0	1	0
	-3	19	-9	0	0
	0	-9	11	0	2
	1	0	0	0	12

b. Make the Norton equivalence of the voltage source and the series resistor:

	2	3	
	19	-9	36
	-9	11	2

c. $V_1 = 12$ V $V_2 = 3.23$ V $V_3 = 2.83$ V

10.4

	1	2	
	0.25	−0.25	3
	−0.25	1.25	2

$V_1 = 17$ V $V_2 = 5$ V

mesh currents $I_1 = 3$ A, $I_2 = 0.5$ A, $I_3 = 2.5$ A, all clockwise. See 1.15

10.5 a.

	a	c	d	
	1.5	−1	−0.5	1
	−1	1.58	−0.33	0
	−0.5	−0.33	1.03	0

b. $V_a = 2.90$ V $V_c = 2.29$ V $V_d = 2.14$ V See 1.22

13.6 a.

	1	2	
	1.83	−5.33	3
	−0.33	5.33	6

(with the Norton equivalence)

b. $V_1 = 6$ V $V_2 = 1.5$ V. See 1.66

10.7 a.

	1	2	3	
	3	−1	−1	2
	−7	8	0	0
	5	−6	2	0

(with the Norton equivalence).

b. $V_1 = 1$ V $V_2 = 0.875$ V $V_3 = 0.125$ See 1.67

10.8 a.

	1	2	3	I	
	1	0	0	1	1
	0	3	0	−7	0
	0	0	4	6	0
	1	−1	0	0	0

b. $V_1 = 0.7$ V $V_2 = 0.7$ V $V_3 = −0.45$ V $I = 0.3$ A See 1.68

10.9 a. Introduce the current I from 2 to 3:

	1	2	3	I	
	2	−1	0	0	1
	−1	2	0	1	0
	0	0	1	−1	0
	6	−7	1	0	0

b. $V_1 = 0.8182$ V $V_2 = 0.6364$ V $V_3 = −0.4546$ V $I = −0.4546$ A

See 1.69

10.10 a. Introduce the current I_a from 2 to 3 and introduce node 4:

	1	2	3	4	I	I_a	
1	3	−2					1
2	−2	2			1		
3			1			−1	
4				1	−1	1	
*		1		−1			
**			−1	1	−3		

b. $V_1 = 0.7143$ V $V_2 = V_4 = 0.5714$ V $V_3 = −0.2857$ V $I_a = −0.2857$ A

See 1.70

10.11

	1	2	3	
	$G_1\left(1 - \dfrac{G_1}{G}\right)$	$-\dfrac{G_1 G_2}{G}$	$-\dfrac{G_1 G_3}{G}$	
	$-\dfrac{G_1 G_2}{G}$	$G_2\left(1 - \dfrac{G_2}{G}\right)$	$-\dfrac{G_2 G_3}{G}$	
	$-\dfrac{G_1 G_3}{G}$	$-\dfrac{G_2 G_3}{G}$	$G_3\left(1 - \dfrac{G_3}{G}\right)$	

$$G = G_1 + G_2 + G_3$$

10.12

	1	2	3	
	G_1	0	$-G_1$	
	0	G_2	$-G_2$	
	$-G_1$	$-G_2$	$G_1 + G_2$	

10.13

	1	2	3	I_1	
	0	0	0	1	
	0	0	0	a	
	0	0	0	$-1-a$	
	1	0	−1	0	

10.14

	a	b	c	d	I_1	
	0	0	0	0	1	
	0	G	0	$-G$	−1	
	0	0	0	0	a	
	0	$-G$	0	G	$-a$	
	1	−1	0	0	0	

10.15

	b	e
b	$G_1(1 - \dfrac{G_1 + G}{G_1 + G_2 + G})$	$\dfrac{-G_1 G_2}{G_1 + G_2 + G}$
c	$G(1 - \dfrac{G_1 + G}{G_1 + G_2 + G})$	$\dfrac{-G G_2}{G_1 + G_2 + G}$
e	$-G_2 \dfrac{G_1 + G}{G_1 + G_2 + G}$	$G_2 \dfrac{G_1 + G}{G_1 + G_2 + G}$

10.16

	b	e
b	G_1	$-G_1$
c	G	$-G$
e	$-G - G_1$	$G_1 + G$

10.17 a.

1	2	3	4	5	I_b	I
2				-1	1	
	1		-1		-1	
		1		-1	2	
	-1		1.5		-2	
-1		-1		2		-1
				1		12
1	-1					

b.

1	3	4	5	I	
3		-1	-1		
2	1	-2	-1		
-3		3.5			
-1	-1		2	-1	
			1		12

c. $V_1 = V_2 = 5.6$ V

$V_3 = 10.4$ V

$V_4 = 4.8$ V

$V_5 = 12$ V

$I = 8$ A

$I_b = 0.8$ A

10.18

1	2	3
2	-2	0
-2	5	-3
0	-3	3

10.19

1	2
2	-2
-2	5

10.20

1	3
2	0
0	3

10.21

1	3
1.2	-1.2
-1.2	1.2

10.22

1	3
a	c
g	i

10.23

k	3
$a + b + d + e$	$c + f$
$g + h$	i

10.24

1	3
$\dfrac{ae - bd}{e}$	$\dfrac{ce - bf}{e}$
$\dfrac{ge - hd}{e}$	$\dfrac{ei - hf}{e}$

10.25 a.

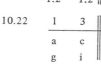

b. $v_C(1) = 6.3636$ V
$v_C(2) = 5.7851$ V
$v_C(3) = 5.2592$ V
$v_C(4) = 4.7811$ V
$v_C(5) = 4.3465$ V

d. $v_C(10) = 2.6988$ V
$v_C(20) = 1.0405$ V
$v_C(30) = 0.4012$ V   See Problem 9.1

10.26 a.

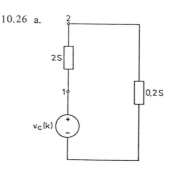

10.27 a.

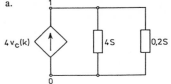

b. $t = \frac{1}{2}k$

t = 1 s v_C = 6.3492 V
t = 2 s v_C = 5.7589 V
t = 3 s v_C = 5.2235 V
t = 4 s v_C = 4.7379 V
t = 5 s v_C = 4.2974 V

d. t = 10 s v_C = 2.6382 V
   t = 20 s v_C = 0.9943 V
   t = 30 s v_C = 0.3747 V   See 9.1

10.28 a.

b. t = 1 s v_C = 6.3333 V
   t = 2 s v_C = 5.7302 V
   t = 3 s v_C = 5.1844 V
   t = 4 s v_C = 4.6901 V
   t = 5 s v_C = 4.2439 V

d. t = 10 s v_C = 2.5730 V
   t = 20 s v_C = 0.9458 V
   t = 30 s v_C = 0.3476 V   See 9.1

10.29 a. $v_C(t+1) = \dfrac{1.4 + 2v_C(t)}{2.2}$

b. $v_C(1)$ = 0.6364 V
   $v_C(2)$ = 1.2149 V
   $v_C(3)$ = 1.7408 V
   $v_C(4)$ = 2.2189 V
   $v_C(5)$ = 2.6536 V   See 9.3

10.30 a. $\begin{bmatrix} v(t+1) \\ i(t+1) \end{bmatrix} = \dfrac{1}{4.2}\begin{bmatrix} 1 & 1 \\ 4 & -0.2 \end{bmatrix}\begin{bmatrix} 1.4 \\ i(t) + 4v(t) \end{bmatrix}$

met v(0) = 0 V en i(0) = 1.4 A.

b. $v_C(1)$ = 0.6667 V
   $v_C(2)$ = 1.2698 V
   $v_C(3)$ = 1.8155 V
   $v_C(4)$ = 2.3093 V
   $v_C(5)$ = 2.7560 V   See 9.3

10.31 t = 0.05 s v_1 = 0.8240 V
          t = 0.1 s v_1 = 1.4828 V   See 9.10

10.32 a. t = 0.1 s v_1 = 0.5238 V v_2 = 0.0476 V
             t = 0.2 s v_1 = 0.9830 V
                         v_2 = 0.1365 V   See 9.24

b. The numerical values are too large due to the small tangent for small values of t.

c. $v_2(0.1)$ = 0.0266 V
   $v_2(0.2)$ = 0.1003 V

10.33 a.

1	2	i	
$\frac{1}{3}$	0	1	$\frac{5}{3}\cos\frac{k+1}{20}$
0	10.5	−1	$10\,v_2$
1	−1	−20	$-20\,i$

(with the Norton equivalence).

b. v_1 = 0.57188 V v_2 = 1.09276 V

See 9.21

10.34

1	2	3	i	i_b	
0.5	−0.5			−1	
−0.5	5.75				$5\,v_2$
		0.5	−1		
1					12
1		−1	−21		$-20\,i$

(i is the inductor current, i_b is the source current).

13.35.

	1	2	i	
1	$0.8 + \frac{C}{h}$		1	$20.8 + \frac{C}{h}v_1$
2		0.5	−1	
*	1	−1	$-\frac{L}{h}$	$-\frac{L}{h}i(k)$

(with the Norton equivalence).

10.37 V = 5.6394 V

10.38 V = 37 mV

10.39 See Figure 10.33.

10.40 a.

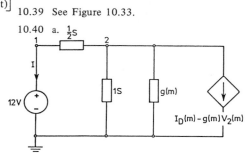

b.

1	2	I	
0.5	−0.5	1	0
−0.5	$1.5 + g(m)$	0	$-I_D(m) + g(m)\,V_2(m)$
1	0	0	12

10.42

	a	e	
a	g	$-g$	$-I_D + gV_a - gV_e$
b	$-ag$	ag	$aI_D - agV_a + agU_e$
c	ag	$-ag$	$-aI_D + agV_a - agV_e$
e	$-g$	g	$I_D - g(V_a - V_e)$

$g = g(m)$

10.43

	b	e	
b	$g(1-a)$	$g(a-1)$	$(a-1)I_D + g(1-a)V_b + g(a-1)V_e$
c	ag	$-ag$	$-aI_D + agV_b - agV_e$
e	$-g$	g	$I_D - gV_b + gV_e$

$g = g(m)$

10.44

	1	2	3	4	I_B	
1	$G_S + G_1 + G_2 + g(1-a)$	$g(a-1)$		$-G_1$		$(a-1)I_D + g(1-a)V_1$ $+ g(a-1)V_2 + G_SV_S$
2	$-g$	$G_e + g$				$I_D - gV_1 + gV_2$
3	ag	$-ag$	G_C	$-G_C$		$-aI_D + agV_1 - agV_2$
4	$-G_1$		$-G_C$	$G_1 + G_C$	-1	
*				1		V_B

$g = g(m)$

10.45 c.

	1	2	3	4	i_1	i_2	
1	1	−1				−1	
2	−1	2					V_2
3			1.25	−0.25			
4			−0.25	0.25	1		
*	1					1	
**			100	−100	−1		

MNA =

10.46 b.

	1	2	i	
1	1,5		1	44
2		0,5	−1	
*	1	−1	−10	−10i

MNA =

c. $i = 11.79$ A for $t = 0.1$ s.

10.47 b.

	1	2	i_1	
1	$2 \cdot 10^{-6}$	$-2 \cdot 10^{-6}$	-1	$2 \cdot 10^{-6}(v_1 - v_2)$
MNA = 2	$-2 \cdot 10^{-6}$	$6 \cdot 10^{-6}$		$3 \cdot 10^{-6} v_2 - 2 \cdot 10^{-6}(v_1 - v_2)$
*	1			2

(i_1 is the voltage source current).

c. $v_2(1) = 0.3333$ V.

d. The exact solution is 0.3275 V; the fault is 1.78%.

10.48 a.

	1	i	
1	1	1	$10 \cos 0.1(k + 1)$
MNA =			
*	1	-20	$-20i(k)$

(In this the Norton equivalence has been taken).

b. $i(1) = 0.4738$ A. Exact solution: 0.4869 A.

10.49 a.

	1	i	
1	50	1	$10 + 50 v_1$
MNA =			
*	1	-2.7	$-2.5 i$

(Two resistors in series have been altered into one resistance of 2.7 Ω).

b. $v_1(1) = 0.1985$ V; $v_1(2) = 0.3942$ V.

10.50 Thank you.

Index